工业和信息化"十三五"人才培养规划教材

网络技术类

Advanced Network Interconnection Technology

高级网络互联技术
项目教程
微课版

崔升广 ● 主编
杨宇 何忠刚 崔凯 徐阳 ● 副主编

人民邮电出版社
北京

图书在版编目（CIP）数据

高级网络互联技术项目教程：微课版 / 崔升广主编. -- 北京：人民邮电出版社，2020.10（2023.6重印）
工业和信息化"十三五"人才培养规划教材. 网络技术类
ISBN 978-7-115-54361-5

Ⅰ．①高… Ⅱ．①崔… Ⅲ．①互联网络－高等学校－教材 Ⅳ．①TP393.4

中国版本图书馆CIP数据核字(2020)第116907号

内 容 提 要

本书以华为网络设备为网络实训环境，以实际项目为导向，共分为13个项目，包括MUX VLAN技术、MSTP、VRRP、RIP与高级配置、OSPF协议与高级配置、IS-IS协议、BGP、路由策略、策略路由、GRE协议、IPSec、MPLS技术和VXLAN技术等。

本书是理论与实践相结合的项目化教材，配有丰富的网络拓扑图、案例和微课视频，实用性强，简单易学，能够帮助学生在训练过程中巩固所学的知识。

本书适合作为高校计算机网络技术专业及其他计算机相关专业的教材，也适合从事网络工程的技术人员参考使用。

◆ 主　　编　崔升广
　 副 主 编　杨　宇　何忠刚　崔　凯　徐　阳
　 责任编辑　郭　雯
　 责任印制　王　郁　马振武

◆ 人民邮电出版社出版发行　北京市丰台区成寿寺路11号
　邮编　100164　电子邮件　315@ptpress.com.cn
　网址　https://www.ptpress.com.cn
　天津翔远印刷有限公司印刷

◆ 开本：787×1092　1/16
　印张：18　　　　　　　　　　2020年10月第1版
　字数：494千字　　　　　　　2023年 6 月天津第 5 次印刷

定价：59.80元

读者服务热线：(010)81055256　印装质量热线：(010)81055316
反盗版热线：(010)81055315
广告经营许可证：京东市监广登字 20170147 号

前言 FOREWORD

随着计算机网络技术的不断发展，计算机网络已经成为人们生活、工作的重要组成部分。建立以网络为核心的工作方式，必将成为未来发展的趋势，培养大批熟练的网络技术人才是当前社会发展的迫切需求。在职业教育中，"网络互联技术"已经成为计算机网络技术专业的一门重要专业基础课程，由于计算机网络技术的普遍应用，网络互联技术受到人们越来越多的重视，越来越多人在从事网络相关领域的工作，计算机相关专业也开设了相关课程。网络互联技术是一门实践性很强的课程，需要一定的理论基础，同时需要大量的实践练习。本书作为重要的专业基础课程教材，特点如下。

1. 以华为网络设备搭建网络实训环境

本书以华为网络设备搭建网络实训环境，在介绍相关理论与技术原理的同时，还提供了大量的网络项目配置案例，以达到理论与实践相结合的目的。

2. 内容安排深入浅出，配备微课视频

编者精心选取了本书的内容，并对教学内容进行了整体规划与设计，使本书在叙述上简明扼要、通俗易懂，既方便教师讲授，又方便学生理解与掌握。本书用大量的网络拓扑图、案例和微课视频讲解相关知识。

3. 与时俱进，内容紧跟技术变化

本书可以让读者学到最前沿和最实用的技术，为以后的工作储备基础知识。

本书由崔升广担任主编，由杨宇、何忠刚、崔凯、徐阳担任副主编。其中，崔升广编写了项目1至项目5、项目10至项目13，杨宇编写了项目6，何忠刚编写了项目7，崔凯编写了项目8，徐阳编写了项目9，全书由崔升广组织编写并统稿。

由于编者水平有限，书中难免存在疏漏和不足之处，恳请广大读者批评指正。

编　者

2020年10月

目录 CONTENTS

项目 1

MUX VLAN 技术 1
- 1.1 项目陈述 1
- 1.2 必备知识 1
  - 1.2.1 MUX VLAN 概述 1
  - 1.2.2 MUX VLAN 基本概念 2
- 1.3 项目实施 3
  - 1.3.1 MUX VLAN 配置 3
  - 1.3.2 二层端口隔离配置 12
  - 1.3.3 三层端口隔离配置 18
- 课后习题 23

项目 2

MSTP 24
- 2.1 项目陈述 24
- 2.2 必备知识 24
  - 2.2.1 MSTP 概述 24
  - 2.2.2 MSTP 基本概念 25
- 2.3 项目实施 30
- 课后习题 35

项目 3

VRRP 36
- 3.1 项目陈述 36
- 3.2 必备知识 36
  - 3.2.1 VRRP 概述 36
  - 3.2.2 VRRP 端口状态 39
  - 3.2.3 VRRP 选举机制 39
- 3.3 项目实施 42
  - 3.3.1 VRRP 配置 42
  - 3.3.2 MSTP 与 VRRP 多备份组实现负载均衡配置 45
- 课后习题 53

项目 4

RIP 与高级配置 54
- 4.1 项目陈述 54

4.2 必备知识 ·· 54
4.3 项目实施 ·· 57
 4.3.1 RIP 路由聚合配置 ·· 57
 4.3.2 RIPv2 报文的认证方式配置 ··· 62
 4.3.3 RIP 与 BFD 联动配置 ··· 66
 4.3.4 RIP 引入外部路由配置 ·· 79
 4.3.5 RIPng 配置 ·· 84
课后习题 ·· 88

项目 5

OSPF 协议与高级配置 ·· 89

5.1 项目陈述 ·· 89
5.2 必备知识 ·· 89
5.3 项目实施 ·· 94
 5.3.1 OSPF 协议路由聚合配置 ··· 94
 5.3.2 OSPF 协议认证方式配置 ··· 98
 5.3.3 OSPF 协议认证方式配置实例 ··· 100
 5.3.4 OSPF 协议的特殊区域虚连接配置 ·· 101
 5.3.5 Stub Area 配置 ·· 104
 5.3.6 NSSA 配置 ··· 110
 5.3.7 OSPF 协议与 BFD 联动配置 ·· 119
 5.3.8 OSPF 协议引入外部路由配置 ·· 126
 5.3.9 OSPFv3 协议配置 ··· 130
课后习题 ·· 133

项目 6

IS-IS 协议 ··· 135

6.1 项目陈述 ·· 135
6.2 必备知识 ·· 135
 6.2.1 IS-IS 协议概述 ·· 135
 6.2.2 IS-IS 协议工作原理 ··· 140
6.3 项目实施 ·· 142
 6.3.1 IS-IS 协议基础配置 ··· 142
 6.3.2 IS-IS 协议基本功能配置实例 ··· 144
 6.3.3 IS-IS 协议与动态 BFD 联动功能配置实例 ··· 147
课后习题 ·· 152

项目 7

BGP ·· 154

7.1 项目陈述 ·· 154
7.2 必备知识 ·· 154

	7.2.1 BGP 概述	154
	7.2.2 BGP 工作原理	155
7.3	项目实施	159
	7.3.1 BGP 基础配置	159
	7.3.2 BGP 基本功能配置实例	160
	7.3.3 BGP 与 BFD 联动功能配置实例	165
课后习题		170

项目 8

路由策略 ······ 171

8.1	项目陈述	171
8.2	必备知识	171
	8.2.1 Route-Policy 概述	171
	8.2.2 Route-Policy 基本原理	171
	8.2.3 配置 Route-Policy	172
8.3	项目实施	173
	8.3.1 Route-Policy 过滤配置实例	173
	8.3.2 路由引入时应用 Route-Policy 配置实例	178
课后习题		183

项目 9

策略路由 ······ 184

9.1	项目陈述	184
9.2	必备知识	184
	9.2.1 PBR 概述	184
	9.2.2 配置本地 PBR	185
	9.2.3 配置端口 PBR	186
9.3	项目实施	187
课后习题		196

项目 10

GRE 协议 ······ 197

10.1	项目陈述	197
10.2	必备知识	197
	10.2.1 GRE 协议概述	197
	10.2.2 GRE 协议基本原理	198
	10.2.3 配置 GRE 协议隧道	199
10.3	项目实施	200
	10.3.1 GRE 协议实现 IPv4 网络互通配置实例	200
	10.3.2 GRE 协议隧道实现 IPv6 网络互通配置实例	205
课后习题		209

项目 11

IPSec ······ 210

11.1 项目陈述 ······ 210
11.2 必备知识 ······ 210
11.2.1 IPSec 概述 ······ 210
11.2.2 配置采用虚拟隧道端口方式建立 IPSec 隧道 ······ 214
11.2.3 配置 IKE 协议 ······ 216
11.3 项目实施 ······ 219
11.3.1 虚拟隧道端口建立 IPSec 隧道配置实例 ······ 219
11.3.2 虚拟隧道端口建立 GRE over IPSec 隧道配置实例 ······ 224
11.3.3 虚拟隧道端口建立 IPSec over GRE 隧道配置实例 ······ 231
课后习题 ······ 238

项目 12

MPLS 技术 ······ 239

12.1 项目陈述 ······ 239
12.2 必备知识 ······ 239
12.2.1 MPLS 技术概述 ······ 239
12.2.2 MPLS 体系结构 ······ 241
12.2.3 MPLS 转发过程 ······ 242
12.3 项目实施 ······ 243
课后习题 ······ 251

项目 13

VXLAN 技术 ······ 253

13.1 项目陈述 ······ 253
13.2 必备知识 ······ 253
13.2.1 VXLAN 技术概述 ······ 253
13.2.2 VXLAN 网关划分 ······ 256
13.2.3 VXLAN 运行机制 ······ 257
13.2.4 配置 VXLAN——静态方式 ······ 259
13.2.5 配置 VXLAN——BGP EVPN 方式 ······ 261
13.3 项目实施 ······ 264
13.3.1 VXLAN 隧道实现同网段用户互通配置实例 ······ 264
13.3.2 VXLAN 三层网关实现不同网段用户互通配置实例 ······ 270
课后习题 ······ 279

项目1
MUX VLAN技术

【学习目标】

- 掌握MUX VLAN应用场景及配置方法。
- 掌握端口隔离应用场景及配置方法。

1.1 项目陈述

我们知道虚拟局域网（Virtual Local Area Network，VLAN）的划分可以隔离广播域。在二层网络中，相同 VLAN 下的用户之间可以相互访问，不同 VLAN 间的用户不可以相互访问。一方面，在实际工作中，同一网段中的用户需要实现不同 VLAN 间的相互通信时，只能通过三层网络来实现。另一方面，可能需要同一 VLAN 的用户间实现相互隔离，这样就需要划分更多的 VLAN 来支持。交换机支持 VLAN 的最大个数为 4094，因此增加了网络的复杂度，同时增加了网络管理员的工作量，如何在二层网络上实现不同 VLAN 间的用户相互访问，又可以使同一 VLAN 的用户间相互隔离呢？使用多路虚拟局域网（Multiplex Virtual Local Area Network，MUX VLAN）就可以轻松解决上述问题。

1.2 必备知识

1.2.1 MUX VLAN 概述

MUX VLAN 提供了一种通过 VLAN 进行网络资源控制的机制。我们知道 VLAN 技术的产生是为了隔离广播域。在企业网络中，企业员工和企业客户可以访问企业的服务器，企业希望企业内部员工之间可以互相交流，而企业客户之间是隔离的，不能互相访问。为了满足所有用户都可访问企业服务器的需求，可通过配置 VLAN 间通信实现。如果企业规模很大，拥有大量的用户，那么就要为不能互相访问的用户分别分配 VLAN，这不但需要耗费大量的 VLAN ID，还增加了网络管理员配置和维护的工作量，而通过 MUX VLAN 提供的二层流量隔离机制可以实现企业内部员工之间的互相访问，并使企业客户之间是隔离的，能够满足企业的需求。

V1-1 MUX VLAN 概述

1.2.2　MUX VLAN 基本概念

1. MUX VLAN 类型

MUX VLAN 分为主 VLAN（Principal VLAN）和从 VLAN（Subordinate VLAN），从 VLAN 又分为隔离型从 VLAN（Separate VLAN）和互通型从 VLAN（Group VLAN）。

（1）主 VLAN：Principal port（主端口）可以和 MUX VLAN 内的所有端口进行通信。

（2）隔离型从 VLAN（简称隔离 VLAN）：Separate port（隔离端口）只能和 Principal port 进行通信，和其他类型的端口实现了完全隔离。每个隔离 VLAN 必须绑定一个主 VLAN。

（3）互通型从 VLAN（简称组 VLAN）：Group port（互通端口）可以和 Principal port 进行通信，同一组内的端口也可互相通信，但不能和其他 Group port 或 Separate port 通信。每个组 VLAN 必须绑定一个主 VLAN。

MUX VLAN 中的主 VLAN、从 VLAN 以及 Principal port、Separate port、Group port 之间的通信关系如图 1.1 所示（注：×代表不可通信）。

图 1.1　MUX VLAN 的通信关系

2. 配置 MUX VLAN 注意事项

（1）如果指定 VLAN 已经用于主 VLAN 或从 VLAN，那么该 VLAN 不能再用于创建 VLANIF 端口，或者不能在 VLAN Mapping（映射）、VLAN Stacking（堆叠）、Super-VLAN、Sub-VLAN 的配置中使用。

（2）禁止端口 MAC 地址学习功能或限制端口 MAC 地址学习数量会影响 MUX VLAN 功能的正常使用。

（3）不能在同一端口上配置 MUX VLAN 和端口安全功能。

（4）不能在同一端口上配置 MUX VLAN 和 MAC 认证功能。

（5）不能在同一端口上配置 MUX VLAN 和 IEEE 802.1x 认证功能。

（6）当同时配置 DHCP Snooping（嗅探）和 MUX VLAN 时，如果 DHCP Server（服务器）在 MUX VLAN 的从 VLAN 侧，而 DHCP Client（客户机）在主 VLAN 侧，则会导致 DHCP Client 无法正常获取 IP 地址，因此，应将 DHCP Server 配置在主 VLAN 侧。

（7）端口使能 MUX VLAN 功能后，该端口不可再配置 VLAN。

1.3 项目实施

1.3.1 MUX VLAN 配置

（1）配置 MUX VLAN，进行网络拓扑连接，如图 1.2 所示。

服务器 Server1 连接在交换机 LSW1 上，服务器 Server2 连接在交换机 LSW2 上，服务器 Server1 与服务器 Server2 同属于 VLAN 100。VLAN 100 为主 VLAN。

主机 PC1 与主机 PC2 连接在交换机 LSW1 上，主机 PC5 与主机 PC6 连接在交换机 LSW2 上，同属于 VLAN 10。VLAN 10 为组 VLAN。

主机 PC3 与主机 PC4 连接在交换机 LSW1 上，同属于 VLAN 20。VLAN 20 为组 VLAN。

主机 PC7 与主机 PC8 连接在交换机 LSW2 上，同属于 VLAN 30。VLAN 30 为隔离 VLAN。

图 1.2 配置 MUX VLAN

V1-2 配置 MUX VLAN——LSW1

V1-3 配置 MUX VLAN——LSW2

V1-4 配置 MUX VLAN——测试结果

（2）对交换机 LSW1 进行相关配置，相关实例代码如下。

```
<Huawei>system-view
[Huawei]sysname LSW1
[LSW1]vlan batch 10 20 30 100
                            //创建 VLAN 10、VLAN 20、VLAN 30、VLAN 100
[LSW1]vlan 100
[LSW1-vlan100]mux-vlan                    //配置 VLAN 100 为主 VLAN
[LSW1-vlan100]subordinate group 10        //配置 VLAN 10 为组 VLAN
[LSW1-vlan100]subordinate group 20        //配置 VLAN 20 为组 VLAN
[LSW1-vlan100]subordinate separate 30     //配置 VLAN 30 为隔离 VLAN
[LSW1-vlan100]quit
[LSW1]interface GigabitEthernet 0/0/1
[LSW1-GigabitEthernet0/0/1]port link-type trunk             //配置端口类型为 Trunk
[LSW1-GigabitEthernet0/0/1]port trunk allow-pass vlan all   //允许所有 VLAN 数据通过
```

```
[LSW1-GigabitEthernet0/0/1]quit
[LSW1]interface GigabitEthernet 0/0/2
[LSW1-GigabitEthernet0/0/2]port link-type access          //配置端口类型为 Access
[LSW1-GigabitEthernet0/0/2]port default vlan 100          //端口加入 VLAN 100
[LSW1-GigabitEthernet0/0/2]port mux-vlan enable           //使能 MUX VLAN 功能
[LSW1-GigabitEthernet0/0/2]quit
[LSW1]interface GigabitEthernet 0/0/3
[LSW1-GigabitEthernet0/0/3]port link-type access
[LSW1-GigabitEthernet0/0/3]port default vlan 10
[LSW1-GigabitEthernet0/0/3]port mux-vlan enable
[LSW1-GigabitEthernet0/0/3]quit
[LSW1]interface GigabitEthernet 0/0/4
[LSW1-GigabitEthernet0/0/4]port link-type access
[LSW1-GigabitEthernet0/0/4]port default vlan 10
[LSW1-GigabitEthernet0/0/4]port mux-vlan enable
[LSW1-GigabitEthernet0/0/4]quit
[LSW1]interface GigabitEthernet 0/0/5
[LSW1-GigabitEthernet0/0/5]port link-type access
[LSW1-GigabitEthernet0/0/5]port default vlan 20
[LSW1-GigabitEthernet0/0/5]port mux-vlan enable
[LSW1-GigabitEthernet0/0/5]quit
[LSW1]interface GigabitEthernet 0/0/6
[LSW1-GigabitEthernet0/0/6]port link-type access
[LSW1-GigabitEthernet0/0/6]port default vlan 20
[LSW1-GigabitEthernet0/0/6]port mux-vlan enable
[LSW1-GigabitEthernet0/0/6]quit
[LSW1]
```

（3）对交换机 LSW2 进行相关配置，相关实例代码如下。

```
<Huawei>system-view
[Huawei]sysname LSW2
[LSW2]vlan batch 10 20 30 100
[LSW2]vlan 100
[LSW2-vlan100]mux-vlan
[LSW2-vlan100]subordinate group 10
[LSW2-vlan100]subordinate group 20
[LSW2-vlan100]subordinate separate 30
[LSW2-vlan100]quit
[LSW2]interface GigabitEthernet 0/0/1
[LSW2-GigabitEthernet0/0/1]port link-type trunk
[LSW2-GigabitEthernet0/0/1]port trunk allow-pass vlan all
[LSW2-GigabitEthernet0/0/1]quit
[LSW2]interface GigabitEthernet 0/0/2
[LSW2-GigabitEthernet0/0/2]port link-type access
[LSW2-GigabitEthernet0/0/2]port default vlan 100
[LSW2-GigabitEthernet0/0/2]port mux-vlan enable
[LSW2-GigabitEthernet0/0/2]quit
[LSW2]interface GigabitEthernet 0/0/3
[LSW2-GigabitEthernet0/0/3]port link-type access
```

[LSW2-GigabitEthernet0/0/3]port default vlan 10
[LSW2-GigabitEthernet0/0/3]port mux-vlan enable
[LSW2-GigabitEthernet0/0/3]quit
[LSW2]interface GigabitEthernet 0/0/4
[LSW2-GigabitEthernet0/0/4]port link-type access
[LSW2-GigabitEthernet0/0/4]port default vlan 10
[LSW2-GigabitEthernet0/0/4]port mux-vlan enable
[LSW2-GigabitEthernet0/0/4]quit
[LSW2]interface GigabitEthernet 0/0/5
[LSW2-GigabitEthernet0/0/5]port link-type access
[LSW2-GigabitEthernet0/0/5]port default vlan 30
[LSW2-GigabitEthernet0/0/5]port mux-vlan enable
[LSW2-GigabitEthernet0/0/5]quit
[LSW2]interface GigabitEthernet 0/0/6
[LSW2-GigabitEthernet0/0/6]port link-type access
[LSW2-GigabitEthernet0/0/6]port default vlan 30
[LSW2-GigabitEthernet0/0/6]port mux-vlan enable
[LSW2-GigabitEthernet0/0/6]quit
[LSW2]

（4）显示交换机 LSW1 的配置信息，主要配置实例代码如下。

```
<LSW1>display current-configuration
#
sysname LSW1
#
vlan batch 10 20 30 100
#
vlan 100
 mux-vlan
 subordinate separate 30
 subordinate group 10 20
#
interface GigabitEthernet0/0/1
 port link-type trunk
 port trunk allow-pass vlan 2 to 4094
#
interface GigabitEthernet0/0/2
 port link-type access
 port default vlan 100
 port mux-vlan enable
#
interface GigabitEthernet0/0/3
 port link-type access
 port default vlan 10
 port mux-vlan enable
#
interface GigabitEthernet0/0/4
 port link-type access
 port default vlan 10
```

```
  port mux-vlan enable
#
interface GigabitEthernet0/0/5
  port link-type access
  port default vlan 20
  port mux-vlan enable
#
interface GigabitEthernet0/0/6
  port link-type access
  port default vlan 20
  port mux-vlan enable
#
user-interface con 0
user-interface vty 0 4
#
return
<LSW1>
```

（5）显示交换机 LSW2 的配置信息，主要配置实例代码如下。

```
[LSW2]display current-configuration
#
sysname LSW2
#
vlan batch 10 20 30 100
#
vlan 100
  mux-vlan
  subordinate separate 30
  subordinate group 10 20
#
interface GigabitEthernet0/0/1
  port link-type trunk
  port trunk allow-pass vlan 2 to 4094
#
interface GigabitEthernet0/0/2
  port link-type access
  port default vlan 100
  port mux-vlan enable
#
interface GigabitEthernet0/0/3
  port link-type access
  port default vlan 10
  port mux-vlan enable
#
interface GigabitEthernet0/0/4
  port link-type access
port default vlan 10
```

```
  port mux-vlan enable
 #
 interface GigabitEthernet0/0/5
  port link-type access
  port default vlan 30
  port mux-vlan enable
 #
 interface GigabitEthernet0/0/6
  port link-type access
  port default vlan 30
  port mux-vlan enable
 #
 user-interface con 0
 user-interface vty 0 4
 #
 return
 [LSW2]
```

（6）显示交换机 LSW1 配置的 VLAN 信息，如图 1.3 所示。

（7）显示交换机 LSW2 配置的 VLAN 信息，如图 1.4 所示。

图 1.3　交换机 LSW1 配置的 VLAN 信息　　　图 1.4　交换机 LSW2 配置的 VLAN 信息

（8）测试相关结果，配置主机 PC1 的 IP 地址为 192.168.11.10，主机 PC2 的 IP 地址为 192.168.11.20，如图 1.5 所示。

配置主机 PC5 的 IP 地址为 192.168.11.50，主机 PC6 的 IP 地址为 192.168.11.60，如图 1.6 所示。

主机 PC1 分别访问主机 PC2、主机 PC5 与主机 PC6，因为它们同属于 VLAN 10，因此组内可以相互访问，测试结果如图 1.7 所示。

（9）测试相关结果，配置主机 PC3 的 IP 地址为 192.168.11.30，主机 PC4 的 IP 地址为 192.168.11.40，如图 1.8 所示。

图 1.5 配置主机 PC1 与主机 PC2 的 IP 地址

图 1.6 配置主机 PC5 与主机 PC6 的 IP 地址

图 1.7 主机 PC1 分别访问主机 PC2、主机 PC5 与主机 PC6 的测试结果

项目 1
MUX VLAN 技术

图 1.8　配置主机 PC3 与主机 PC4 的 IP 地址

主机 PC1 访问主机 PC3 时，因为它们属于不同的组 VLAN，因此组与组之间不可以相互访问，测试结果如图 1.9 所示。

图 1.9　主机 PC1 访问主机 PC3 的测试结果

（10）测试相关结果，配置主机 PC7 的 IP 地址为 192.168.11.70，主机 PC8 的 IP 地址为 192.168.11.80，如图 1.10 所示。

图 1.10　配置主机 PC7 与主机 PC8 的 IP 地址

主机 PC7 与主机 PC8 同属于一个隔离 VLAN。在隔离 VLAN 中，组内的主机不能相互访问，组间也不能相互访问。主机 PC7 访问主机 PC8 的测试结果如图 1.11 所示。

9

图 1.11　主机 PC7 访问主机 PC8 的测试结果

主机 PC7 分别访问主机 PC1、主机 PC3 与主机 PC5，组间不可以相互访问，测试结果如图 1.12 所示。

图 1.12　主机 PC7 分别访问主机 PC1、主机 PC3 与主机 PC5 的测试结果

（11）配置服务器 Server1 的 IP 地址为 192.168.11.100，服务器 Server2 的 IP 地址为 192.168.11.200，如图 1.13 所示。

服务器 Server1 连接在交换机 LSW1 上，服务器 Server2 连接在交换机 LSW2 上，服务器 Server1 与服务器 Server2 同属于 VLAN 100，VLAN 100 为主 VLAN。

组 VLAN 中的主机可以与主 VLAN 中的服务器进行通信，组 VLAN 中的主机 PC1 访问服务器 Server1 与服务器 Server2 的测试结果如图 1.14 所示。

图 1.13　配置服务器 Server1 与服务器 Server2 的 IP 地址

图 1.14　组 VLAN 中的主机 PC1 访问服务器 Server1 与服务器 Server2 的测试结果

隔离 VLAN 中的主机 PC7 访问服务器 Server1 的测试结果如图 1.15 所示。

图 1.15　隔离 VLAN 中的主机 PC7 访问服务器 Server1 的测试结果

1.3.2 二层端口隔离配置

1. 端口隔离应用场景

端口隔离用于实现报文之间的二层隔离,将不同的端口加入不同的 VLAN 也可以实现隔离,但这会浪费有限的 VLAN 资源。利用端口隔离特性,可以实现同一 VLAN 内端口之间的隔离,用户只需要将端口加入到隔离组中,就可以实现隔离组内端口之间二层数据的隔离,端口隔离功能为用户提供了更安全、更灵活的组网方案。端口隔离应用场景如图 1.16 所示。

图 1.16 端口隔离应用场景

2. 端口隔离基本概念

端口隔离分为二层隔离三层互通和二层三层均隔离两种模式。

如果用户希望隔离同一 VLAN 内的广播报文,但是不同端口下的用户还可以进行三层通信,则可以将隔离模式设置为二层隔离三层互通。

如果用户希望完全隔离同一 VLAN 不同端口下的用户,则可以将隔离模式配置为二层三层均隔离。

如果没有特殊需求,建议用户不要将上行口和下行口加入到同一端口隔离组中,否则上行口和下行口之间将不能相互通信。

如图 1.17 所示,同一端口隔离组内的用户不能进行二层通信,但是不同端口隔离组内的用户可以正常通信;未划分端口隔离的用户也能与端口隔离组内的用户正常通信。

3. 二层端口隔离配置

如图 1.18 所示,同一项目组的员工都被划分到 VLAN 10 中,其中,企业内部的员工允许相互通信,企业外部的员工不允许相互通信,外部员工与内部员工之间允许通信。交换机 LSW1 与交换机 LSW2 为二层交换机,交换机 LSW3 与交换机 LSW4 为三层交换机,交换机 LSW3 中的 VLAN 10 的 IP 地址为 192.168.10.100/24;主机 PC1~主机 PC4 为内部员工,主机 PC5~主机 PC8 为外部员工;主机 PC5 与主机 PC6 在隔离组 Group 1 中,主机 PC7 与主机 PC8 在隔离组 Group 2 中。配置相关端口与 IP 地址,进行端口隔离配置。

图 1.17 二层端口隔离应用场景

（1）配置命令。port-isolate enable 命令用来使能端口隔离功能，默认将端口划入隔离组 Group 1。

如果希望创建新的 Group，则使用命令 port-isolate enable group X（要创建的隔离组组号），X 值为 1～64。

可以在系统视图下使用 port-isolate mode all 命令配置隔离模式为二层三层均隔离。

（2）查看命令。使用 display port-isolate group all 命令可以查看创建的所有隔离组端口情况。

使用 display port-isolate group X（组号）命令可以查看特定隔离组端口情况。

图 1.18 二层端口隔离配置

（3）对交换机 LSW1 进行相关配置，相关实例代码如下。

V1-8 二层端口隔离
配置——测试结果

```
< Huawei >system-view
[Huawei]sysname LSW1
[LSW1]vlan 10
[LSW1-vlan10]quit
[LSW1]interface Ethernet 0/0/1
[LSW1-Ethernet0/0/1]port link-type trunk                //配置端口类型为 Trunk
[LSW1-Ethernet0/0/1]port trunk allow-pass vlan all      //允许所有 VLAN 数据通过
[LSW1-Ethernet0/0/1]quit
[LSW1]interface Ethernet 0/0/2
[LSW1-Ethernet0/0/2]port link-type access
[LSW1-Ethernet0/0/2]port default vlan 10
[LSW1-Ethernet0/0/2]quit
[LSW1]interface Ethernet 0/0/3
[LSW1-Ethernet0/0/3]port link-type access
[LSW1-Ethernet0/0/3]port default vlan 10
[LSW1-Ethernet0/0/3]quit
[LSW1]interface Ethernet 0/0/4
[LSW1-Ethernet0/0/4]port link-type access
[LSW1-Ethernet0/0/4]port default vlan 10
[LSW1-Ethernet0/0/4]quit
[LSW1]interface Ethernet 0/0/5
[LSW1-Ethernet0/0/5]port link-type access
[LSW1-Ethernet0/0/5]port default vlan 10
[LSW1-Ethernet0/0/5]quit
[LSW1]
```

（4）对交换机 LSW2 进行相关配置，相关实例代码如下。

```
< Huawei >system-view
[Huawei]sysname LSW2
[LSW2]vlan 10
[LSW2-vlan10]quit
[LSW2]interface Ethernet 0/0/1
[LSW2-Ethernet0/0/1]port link-type trunk
[LSW2-Ethernet0/0/1]port trunk allow-pass vlan all
[LSW2-Ethernet0/0/1]quit
[LSW2]interface Ethernet 0/0/2
[LSW2-Ethernet0/0/2]port link-type access
[LSW2-Ethernet0/0/2]port default vlan 10
[LSW2-Ethernet0/0/2]port-isolate enable                 //配置隔离端口，默认将端口划入隔离组 Group 1
[LSW2-Ethernet0/0/2]quit
[LSW2]interface Ethernet 0/0/3
[LSW2-Ethernet0/0/3]port link-type access
[LSW2-Ethernet0/0/3]port default vlan 10
[LSW2-Ethernet0/0/3]port-isolate enable
[LSW2-Ethernet0/0/3]quit
[LSW2]interface Ethernet 0/0/4
[LSW2-Ethernet0/0/4]port link-type access
```

```
[LSW2-Ethernet0/0/4]port default vlan 10
[LSW2-Ethernet0/0/4]port-isolate enable group 2    //配置隔离端口，将端口划入隔离组 Group 2
[LSW2-Ethernet0/0/4]quit
[LSW2]interface Ethernet 0/0/5
[LSW2-Ethernet0/0/5]port link-type access
[LSW2-Ethernet0/0/5]port default vlan 10
[LSW2-Ethernet0/0/5]port-isolate enable group 2
[LSW2-Ethernet0/0/5]quit
[LSW2]
```

（5）显示交换机 LSW2 的配置信息，主要配置实例代码如下。

```
[LSW2]display current-configuration
#
sysname LSW2
#
vlan batch 10
#
interface Ethernet0/0/1
 port link-type trunk
 port trunk allow-pass vlan 2 to 4094
#
interface Ethernet0/0/2
 port link-type access
 port default vlan 10
 port-isolate enable group 1
#
interface Ethernet0/0/3
 port link-type access
 port default vlan 10
 port-isolate enable group 1
#
interface Ethernet0/0/4
 port link-type access
 port default vlan 10
 port-isolate enable group 2
#
interface Ethernet0/0/5
 port link-type access
 port default vlan 10
 port-isolate enable group 2
#
user-interface con 0
user-interface vty 0 4
#
return
[LSW2]
```

（6）对交换机 LSW3 进行相关配置，相关实例代码如下。

```
<Huawei>system-view
```

```
[Huawei]sysname LSW3
[LSW3]vlan 10
[LSW3]interface Vlanif 10
[LSW3-Vlanif10]ip address 192.168.10.100 255.255.255.0    //配置 VLANIF 10 的 IP 地址
[LSW3-Vlanif10]quit
[LSW3]interface GigabitEthernet 0/0/1
[LSW3-GigabitEthernet0/0/1]port link-type trunk
[LSW3-GigabitEthernet0/0/1]port trunk allow-pass vlan all
[LSW3-GigabitEthernet0/0/1]quit
[LSW3]interface GigabitEthernet 0/0/2
[LSW3-GigabitEthernet0/0/2]port link-type trunk
[LSW3-GigabitEthernet0/0/2]port trunk allow-pass vlan all
[LSW3-GigabitEthernet0/0/2]quit
[LSW3]interface GigabitEthernet 0/0/3
[LSW3-GigabitEthernet0/0/3]port link-type trunk
[LSW3-GigabitEthernet0/0/3]port trunk allow-pass vlan all
[LSW3-GigabitEthernet0/0/3]quit
[LSW3]
```

（7）配置主机 PC1 的 IP 地址为 192.168.10.1，主机 PC2 的 IP 地址为 192.168.10.2，如图 1.19 所示。

图 1.19　配置主机 PC1 与主机 PC2 的 IP 地址

主机 PC1 访问主机 PC2，VLAN 10 内部员工之间可以相互访问，测试结果如图 1.20 所示。

图 1.20　主机 PC1 访问主机 PC2 的测试结果

（8）配置主机 PC5 的 IP 地址为 192.168.10.5，主机 PC6 的 IP 地址为 192.168.10.6，如图 1.21 所示。

图 1.21　配置主机 PC5 与主机 PC6 的 IP 地址

主机 PC5 访问主机 PC6，VLAN 10 外部员工之间不可以相互访问，主机 PC5 与主机 PC6 属于隔离组 Group 1，测试结果如图 1.22 所示。

图 1.22　主机 PC5 访问主机 PC6 的测试结果

（9）配置主机 PC7 的 IP 地址为 192.168.10.7，主机 PC8 的 IP 地址为 192.168.10.8，如图 1.23 所示。

图 1.23　配置主机 PC7 与主机 PC8 的 IP 地址

主机 PC7 访问主机 PC8，VLAN 10 外部员工之间不可以相互访问，主机 PC7 与主机 PC8 属于隔离组 Group 2，测试结果如图 1.24 所示。

（10）主机 PC1 的 IP 地址为 192.168.10.1，主机 PC5 的 IP 地址为 192.168.10.5。主机 PC1 访问主机 PC5，属于内部员工访问外部员工，同属于 VLAN 10，相互之间可以访问，测试结果如图 1.25 所示。

图 1.24　主机 PC7 访问主机 PC8 的测试结果　　　图 1.25　主机 PC1 访问主机 PC5 的测试结果

（11）主机 PC5 的 IP 地址为 192.168.10.5，主机 PC7 的 IP 地址为 192.168.10.7。主机 PC5 访问主机 PC7，属于外部员工不同隔离组之间的相互访问，隔离组 Group 1 与隔离组 Group 2 相互之间可以访问，测试结果如图 1.26 所示。

（12）使用 display port-isolate group all 命令可以查看交换机 LSW2 上创建的所有隔离组端口情况，使用 display port-isolate group X（组号）命令可以查看交换机 LSW2 特定隔离组端口情况，如图 1.27 所示。

图 1.26　主机 PC5 访问主机 PC7 的测试结果　　　图 1.27　查看交换机 LSW2 创建的隔离组端口情况

1.3.3　三层端口隔离配置

如图 1.28 所示，同项目组的内部员工被划分到 VLAN 20 中，外部员工被划分到 VLAN 10 中；交换机 LSW1 为三层交换机，交换机 LSW1 中的 VLAN 10 的 IP 地址为 192.168.10.100/24，VLAN 20 的 IP 地址为 192.168.20.100/24；主机 PC3 与主机 PC4 为内部员工；主机 PC1 与主机 PC2 为外部员工，在隔离组 Group 1 中；主机 PC5 与主机 PC6 为外部员工，在隔离组 Group 2 中。配置相关端口与 IP 地址，进行三层端口隔离配置。

V1-9 三层端口隔离配置——LSW1

项目 1
MUX VLAN 技术

图 1.28 三层端口隔离配置

（1）对交换机 LSW1 进行相关配置，相关实例代码如下。

V1-10 三层端口
隔离配置
——测试结果

```
<Huawei>system-view
[Huawei]sysname LSW1
[LSW1]vlan batch 10 20                              //创建 VLAN 10、VLAN 20
[LSW1]interface Vlanif 10
[LSW1-Vlanif10]ip address 192.168.10.100 24         //配置 VLAN 10 的 IP 地址
[LSW1-Vlanif10]quit
[LSW1]interface Vlanif 20
[LSW1-Vlanif20]ip address 192.168.20.100 24         //配置 VLAN 20 的 IP 地址
[LSW1-Vlanif20]quit
[LSW1]port-isolate mode all                         //配置隔离模式为二层三层均隔离
[LSW1]interface GigabitEthernet 0/0/1
[LSW1-GigabitEthernet0/0/1]port link-type access
[LSW1-GigabitEthernet0/0/1]port default vlan 10
[LSW1-GigabitEthernet0/0/1]port-isolate enable //配置为隔离端口，默认将端口划入隔离组 Group 1
[LSW1-GigabitEthernet0/0/1]quit
[LSW1]interface GigabitEthernet 0/0/2
[LSW1-GigabitEthernet0/0/2]port link-type access
[LSW1-GigabitEthernet0/0/2]port default vlan 10
[LSW1-GigabitEthernet0/0/2]port-isolate enable //配置为隔离端口，默认将端口划入隔离组 Group 1
[LSW1-GigabitEthernet0/0/2]quit
[LSW1]interface GigabitEthernet 0/0/3
[LSW1-GigabitEthernet0/0/3]port link-type access
[LSW1-GigabitEthernet0/0/3]port default vlan 20
[LSW1-GigabitEthernet0/0/3]quit
[LSW1]interface GigabitEthernet 0/0/4
[LSW1-GigabitEthernet0/0/4]port link-type access
[LSW1-GigabitEthernet0/0/4]port default vlan 20
[LSW1-GigabitEthernet0/0/4]quit
[LSW1]interface GigabitEthernet 0/0/5
```

```
[LSW1-GigabitEthernet0/0/5]port link-type access
[LSW1-GigabitEthernet0/0/5]port default vlan 10
[LSW1-GigabitEthernet0/0/5]port-isolate enable group 2
                                    //配置为隔离端口,将端口划入隔离组 Group 2
[LSW1-GigabitEthernet0/0/5]quit
[LSW1]interface GigabitEthernet 0/0/6
[LSW1-GigabitEthernet0/0/6]port link-type access
[LSW1-GigabitEthernet0/0/6]port default vlan 10
[LSW1-GigabitEthernet0/0/6]port-isolate enable group 2
                                    //配置为隔离端口,将端口划入隔离组 Group 2
[LSW1-GigabitEthernet0/0/6]quit
[LSW1]
```

(2)显示交换机 LSW1 的配置信息,主要配置实例代码如下。

```
[LSW1]display current-configuration
#
sysname LSW1
#
vlan batch 10 20
#
port-isolate mode all
#
interface Vlanif10
 ip address 192.168.10.100 255.255.255.0
#
interface Vlanif20
 ip address 192.168.20.100 255.255.255.0
#
interface GigabitEthernet0/0/1
 port link-type access
 port default vlan 10
 port-isolate enable group 1
#
interface GigabitEthernet0/0/2
 port link-type access
 port default vlan 10
 port-isolate enable group 1
#
interface GigabitEthernet0/0/3
 port link-type access
 port default vlan 20
#
interface GigabitEthernet0/0/4
 port link-type access
 port default vlan 20
#
interface GigabitEthernet0/0/5
 port link-type access
 port default vlan 10
```

```
  port-isolate enable group 2
#
interface GigabitEthernet0/0/6
 port link-type access
 port default vlan 10
 port-isolate enable group 2
#
user-interface con 0
user-interface vty 0 4
#
return
[LSW1]
```

（3）配置主机 PC1 的 IP 地址为 192.168.10.1，主机 PC2 的 IP 地址为 192.168.10.2，网关均为 192.168.10.100，如图 1.29 所示。

图 1.29　配置主机 PC1 与主机 PC2 的 IP 地址和网关

主机 PC1 访问主机 PC2，主机 PC1 与主机 PC2 同属于 VLAN 10（外部员工）的隔离组 Group 1。因隔离模式为二层三层均隔离，主机 PC1 与主机 PC2 之间不可以相互访问，测试结果如图 1.30 所示。

图 1.30　主机 PC1 访问主机 PC2 的测试结果

（4）配置主机 PC3 的 IP 地址为 192.168.20.3，主机 PC4 的 IP 地址为 192.168.20.4，网关均为 192.168.20.100/24，如图 1.31 所示。

图 1.31　配置主机 PC3 与主机 PC4 的 IP 地址和网关

主机 PC1 访问主机 PC3，主机 PC1 属于 VLAN 10（外部员工），属于隔离组 Group 1，主机 PC3 属于 VLAN 20（内部员工）。主机 PC1 与主机 PC3 之间可以相互访问，测试结果如图 1.32 所示。

主机 PC1 访问主机 PC5，主机 PC1 属于 VLAN 10（外部员工），属于隔离组 Group 1，主机 PC5 属于 VLAN 10（外部员工），属于隔离组 Group 2。主机 PC1 与主机 PC5 之间可以相互访问，测试结果如图 1.33 所示。

图 1.32　主机 PC1 访问主机 PC3 的测试结果

图 1.33　主机 PC1 访问主机 PC5 的测试结果

课后习题

1. 选择题

（1）在 MUX VLAN 中，可以与所有 VLAN 进行通信的是（　　）。
　　A. Group VLAN　　　　　　B. Principal VLAN
　　C. Separate VLAN　　　　　D. Subordinate VLAN

（2）在 MUX VLAN 中，下列说法错误的是（　　）。
　　A. MUX VLAN 分为主 VLAN 和从 VLAN，从 VLAN 又分为隔离型从 VLAN 和互通型从 VLAN
　　B. Principal port 可以和 MUX VLAN 内的所有端口进行通信
　　C. Separate port 只能和 Principal port 进行通信，Separate port 与其他类型的端口完全隔离
　　D. Group port 可以和 Principal port 进行通信，同一组内的端口可互相通信，同时能与其他 Group port 或 Separate port 通信

（3）【多选】下列关于二层隔离端口描述正确的是（　　）。
　　A. 划分到相同 VLAN 中的相同隔离组之间的主机可以相互访问
　　B. 划分到相同 VLAN 中的相同隔离组之间的主机不可以相互访问
　　C. 划分到相同 VLAN 中的不同隔离组之间的主机可以相互访问
　　D. 划分到相同 VLAN 中的不同隔离组之间的主机不可以相互访问

（4）下列关于三层隔离端口描述错误的是（　　）。
　　A. 划分到相同 VLAN 中的内部员工之间的主机可以相互访问
　　B. 划分到相同 VLAN 中的外部员工不同隔离组之间的主机可以相互访问
　　C. 划分到相同 VLAN 中的外部员工相同隔离组之间的主机可以相互访问
　　D. 划分到不同 VLAN 中的内部员工与外部员工隔离组之间的主机可以相互访问

2. 简答题

（1）简述 MUX VLAN 的功能及应用场景。
（2）简述配置 MUX VLAN 的注意事项。
（3）简述 MUX VLAN 的配置步骤与配置命令。
（4）简述端口隔离的功能及应用场景。
（5）简述端口隔离的配置步骤与配置命令。

项目2 MSTP

【学习目标】

- 了解MSTP的基本概念及其应用场景。
- 掌握MSTP的配置方法。

2.1 项目陈述

生成树协议（Spanning Tree Protocol，STP）实现了链路的冗余，增强了网络的稳定性，但网络的收敛时间较长，大约需要50s。快速生成树协议（Rapid Spanning Tree Protocol，RSTP）在STP的基础上进行了改进，实现了网络拓扑的快速收敛，大约需要1s，网络收敛速度问题已经得到了解决。随着网络技术的发展，VLAN技术的应用对可靠性、服务质量、传送效率、业务处理灵活性、可管理性等网络服务提出了更高的要求，新的问题也随之而来，网络中的根交换机负载明显过重，而其他非根交换机负载较少，负载明显不均衡，需要新的技术来解决这些问题。

2.2 必备知识

2.2.1 MSTP概述

多生成树（Multiple Spanning Tree，MST）使用了修正的RSTP，故被称为多生成树协议（Multiple Spanning Tree Protocol，MSTP），多生成树是在IEEE 802.1w标准的快速生成树算法上扩展而得到的。

RSTP在STP的基础上进行了改进，实现了网络拓扑的快速收敛。但由于局域网内所有的VLAN共享一棵生成树，因此被阻塞后链路将不承载任何流量，无法在VLAN间实现数据流量的负载均衡，从而造成带宽浪费。为了弥补STP和RSTP的缺陷，电气和电子工程师协会（Institute of Electrical and Electronics Engineers，IEEE）在2002年发布的IEEE 802.1s标准中定义了MSTP。MSTP兼容STP和RSTP，既可以快速收敛，又提供了数据转发的多个冗余路径，在数据转发过程中实现了VLAN数据的负载均衡。

采用MST能够通过干道（Trunk）建立多棵生成树，关联多个VLAN到相关的生成树进程，每个生成树进程具备独立于其他进程的拓扑结构；MST提供了多个数据转发路径实现负载均衡，提高了网络容错能力，因为一个进程（转发路径）的故障不会影响其他进程。一个生成树进程只能存在于具备一致的 VLAN 进程分配的桥中，必须用同样的 MST 配置信息来配置一组桥，以使桥能被分配到一组生成树进程中，则具备同样的 MST 配置信息的互连的桥构成多生成树域（Multiple Spanning Tree Region，MST Region）。

2.2.2 MSTP 基本概念

1. MST Region

MST Region 由交换网络中的多台交换设备以及它们之间的网段构成。同一个 MST Region 内的设备具有如下特点：都启动了 MSTP；具有相同的域名；具有相同的 VLAN 到生成树实例的映射；具有相同的 MSTP 修订级别配置。

实例（Instance）是针对一组 VLAN 的一个独立计算的 STP，通过将多个 VLAN 捆绑到一个实例，相对于每个 VLAN 独立运算来说，可以节省通信开销和资源占用率。MSTP 各个实例的计算过程相互独立，使用多个实例可以实现物理链路的负载均衡。当把多个拓扑结构相同的 VLAN 映射到一个实例之后，它们在端口上的转发状态取决于该端口在对应 MSTP 实例中的状态。

2. CIST/总根/CST/IST/主桥/SST

公共和内部生成树（Common and Internal Spanning Tree，CIST）是通过 STP 或 RSTP 计算生成的，连接一个交换网络内所有交换设备的单生成树。

总根是整个网络中优先级最高的网桥，即为 CIST 的根桥。

公共生成树（Common Spanning Tree，CST）是通过 STP 或 RSTP 计算生成，连接交换网络内所有 MST Region 的一棵生成树。

内部生成树（Internal Spanning Tree，IST）是各 MST Region 内的一棵生成树，IST 是 CIST 在 MST Region 中的一个片段。MST Region 内每棵生成树都对应一个实例号，IST 的实例号为 0。实例 0 无论有没有配置都是存在的，没有映射到其他实例的 VLAN 默认都会映射到实例 0（即 IST）。

主桥（Master Bridge）就是 IST Master，它是域内距离总根最近的交换设备。如果总根在 MST Region 中，则总根为该域的主桥。

单生成树（Single Spanning Tree，SST）的产生分为两种情况：一种是运行 STP 或 RSTP 的交换设备只能属于一棵生成树；另一种是 MST Region 中只有一个交换设备，则该交换设备构成单生成树。

3. MSTI/MSTI 域根

一个 MST Region 内可以存在多棵生成树，每棵生成树都称为一个多生成树实例（Multiple Spanning Tree Instance，MSTI）。MSTI 域根即 MSTI 的根，一个 MSTI Region 中不同的 MSTI 有各自的域根。MSTI 之间彼此独立，MSTI 可以与一个或者多个 VLAN 对应，但一个 VLAN 只能与一个 MSTI 对应。每一个 MSTI 对应一个实例号，实例号从 1 开始，以区分实例号为 0 的 IST。MSTI 域根是每个 MSTI 上优先级最高的网桥，MST Region 内每个 MSTI 都可以指定不同的根。

4. MSTP 端口角色

MSTP 在 RSTP 的基础上新增了 2 种端口，MSTP 的端口角色共有 7 种：主端口（Master Port）、根端口（Root Port）、指定端口（Designated Port）、替代端口（Alternate Port）、备份

端口（Backup Port）、边缘端口和域边缘端口。

主端口是 MST Region 和总根相连的最短路径的端口，是域中报文发往总根的必经之路。主端口是特殊域边缘端口，主端口在 CIST 上的角色是根端口，在其他各实例上的角色都是主端口。

域边缘端口指 MST Region 内网桥与其他 MST Region 或者 STP/RSTP 网桥相连的端口。

5．MSTP 快速收敛

MSTP 快速收敛的方式有两种：一种是普通 P/A 快速收敛，与 RSTP 相同；另一种是增强型 P/A 快速收敛。在 MSTP 中，P/A 机制的工作过程如下。

（1）上游设备发送 Proposal 报文，请求进行快速迁移。下游设备接收到此报文后，将与上游设备相连的端口设置为根端口，并阻塞所有非边缘端口。

（2）上游设备继续发送 Agreement 报文。下游设备接收到此报文后，根端口转为 Forwarding 状态。

（3）下游设备回应 Agreement 报文。上游设备接收到此报文后，将与下游设备相连的端口设置为指定端口，指定端口进入 Forwarding 状态。

6．MSTP 的特点

（1）MSTP 将环路网络修剪为一个无环的树形网络，避免报文在环路网络中的增生和无限循环，同时提供了用于数据转发的多个冗余路径，在数据转发过程中保证负载均衡，如图 2.1 所示。

V2-1 MSTP 的特点

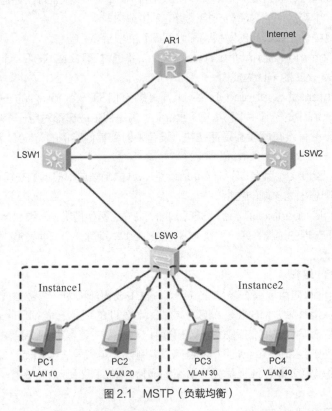

图 2.1　MSTP（负载均衡）

（2）MSTP 设置了 VLAN 映射表（即 VLAN 和生成树的对应关系表），把 VLAN 和生成树联系起来；通过增加"实例"的概念，将多个 VLAN 捆绑到一个实例中，以节省通信开销和资源占用率，如图 2.2 和图 2.3 所示。

项目 2
MSTP

图 2.2　MSTP（实例 Instance1 流量）　　　图 2.3　MSTP（实例 Instance2 流量）

（3）MSTP 将一个交换网络划分成多个域，每个域内形成多棵生成树，各生成树之间彼此独立。

（4）MSTP 兼容 STP 和 RSTP。

7. MSTP 配置命令

MSTP 的配置命令，包括启动/关闭生成树协议、配置生成树协议模式、配置区域属性、配置交换机优先级、配置端口优先级等，下面分别进行介绍。

（1）启动/关闭生成树协议。

```
<Huawei>system-view
[Huawei]stp enable              //启动生成树协议
[Huawei]undo stp enable         //关闭生成树协议
Warning: The global STP state will be changed. Continue? [Y/N]y
Info: This operation may take a few seconds. Please wait for a moment...done.
```

（2）配置生成树协议模式。

```
<Huawei>system-view
[Huawei]stp enable              //启动生成树协议
[Huawei]stp mode ?
  mstp    Multiple Spanning Tree Protocol (MSTP) mode
  rstp    Rapid Spanning Tree Protocol (RSTP) mode
  stp     Spanning Tree Protocol (STP) mode
[Huawei]stp mode mstp           //配置为多生成树协议模式
```

交换机有 3 种生成树模式，即 STP、RSTP、MSTP，默认情况下为 MSTP。将交换机生成树协议恢复为默认模式的代码如下。

```
<Huawei>system-view
[Huawei]undo stp mode           //恢复生成树协议默认模式
```

（3）配置区域属性。

```
<Huawei>system-view
[Huawei]stp mode mstp
```

27

```
[Huawei]stp region-configuration                        //配置 MSTP 区域
[Huawei-mst-region]region-name RG1                      //区域名称为 RG1
[Huawei-mst-region]instance ?
   INTEGER<0-48>   Identifier of spanning tree instance     //实例号的范围为 0~48
[Huawei-mst-region]instance 1 vlan 10 20                //创建实例 1，关联 VLAN 10、VLAN 20
[Huawei-mst-region]instance 2 vlan 30 40                //创建实例 2，关联 VLAN 30、VLAN 40
[Huawei-mst-region]active  region-configuration         //激活区域配置
[Huawei-mst-region]quit
[Huawei]stp instance 1 priority ?
   INTEGER<0-61440>   Bridge priority, in steps of 4096
//实例区域优先级取值为 0~61440，按 4096 的倍数递增，默认值为 32768
[Huawei]stp instance 1 priority 4096    //配置优先级，使交换机 Huawei 为实例 1 的根桥
[Huawei]
```

（4）配置交换机优先级。

```
<Huawei>system-view
[Huawei]stp priority ?
   INTEGER<0-61440>   Bridge priority, in steps of 4096
//交换机的优先级取值为 0~61440，按 4096 的倍数递增，默认值为 32768
[Huawei]stp priority 4096       //配置交换机的优先级为 4096
```

（5）配置交换机端口优先级。

```
<Huawei>system-view
Enter system view, return user view with Ctrl+Z.
[Huawei]stp enable
[Huawei]stp mode mstp
[Huawei]interface GigabitEthernet 0/0/1
[Huawei-GigabitEthernet0/0/1]stp instance 1 port priority ?
   INTEGER<0-240>   Port priority, in steps of 16
[Huawei-GigabitEthernet0/0/1]stp instance 1 port priority 32
//配置实例 1 端口优先级为 32，端口优先级取值为 0~240，按 16 的倍数递增，默认值为 128
[Huawei-GigabitEthernet0/0/1]stp port priority ?
   INTEGER<0-240>   Port priority, in steps of 16
[Huawei-GigabitEthernet0/0/1]stp port priority 16       //默认为实例 0 端口优先级
[Huawei-GigabitEthernet0/0/1]quit
[Huawei]
```

（6）显示 MSTP 配置信息。

```
[Huawei]display stp ?
   brief                  Brief information
   error                  Information of error packet
   instance               Spanning tree instance
   interface              Specify interface
   process                The MSTP process
   region-configuration   Region configuration
   slot                   Specify the slot
   tc-bpdu                Number of TC/TCN BPDUs sent from and received by the port
   topology-change        Display information about topology change
```

```
  vlan                    Virtual LAN
  |                       Matching output
  <cr>
[Huawei]display stp brief        //显示生成树协议配置信息
 MSTID   Port                    Role    STP State     Protection
   0     GigabitEthernet0/0/1    DESI    FORWARDING    NONE
   0     GigabitEthernet0/0/2    DESI    FORWARDING    NONE
   0     GigabitEthernet0/0/3    ROOT    FORWARDING    NONE
[Huawei]display stp interface GigabitEthernet 0/0/2 brief
 MSTID   Port                    Role    STP State     Protection
   0     GigabitEthernet0/0/2    DESI    FORWARDING    NONE
[Huawei]
[Huawei]display stp instance 0 ?
  brief              Brief information
  interface          Specify interface
  slot               Specify the slot
  tc-bpdu            Number of TC/TCN BPDUs sent from and received by the port
  topology-change    Display information about topology change
  |                  Matching output
  <cr>
[Huawei]display stp instance 0 brief     //显示实例配置信息
 MSTID   Port                    Role    STP State     Protection
   0     GigabitEthernet0/0/1    DESI    FORWARDING    NONE
   0     GigabitEthernet0/0/2    DESI    FORWARDING    NONE
   0     GigabitEthernet0/0/3    ROOT    FORWARDING    NONE
[Huawei]
[Huawei]display stp interface GigabitEthernet 0/0/2 ?
  brief      Brief information
  tc-bpdu    Number of TC/TCN BPDUs sent from and received by the port
  |          Matching output
  <cr>
[Huawei]display stp interface GigabitEthernet 0/0/2 brief   //显示端口配置信息
 MSTID   Port                    Role    STP State     Protection
   0     GigabitEthernet0/0/2    DESI    FORWARDING    NONE
[Huawei]
[Huawei]display stp vlan 10 ?
  |          Matching output
  <cr>
[Huawei]display stp vlan 10      //显示 VLAN 配置信息
 ProcessId   InstanceId   Port                    Role    State
-------------------------------------------------------------------
     0           0        GigabitEthernet0/0/1    DESI    FORWARDING
[Huawei]
```

2.3 项目实施

(1) 配置 MSTP,交换机 LSW1 为实例 1 的根桥,实例 1 关联 VLAN 10 与 VLAN 30;交换机 LSW2 为实例 2 的根桥,实例 2 关联 VLAN 20 与 VLAN 40,如图 2.4 所示,进行网络拓扑连接。

V2-2 MSTP 配置
——LSW1

V2-3 MSTP 配置
——LSW2

图 2.4 MSTP 配置

(2) 配置交换机 LSW1,相关实例代码如下。

```
<Huawei>system-view
[Huawei]sysname LSW1
Enter system view, return user view with Ctrl+Z.
[LSW1]vlan batch 10 20 30 40    //创建 VLAN 10、VLAN 20、VLAN 30、VLAN 40
[LSW1]port-group 1              //创建端口组,进行统一设置
[LSW1-port-group-1]group-member Ethernet 0/0/1 to Ethernet 0/0/3
[LSW1-Ethernet0/0/1]port link-type trunk      //配置端口类型
[LSW1-port-group-1]port trunk allow-pass vlan all  //允许所有 VLAN 数据通过
[LSW1-port-group-1]quit
[LSW1]stp mode mstp                           //配置 MSTP
[LSW1]stp region-configuration                //配置 MSTP 区域
[LSW1-mst-region]region-name RG1              //区域名称为 RG1
[LSW1-mst-region]instance 1 vlan 10 30        //创建实例 1,关联 VLAN 10、VLAN 30
[LSW1-mst-region]instance 2 vlan 20 40        //创建实例 2,关联 VLAN 20、VLAN 40
[LSW1-mst-region]active  region-configuration //激活区域配置
[LSW1-mst-region]quit
[LSW1]stp instance 1 priority 4096    //配置优先级,使交换机 LSW1 为实例 1 的根桥
[LSW1]stp instance 2 priority 8192    //配置优先级,使交换机 LSW2 为实例 2 的根桥
[LSW1]
```

V2-4 MSTP 配置
——LSW3

V2-5 MSTP 配置
——LSW4

(3) 配置交换机 LSW2,相关实例代码如下。

```
<Huawei>system-view
```

```
Enter system view, return user view with Ctrl+Z.
[Huawei]sysname LSW2
[LSW2]vlan batch 10 20 30 40          //创建 VLAN 10、VLAN 20、VLAN 30、VLAN 40
[LSW2]port-group 1                    //创建端口组，进行统一设置
[LSW2-port-group-1]group-member Ethernet 0/0/1 to Ethernet 0/0/3
[LSW2-Ethernet0/0/1]port link-type trunk           //配置端口类型
[LSW2-port-group-1]port trunk allow-pass vlan all  //允许所有 VLAN 数据通过
[LSW2-port-group-1]quit
[LSW2]stp mode mstp                   //配置 MSTP
[LSW2]stp region-configuration        //配置 MSTP 区域
[LSW2-mst-region]region-name RG1      //区域名称为 RG1
[LSW2-mst-region]instance 1 vlan 10 30  //创建实例 1，关联 VLAN 10、VLAN 30
[LSW2-mst-region]instance 2 vlan 20 40  //创建实例 2，关联 VLAN 20、VLAN 40
[LSW2-mst-region]active  region-configuration  //激活区域配置
[LSW2-mst-region]quit
[LSW2]stp instance 1 priority 8192    //配置优先级，使交换机 LSW1 为实例 1 的根桥
[LSW2]stp instance 2 priority 4096    //配置优先级，使交换机 LSW2 为实例 2 的根桥
[LSW2]
```

（4）显示交换机 LSW1、LSW2 的配置信息，以交换机 LSW1 为例，主要配置实例代码如下。

```
<LSW1>display current-configuration
#
sysname LSW1
#
vlan batch 10 20 30 40
#
stp instance 1 priority 4096
stp instance 2 priority 8192
#
stp region-configuration
 region-name RG1
 instance 1 vlan 10 30
 instance 2 vlan 20 40
 active region-configuration
#
interface Ethernet0/0/1
 port link-type trunk
 port trunk allow-pass vlan 2 to 4094
#
interface Ethernet0/0/2
 port link-type trunk
 port trunk allow-pass vlan 2 to 4094
#
interface Ethernet0/0/3
port link-type trunk
 port trunk allow-pass vlan 2 to 4094
#
user-interface con 0
user-interface vty 0 4
```

```
#
port-group 1
 group-member Ethernet0/0/1
 group-member Ethernet0/0/2
 group-member Ethernet0/0/3
#
return
<LSW1>
```

（5）配置交换机 LSW3，相关实例代码如下。

```
<Huawei>system-view
Enter system view, return user view with Ctrl+Z.
[Huawei]sysname LSW3
[LSW3]vlan batch 10 20 30 40
[LSW3]port-group 1
[LSW3-port-group-1]group-member Ethernet 0/0/1 to Ethernet 0/0/3
[LSW3-Ethernet0/0/1]port link-type trunk
[LSW3-port-group-1]port trunk allow-pass vlan all
[LSW3-port-group-1]quit
[LSW3]stp mode mstp
[LSW3]stp region-configuration
[LSW3-mst-region]region-name RG1
[LSW3-mst-region]instance 1 vlan 10 30
[LSW3-mst-region]instance 2 vlan 20 40
[LSW3-mst-region]active region-configuration
[LSW3-mst-region]quit
[LSW3]port-group 2
[LSW3-port-group-2]group-member Ethernet 0/0/4 to Ethernet 0/0/5
[LSW3-port-group-2]port link-type access
[LSW3-port-group-2]stp edged-port enable      //配置为边缘端口
[LSW3-port-group-2]quit
[LSW3]interface Ethernet 0/0/4
[LSW3-Ethernet0/0/4]port default vlan 10
[LSW3-Ethernet0/0/4]quit
[LSW3]interface Ethernet 0/0/5
[LSW3-Ethernet0/0/5]port default vlan 20
[LSW3-Ethernet0/0/5]quit
[LSW3]
```

（6）配置交换机 LSW4，相关实例代码如下。

```
<Huawei>system-view
Enter system view, return user view with Ctrl+Z.
[Huawei]sysname LSW4
[LSW4]vlan batch 10 20 30 40
[LSW4]port-group 1
[LSW4-port-group-1]group-member Ethernet 0/0/1 to Ethernet 0/0/3
[LSW4-Ethernet0/0/1]port link-type trunk
```

[LSW4-port-group-1]port trunk allow-pass vlan all
[LSW4-port-group-1]quit
[LSW4]stp mode mstp
[LSW4]stp region-configuration
[LSW4-mst-region]region-name RG1
[LSW4-mst-region]instance 1 vlan 10 30
[LSW4-mst-region]instance 2 vlan 20 40
[LSW4-mst-region]active region-configuration
[LSW4-mst-region]quit
[LSW4]port-group 2
[LSW4-port-group-2]group-member Ethernet 0/0/4 to Ethernet 0/0/5
[LSW4-port-group-2]port link-type access
[LSW4-port-group-2]stp edged-port enable   //配置为边缘端口
[LSW4-port-group-2]quit
[LSW4]interface Ethernet 0/0/4
[LSW4-Ethernet0/0/4]port default vlan 30
[LSW4-Ethernet0/0/4]quit
[LSW4]interface Ethernet 0/0/5
[LSW4-Ethernet0/0/5]port default vlan 40
[LSW4-Ethernet0/0/5]quit
[LSW4]

（7）显示交换机 LSW3、LSW4 的配置信息，以交换机 LSW3 为例，主要配置实例代码如下。

```
<LSW3>display current-configuration
#
sysname LSW3
#
vlan batch 10 20 30 40
#
stp region-configuration
 region-name RG1
 instance 1 vlan 10 30
 instance 2 vlan 20 40
 active region-configuration
#
interface Ethernet0/0/1
 port link-type trunk
 port trunk allow-pass vlan 2 to 4094
#
interface Ethernet0/0/2
 port link-type trunk
 port trunk allow-pass vlan 2 to 4094
#
interface Ethernet0/0/3
 port link-type trunk
```

```
    port trunk allow-pass vlan 2 to 4094
#
interface Ethernet0/0/4
 port link-type access
 port default vlan 10
 stp edged-port enable
#
interface Ethernet0/0/5
 port link-type access
 port default vlan 20
 stp edged-port enable
#
user-interface con 0
user-interface vty 0 4
#
port-group 1
 group-member Ethernet0/0/1
 group-member Ethernet0/0/2
 group-member Ethernet0/0/3
#
port-group 2
 group-member Ethernet0/0/4
 group-member Ethernet0/0/5
#
return
<LSW3>
```

（8）查看 MSTP 运行状态，使用 display stp instance 1 brief 命令，可以看到实例 1 的端口角色及端口运行状态，如图 2.5 所示。

（9）查看 MSTP 运行状态，使用 display stp instance 2 brief 命令，可以看到实例 2 的端口角色及端口运行状态，如图 2.6 所示。

图 2.5　实例 1 端口角色及端口运行状态　　　图 2.6　实例 2 端口角色及端口运行状态

（10）查看交换机 LSW1 的 MSTP 运行状态，如图 2.7 所示。

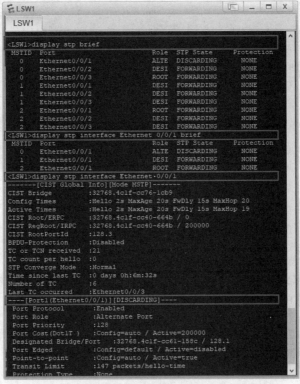

图 2.7 查看交换机 LSW1 的 MSTP 运行状态

课后习题

1. 选择题

（1）在生成树协议中，交换机端口默认优先级为（　　）。
　　A. 16　　　　　B. 32　　　　　C. 64　　　　　D. 128

（2）在生成树协议中，交换机默认优先级为（　　）。
　　A. 65535　　　B. 8192　　　　C. 32768　　　D. 4096

（3）关于 MSTP 的描述，错误的是（　　）。
　　A. MSTP 兼容 RSTP 和 STP
　　B. 一个 MSTI 可以与一个或多个 VLAN 对应
　　C. 一个 MST Region 内只能有一个生成树实例
　　D. 每个生成树实例可以独立地运行快速生成树算法

（4）定义 MSTP 的是（　　）。
　　A. IEEE 802.1w　　　　　　　　B. IEEE 802.1s
　　C. IEEE 802.1d　　　　　　　　D. IEEE 802.1q

2. 简答题

（1）简述生成树协议的主要作用以及 STP、RSTP 的局限性。
（2）MSTP 的出现主要解决了什么问题？

项目 3
VRRP

【学习目标】

- 了解VRRP的基本概念及其应用场景。
- 掌握VRRP的配置方法。
- 掌握配置MSTP与VRRP多备份组实现负载均衡的方法。

3.1 项目陈述

我们知道在 TCP/IP 网络进行跨网段的数据转发时，必须通过网关，但是，默认网关一旦出现故障，就会导致使用该网关所连接的主机因故障而无法进行正常通信。通常的解决方法是配置多个默认网关以实现网关的冗余，这需要管理员手动配置网关地址，在实际工作中的效果并不理想。为了有效地解决这样的问题呢，虚拟路由冗余协议（Virtual Router Redundancy Protocol，VRRP）应运而生，它同时提供了负载均衡的解决方案。

3.2 必备知识

3.2.1 VRRP 概述

VRRP 是由互联网工程任务组（Internet Engineering Task Force，IETF）在 1998 年推出的 RFC2338 协议标准中提出的，用于解决局域网中配置的静态网关出现单点失效现象的路由协议。VRRP 广泛应用在边缘网络中，它的设计目标是在特定情况下，使 IP 数据流量的失败转移不引起混乱，允许主机使用单路由器，以及即使在实际第一跳路由器使用失败的情况下也能够维护路由器间的连通性。

VRRP 是一种选择协议，它可以把一个虚拟路由器的数据动态地分配到局域网上的 VRRP 路由器中的一台。控制虚拟路由器 IP 地址的 VRRP 路由器称为主控路由器，它负责转发数据包到虚拟 IP 地址。一旦主控路由器不能工作，VRRP 提供了动态的故障转移机制，它允许虚拟路由器的 IP 地址作为终端主机的默认第一跳路由器，它是一种局域网接入设备备份协议。一个局域网络内的所有主机都配置了默认网关，如图 3.1 所示，路由器 AR1 的 GE 0/0/0 的端口的 IP 地址为 192.168.1.254，主机发送的目的地址不在本网段的报文将被通过默认网关发往三层交换机，从而实现主机和外部网络的通信。

图 3.1　配置默认网关

当默认网关的路由器出现故障时，所有使用该路由器作为默认网关的主机都会因为默认网关故障而无法进行通信，这个问题的解决方案有两种，下面分别进行介绍。

一种方案是配置多个默认网关，这种方案可以实现网关的冗余，如果第一个默认网关出现故障，则第二个默认网关可以转发数据。但在实际工作中效果并不理想，其原因在于不能真正做到网关及时切换转发，需要人工参与配置。如图 3.2 所示，主机 PC1 设置了双网关（192.168.1.254 及 192.168.2.254），访问 Internet 时，如果路由器 AR1 正常工作，那么主机 PC1 就会把数据包发送给网关地址 192.168.1.254，再由路由器 AR1 转发到 Internet。当路由器 AR1 出现故障，无法通信时，主机 PC1 因不知道路由器故障，会继续将数据包发送给网关地址 192.168.1.254，也就是说，在没有人工参与的情况下，主机 PC1 不会自动切换到另一台路由器 AR2 的网关地址 192.168.2.254，即无法实现默认网关冗余数据通信。

图 3.2　配置多个默认网关

另一种方案是在路由器 AR1 与路由器 AR2 上启动 VRRP。当路由器 AR1 正常工作时，由其转发数据进行通信，当路由器 AR1 出现故障时，系统自动切换到路由器 AR2 上，由路由器 AR2 代替路由器 AR1 的工作，实现网关的冗余，使数据通信不间断。

VRRP 是一种路由容错协议，也称备份路由协议。一个局域网络内的所有主机都配置了默认路由，当网络内主机发出的目的地址不在本网段时，报文将被通过默认路由发往外部路由器，从而实现主机与外部网络的通信。当默认路由器故障（即端口关闭）之后，内部主机将无法与外部通信，如果路由器设置了 VRRP，则此时虚拟路由将启用备份路由器，从而实现通信不间断。

在 VRRP 中，有两组重要的概念：VRRP 路由器和虚拟路由器、主控路由器（Master）和备份路由器（Backup）。VRRP 路由器是指运行 VRRP 的路由器，是物理实体；虚拟路由器是指 VRRP 创建的路由器，是逻辑概念。一组 VRRP 路由器协同工作，共同构成一台虚拟路由器，该虚拟路由器对外表现为一个具有唯一固定的 IP 地址和 MAC 地址的逻辑路由器。处于同一个 VRRP 组中的路由器具有两种互斥的角色：主控路由器和备份路由器。一个 VRRP 组中有且只有一台处于主控角色的路由器，可以有一台或者多台处于备份角色的路由器。VRRP 从路由器组中选出一台作为主控路由器，负责通过地址解析协议（Address Resolution Protocol，ARP）解析和转发 IP 数据包，组中的其他路由器作为备份路由器处于待命状态。当主控路由器发生故障时，其中的一台备份路由器能在瞬间的时延后升级为主控路由器，由于此切换非常迅速且不改变 IP 地址和 MAC 地址，故对终端使用者而言系统是透明的。

如图 3.3 所示，路由器 AR1、路由器 AR2 组成一组 VRRP 路由器，路由器 AR1、路由器 AR2 的虚拟路由器 IP 地址统一设置为 192.168.1.254，网络中所有主机的默认网关地址都为虚拟路由器的 IP 地址。路由器 AR1 为主控路由器，负责对发送到该虚拟 IP 地址的数据包进行转发；路由器 AR2 为备份路由器，当主控路由器 AR1 发生故障时，备份路由器 AR2 将替换主控路由器 AR1 并进行数据转发，当主控路由器 AR1 恢复正常工作后，路由器 AR1 将再次成为主控路由器。

图 3.3　一组 VRRP 路由器

每一个 VRRP 组中的路由器都有唯一的 VRID 标志（值为 1~255），它决定了运行 VRRP 的路由器属于哪一个 VRRP 组，VRRP 组中的虚拟路由器对外表现为唯一的虚拟 MAC 地址，地址的格式为 00-00-5E-00-01[VRID]。主控路由器负责对发送到虚拟路由器 IP 地址的 ARP 请求做出响应并以该虚拟 MAC 地址做出应答，因此，无论如何切换，都能保证终端主机设备是唯一的 IP 地址和 MAC 地址，避免发生网络中网关设备出现故障时导致终端设备无法通信的情况。

3.2.2 VRRP 端口状态

VRRP 有 3 种状态：Initialize（初始状态）、Master（主用状态）和 Backup（备用状态），下面分别进行介绍。

1. Initialize

路由器启动时，如果路由器的优先级是 255（最高优先级，当且仅当配置的 VRRP 虚拟 IP 地址和端口 IP 地址相同时，即所谓的 IP 地址拥有者），要想发送 VRRP 通告信息，并发送广播 ARP 信息通告路由器 IP 地址对应的 MAC 地址为路由虚拟 MAC 地址，则设置通告信息定时器准备定时发送 VRRP 通告信息，转为 Master 状态；否则进入 Backup 状态，设置定时器定时检查是否收到 Master 的通告信息。

2. Master

Master 状态即 VRRP 路由器中真正起作用的路由器，可实现如下功能。

设置定时通告定时器；用 VRRP 虚拟 MAC 地址响应路由器 IP 地址的 ARP 请求；转发目的 MAC 地址是 VRRP 虚拟 MAC 地址的数据包；如果是虚拟路由器 IP 地址的拥有者，则接收目的地址是虚拟路由器 IP 地址的数据包，否则丢弃；当收到 shutdown 事件时，删除定时通告定时器，发送优先值为 0 的通告包，转为初始状态；如果定时通告定时器超时，则发送 VRRP 通告信息。收到 VRRP 通告信息时，如果优先值为 0，则发送 VRRP 通告信息；否则判断数据的优先级是否高于本机，若相等且实际 IP 地址大于本地实际 IP 地址，则设置定时通告定时器，复位主机超时定时器，转为 Backup 状态，否则丢弃该通告包。

3. Backup

Backup 状态下的路由器为 Master 的备份，要实现以下功能。

设置主机超时定时器；不能响应针对虚拟路由器 IP 地址的 ARP 请求信息；丢弃所有目的 MAC 地址是虚拟路由器 MAC 地址的数据包；不接收目的地址是虚拟路由器 IP 地址的所有数据包；当收到 shutdown 事件时，删除主机超时定时器，转为初始状态；当主机超时定时器超时的时候，发送 VRRP 通告信息，广播 ARP 地址信息，转为 Master 状态。收到 VRRP 通告信息时，如果优先值为 0，则表示进入 Master 选举；否则判断数据的优先级是否高于本机，如果高，则承认 Master 有效，复位主机超时定时器，否则丢弃该通告包。

3.2.3 VRRP 选举机制

VRRP 使用选举机制来确定路由器的状态（Master 或 Backup）。运行 VRRP 的一组路由器对外组成了一个虚拟路由器，其中一台路由器处于 Master 状态，其他的路由器处于 Backup 状态。

运行 VRRP 的路由器都会发送和接收 VRRP 通告消息，通告消息中包含了自身的 VRRP 优先级信息。VRRP 通过比较路由器的优先级进行选举，优先级高的路由器将成为主控路由器，其他路由器都为备份路由器。

虚拟路由器和 VRRP 路由器都有自己的 IP 地址（虚拟路由器的 IP 地址可以和 VRRP 备份组内的某个路由器的端口地址相同）。如果 VRRP 组中存在 IP 地址拥有者，即虚拟 IP 地址与某台

VRRP 路由器的 IP 地址相同时，IP 地址拥有者将成为主控路由器，并拥有最高优先级 255；如果 VRRP 组中不存在 IP 地址拥有者，则 VRRP 路由器将通过比较优先级来确定主控路由器。路由器可配置的优先级值为 1~254，默认情况下，VRRP 路由器的优先级为 100。当优先级相同时，VRRP 将通过比较 IP 地址来进行选举，IP 地址大的路由器将成为主控路由器。

如图 3.4 所示，路由器 AR1 的 VRRP 的优先级为 120，路由器 AR2 的 VRRP 的优先级为 100（默认优先级），因此，路由器 AR1 将成为该组的主控路由器，路由器 AR2 将成为该组的备份路由器。

V3-1 VRRP 选举

图 3.4 VRRP 选举

当路由器 AR1 与路由器 AR2 的优先级相同时，VRRP 组内路由器将比较端口的 IP 地址，端口 IP 地址大的优先成为主控路由器。假设在图 3.4 中，路由器 AR1 与路由器 AR2 都没有设置优先级，优先级都是默认值 100，则比较端口的 IP 地址，路由器 AR2 的端口 IP 地址为 192.168.1.252，而路由器 AR1 的端口 IP 地址为 192.168.1.251，因此，路由器 AR2 成为主控路由器，路由器 AR1 成为备份路由器。

1. 创建 VRRP 组，并配置虚拟 IP 地址

在端口模式下，使用如下命令创建 VRRP 组，并配置虚拟 IP 地址。

```
[Huawei-GigabitEthernet0/0/0]vrrp vrid ?
   INTEGER<1-255>   Virtual router identifier
[Huawei-GigabitEthernet0/0/0]vrrp vrid 10 virtual-ip 192.168.1.254
```

其中，vrid 为 VRRP 组的编号，取值为 1~255。属于同一个 VRRP 组的路由器必须配置相同的 vrid 才能正常工作。

virtual-ip 为 VRRP 组的虚拟 IP 地址，虚拟 IP 地址可以是该子网中未使用的地址，也可以是该路由器的端口 IP 地址，虚拟 IP 地址必须与端口 IP 地址位于同一个子网中。

默认情况下，端口没有启用 VRRP 功能，即没有 VRRP 组，使用 undo 命令，可以禁止端口上的 VRRP 功能并取消虚拟 IP 地址的设置，配置命令如下。

```
[Huawei-GigabitEthernet0/0/0]undo vrrp vrid 10 virtual-ip 192.168.1.254
```

2. 配置 VRRP 优先级

在端口模式下，使用如下命令配置 VRRP 优先级。

[Huawei-GigabitEthernet0/0/0]vrrp vrid 10 priority ?
  INTEGER<1-254>　The level of priority(default is 100)
[Huawei-GigabitEthernet0/0/0]vrrp vrid 10 priority 120

优先级的配置是基于端口和 VRRP 组的，也就是说，对于不同的端口和不同的 VRRP 组，可以分配不同的优先级。路由器可配置的优先级值为 1~254，默认情况下，VRRP 路由器的优先级为 100，数值越大，优先级越高。当虚拟地址与某台 VRRP 路由器的地址相同时，IP 地址拥有者将成为主控路由器，并拥有最高优先级 255。

3. 配置 VRRP 抢占模式

在 VRRP 运行过程中，主控路由器定期发送 VRRP 通告信息，备份路由器将监听主控路由器的通告信息，当主控路由器失效时，备份路由器将接替后者的角色。路由器在运行 VRRP 时，VRRP 组中默认采用立即抢占方式，修改主控路由器 Master 设备的延迟抢占时间，以避免在网络环境不稳定时，由于双方频繁抢占而导致通信中断的问题，配置命令如下。

[Huawei-GigabitEthernet0/0/0]vrrp vrid 10 preempt-mode timer delay ?
  INTEGER<0-3600>　Value of timer, in seconds(default is 0)
[Huawei-GigabitEthernet0/0/0]vrrp vrid 10 preempt-mode timer delay 20

配置 VRRP 抢占模式延迟时间值为 0~3600s，默认时间为 0s，即立即抢占方式。可以使用 undo 命令，禁止抢占模式延迟，配置命令如下。

[Huawei-GigabitEthernet0/0/0]undo vrrp vrid 10 preempt-mode timer delay 20

4. 端口跟踪配置

如图 3.4 所示，当路由器 AR1 的端口 GE 0/0/1 连接的链路出现故障时，路由器 AR1 仍作为主控路由器从 GE 0/0/0 接收局域网内的主机发送的报文，然而，路由器 AR1 无法对报文进行转发，使得网络无法正常通信。为了解决这个问题，可以使用 VRRP 端口跟踪机制，端口跟踪能够使 VRRP 根据路由器其他端口的状态自动调整该路由器的 VRRP 优先级，当被跟踪的端口不可用时，对应路由器 VRRP 优先级将降低，从而使得该路由器不再是主控路由器，从而使备份路由器有机会被选举为新的主控路由器，配置命令如下。

[Huawei-GigabitEthernet0/0/0]vrrp vrid 10 track interface GigabitEthernet 0/0/1 ?
  increased　　Increase priority
  reduced　　　Reduce priority
  <cr>　　　　　Please press ENTER to execute command
[Huawei-GigabitEthernet0/0/0]vrrp vrid 10 track interface GigabitEthernet 0/0/1 reduced 30

5. VRRP 定时器配置

VRRP 路由器使用通告报文来进行选举与状态监测，当选举结束后，主控路由器定期发送通告报文，备份路由器将进行监听。可以在端口配置模式下，使用命令修改 VRRP 通告报文的发送时间间隔，配置命令如下。

[Huawei-GigabitEthernet0/0/0]vrrp vrid 10 timer advertise ?
  INTEGER<1-255>　Value of timer, in seconds(default is 1)
[Huawei-GigabitEthernet0/0/0]vrrp vrid 10 timer advertise 10

发送通告报文的单位为秒，取值为 1~255，默认时间为 1s。

6. 查看 VRRP 运行状态

查看路由器 AR1 的 VRRP 运行状态。

```
<AR1>display vrrp brief
Total:1      Master:1       Backup:0       Non-active:0
VRID   State        Interface             Type      Virtual IP
----------------------------------------------------------------
10     Master       GE0/0/0               Normal    192.168.1.254
< AR1>
```

路由器 AR1 的 VRRP 组的端口优先级配置为 120，而路由器 AR2 的 VRRP 组的端口优先级配置为默认值，即 100，所以路由器 AR1 被选举为主控路由器，而路由器 AR2 被选举为备份路由器。

查看路由器 AR2 的 VRRP 运行状态。

```
<AR2>display vrrp brief
Total:1      Master:0       Backup:1       Non-active:0
VRID   State        Interface             Type      Virtual IP
----------------------------------------------------------------
10     Backup       GE0/0/0               Normal    192.168.1.254
[AR2]
```

3.3 项目实施

3.3.1 VRRP 配置

VRRP 技术是解决网络中主机配置单网关时容易出现单点故障问题的一种技术，即将多台路由器配置到一个 VRRP 组中，每一个 VRRP 组虚拟出一台虚拟路由器，作为网络中主机的网关。在一个 VRRP 组中，选举出一台优先级最高的路由器作为主控路由器，虚拟路由器的转发工作由主控路由器承担，当主控路由器因出现故障而宕机时，备份路由器将被选举为主控路由器，承担虚拟路由器的转发工作，从而保证网络的稳定性，如图 3.5 所示，进行网络拓扑连接。

V3-2 配置 VRRP 单备份组

图 3.5　配置 VRRP 单备份组

（1）配置路由器 AR1，相关配置实例代码如下。

```
<Huawei>system-view
[Huawei]sysname AR1
[AR1]interface GigabitEthernet 0/0/1
[AR1-GigabitEthernet0/0/1]ip address 192.168.11.1 24
[AR1-GigabitEthernet0/0/1]vrrp vrid   1 virtual-ip 192.168.11.254
                                                //设置虚拟网关地址
[AR1-GigabitEthernet0/0/1]vrrp vrid 1 priority 120    //设置优先级
[AR1-GigabitEthernet0/0/1]interface    GigabitEthernet0/0/0
[AR1-GigabitEthernet0/0/0]ip address 192.168.12.1 30
[AR1-GigabitEthernet0/0/0]quit
[AR1]interface LoopBack 1                //回环端口
[AR1-LoopBack1]ip address 10.10.10.1 32    //回环地址
[AR1]
```

V3-3 配置 VRRP 单备份组——测试结果

（2）显示路由器 AR1 的配置信息，主要配置实例代码如下。

```
<AR1>display current-configuration
#
 sysname AR1
#
interface GigabitEthernet0/0/0
 ip address 192.168.12.1 255.255.255.252
#
interface GigabitEthernet0/0/1
 ip address 192.168.11.1 255.255.255.0
 vrrp vrid 1 virtual-ip 192.168.11.254
 vrrp vrid 1 priority 120
#
interface LoopBack1
 ip address 10.10.10.1 255.255.255.255
#
return
<AR1>
```

（3）配置路由器 AR2，相关配置实例代码如下。

```
<Huawei>system-view
Enter system view, return user view with Ctrl+Z.
[Huawei]sysname AR2
[AR2]interface GigabitEthernet 0/0/1
[AR2-GigabitEthernet0/0/1]ip address 192.168.11.2 24
```

```
[AR2-GigabitEthernet0/0/1]vrrp vrid   1 virtual-ip 192.168.11.254
[AR2-GigabitEthernet0/0/1]interface   GigabitEthernet0/0/0
[AR2-GigabitEthernet0/0/1]ip address 192.168.13.1 30
[AR2-GigabitEthernet0/0/0]quit
[AR2]interface LoopBack 1
[AR2-LoopBack1]ip address 20.20.20.1 32
[AR2]
```

（4）显示路由器 AR2 的配置信息，主要配置实例代码如下。

```
<AR2>display current-configuration
#
 sysname AR2
#
interface GigabitEthernet0/0/0
 ip address 192.168.13.1 255.255.255.252
#
interface GigabitEthernet0/0/1
 ip address 192.168.11.2 255.255.255.0
 vrrp vrid 1 virtual-ip 192.168.11.254
#
interface GigabitEthernet0/0/2
#
interface LoopBack1
 ip address 20.20.20.1 255.255.255.255
#
return
<AR2>
```

（5）在路由器 AR1 上验证配置，使用 display vrrp brief 命令，如图 3.6 所示。

```
<AR1>display vrrp brief
Total:1     Master:1    Backup:0    Non-active:0
VRID  State       Interface                Type        Virtual IP
----------------------------------------------------------------
1     Master      GE0/0/1                  Normal      192.168.11.254
<AR1>
```

图 3.6　路由器 AR1 验证配置

（6）在路由器 AR2 上验证配置，使用 display vrrp brief 命令，如图 3.7 所示。

```
<AR2>display vrrp brief
Total:1     Master:0    Backup:1    Non-active:0
VRID  State       Interface                Type        Virtual IP
----------------------------------------------------------------
1     Backup      GE0/0/1                  Normal      192.168.11.254
<AR2>
```

图 3.7　路由器 AR2 验证配置

（7）在主机 PC1 上测试 VRRP 验证结果，如图 3.8 所示。

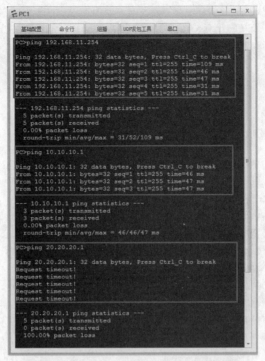

图 3.8　在主机 PC1 上测试 VRRP 验证结果

3.3.2　MSTP 与 VRRP 多备份组实现负载均衡配置

配置 MSTP 与 VRRP 多备份组，实现负载均衡。配置 VLAN 10 与 VLAN 30 属于 MSTP 实例 1，VLAN 10 与 VLAN 30 属于 VRRP 主控交换机 LSW1，VRRP 备份交换机属于 LSW2；VLAN 20 与 VLAN 40 属于 MSTP 实例 2，VLAN 20 与 VLAN 40 属于 VRRP 主控交换机 LSW2，VRRP 备份交换机属于 LSW1；交换机 LSW1 与交换机 LSW2 之间做链路聚合，增加带宽并提高可靠性，如图 3.9 所示。

图 3.9　配置 MSTP 与 VRRP 多备份组

V3-4 配置 MSTP 与 VRRP 多备份组 ——LSW1

V3-5 配置 MSTP 与 VRRP 多备份组 ——LSW2

V3-6 配置 MSTP 与 VRRP 多备份组 ——LSW3-LSW4

V3-7 配置 MSTP 与 VRRP 多备份组 ——测试结果

（1）配置交换机 LSW1，相关配置实例代码如下。

```
<Huawei>system-view
Enter system view, return user view with Ctrl+Z.
[Huawei]sysname LSW1
[LSW1]vlan batch 10 20 30 40
[LSW1]port-group group-member GigabitEthernet 0/0/1 to GigabitEthernet 0/0/2
[LSW1-port-group]port link-type trunk
[LSW1-port-group]port trunk allow-pass vlan all
[LSW1-port-group]quit
[LSW1]port-group group-member GigabitEthernet 0/0/3 to GigabitEthernet 0/0/4
[LSW1-port-group]port link-type hybrid
[LSW1-port-group]quit
[LSW1]interface Eth-Trunk 1                           //配置链路聚合
[LSW1-Eth-Trunk1]trunkport GigabitEthernet 0/0/3 to 0/0/4
[LSW1-Eth-Trunk1]port link-type hybrid
[LSW1-Eth-Trunk1]quit
[LSW1]interface Vlanif 10
[LSW1-Vlanif10]ip address 192.168.10.1 24
[LSW1-Vlanif10]vrrp vrid 10 virtual-ip 192.168.10.254 //设置虚拟网关地址
[LSW1-Vlanif10]vrrp vrid 10 priority 120              //设置优先级
[LSW1-Vlanif10]vrrp vrid 11 virtual-ip 192.168.10.253
[LSW1-Vlanif10]quit
[LSW1]interface Vlanif 20
[LSW1-Vlanif20]ip address 192.168.20.1 24
[LSW1-Vlanif20]vrrp vrid 20 virtual-ip 192.168.20.254
[LSW1-Vlanif20]vrrp vrid 20 priority 120
[LSW1-Vlanif20]vrrp vrid 21 virtual-ip 192.168.20.253
[LSW1-Vlanif20]quit
[LSW1]interface Vlanif 30
[LSW1-Vlanif30]ip address 192.168.30.1 24
[LSW1-Vlanif30]vrrp vrid 30 virtual-ip 192.168.30.254
[LSW1-Vlanif30]vrrp vrid 30 priority 120
[LSW1-Vlanif30]vrrp vrid 31 virtual-ip 192.168.30.253
[LSW1-Vlanif30]quit
[LSW1]interface Vlanif 40
[LSW1-Vlanif40]ip address 192.168.40.1 24
[LSW1-Vlanif40]vrrp vrid 40 virtual-ip 192.168.40.254
[LSW1-Vlanif40]vrrp vrid 40 priority 120
[LSW1-Vlanif40]vrrp vrid 41 virtual-ip 192.168.40.253
```

[LSW1-Vlanif40]quit
[LSW1]stp mode mstp
[LSW1]stp region-configuration
[LSW1-mst-region]region-name RG1
[LSW1-mst-region]instance 1 vlan 10 30
[LSW1-mst-region]instance 2 vlan 20 40
[LSW1-mst-region]active region-configuration
[LSW1-mst-region]quit
[LSW1]stp instance 1 priority 4096
[LSW1]stp instance 2 priority 8192
[LSW1]

（2）配置交换机 LSW2，相关配置实例代码如下。

<Huawei>system-view
Enter system view, return user view with Ctrl+Z.
[Huawei]sysname LSW2
[LSW2]vlan batch 10 20 30 40
[LSW2]port-group group-member GigabitEthernet 0/0/1 to GigabitEthernet 0/0/2
[LSW2-port-group]port link-type trunk
[LSW2-port-group]port trunk allow-pass vlan all
[LSW2-port-group]quit
[LSW2]port-group group-member GigabitEthernet 0/0/3 to GigabitEthernet 0/0/4
[LSW2-port-group]port link-type hybrid
[LSW2-port-group]quit
[LSW2]interface Eth-Trunk 1
[LSW2-Eth-Trunk1]trunkport GigabitEthernet 0/0/3 to 0/0/4
[LSW2-Eth-Trunk1]port link-type hybrid
[LSW2-Eth-Trunk1]quit
[LSW2]interface Vlanif 10
[LSW2-Vlanif10]ip address 192.168.10.2 24
[LSW2-Vlanif10]vrrp vrid 10 virtual-ip 192.168.10.254
[LSW2-Vlanif10]vrrp vrid 11 virtual-ip 192.168.10.253
[LSW2-Vlanif10]vrrp vrid 11 priority 120
[LSW2-Vlanif10]quit
[LSW2]interface Vlanif 20
[LSW2-Vlanif20]ip address 192.168.20.2 24
[LSW2-Vlanif20]vrrp vrid 20 virtual-ip 192.168.20.254
[LSW2-Vlanif20]vrrp vrid 21 virtual-ip 192.168.20.253
[LSW2-Vlanif20]vrrp vrid 21 priority 120
[LSW2-Vlanif20]quit
[LSW2]interface Vlanif 30
[LSW2-Vlanif30]ip address 192.168.30.2 24
[LSW2-Vlanif30]vrrp vrid 30 virtual-ip 192.168.30.254
[LSW2-Vlanif30]vrrp vrid 31 virtual-ip 192.168.30.253
[LSW2-Vlanif30]vrrp vrid 31 priority 120
[LSW2-Vlanif30]quit
[LSW2]interface Vlanif 40
[LSW2-Vlanif40]ip address 192.168.40.2 24
[LSW2-Vlanif40]vrrp vrid 40 virtual-ip 192.168.40.254

```
[LSW2-Vlanif40]vrrp vrid 41 virtual-ip 192.168.40.253
[LSW2-Vlanif40]vrrp vrid 41 priority 120
[LSW2-Vlanif40]quit
[LSW2]stp mode mstp
[LSW2]stp region-configuration
[LSW2-mst-region]region-name RG1
[LSW2-mst-region]instance 1 vlan 10 30
[LSW2-mst-region]instance 2 vlan 20 40
[LSW2-mst-region]active region-configuration
[LSW2-mst-region]quit
[LSW2]stp instance 1 priority 8192
[LSW2]stp instance 2 priority 4096
[LSW2]
```

（3）显示交换机 LSW1、LSW2 的配置信息，以交换机 LSW1 为例，其主要配置实例代码如下。

```
<LSW1>display current-configuration
#
sysname LSW1
#
vlan batch 10 20 30 40
#
stp instance 1 priority 4096
stp instance 2 priority 8192
#
stp region-configuration
 region-name RG1
 instance 1 vlan 10 30
 instance 2 vlan 20 40
 active region-configuration
#
interface Vlanif10
 ip address 192.168.10.1 255.255.255.0
 vrrp vrid 10 virtual-ip 192.168.10.254
 vrrp vrid 10 priority 120
 vrrp vrid 11 virtual-ip 192.168.10.253
#
interface Vlanif20
 ip address 192.168.20.1 255.255.255.0
 vrrp vrid 20 virtual-ip 192.168.20.254
 vrrp vrid 20 priority 120
 vrrp vrid 21 virtual-ip 192.168.20.253
#
interface Vlanif30
 ip address 192.168.30.1 255.255.255.0
 vrrp vrid 30 virtual-ip 192.168.30.254
 vrrp vrid 30 priority 120
 vrrp vrid 31 virtual-ip 192.168.30.253
```

```
#
interface Vlanif40
 ip address 192.168.40.1 255.255.255.0
 vrrp vrid 40 virtual-ip 192.168.40.254
 vrrp vrid 40 priority 120
 vrrp vrid 41 virtual-ip 192.168.40.253
#
interface MEth0/0/1
#
interface Eth-Trunk1
#
interface GigabitEthernet0/0/1
 port link-type trunk
 port trunk allow-pass vlan 2 to 4094
#
interface GigabitEthernet0/0/2
 port link-type trunk
 port trunk allow-pass vlan 2 to 4094
#
interface GigabitEthernet0/0/3
 eth-trunk 1
#
interface GigabitEthernet0/0/4
 eth-trunk 1
#
user-interface con 0
user-interface vty 0 4
#
<LSW1>
```

（4）配置交换机 LSW3，相关配置实例代码如下。

```
<Huawei>system-view
Enter system view, return user view with Ctrl+Z.
[Huawei]sysname LSW3
[LSW3]vlan batch 10 20 30 40
[LSW3]port-group 1
[LSW3]port-group group-member Ethernet 0/0/1 to Ethernet 0/0/2
[LSW3-port-group]port link-type trunk
[LSW3-port-group]port trunk allow-pass vlan all
[LSW3-port-group]quit
[LSW3]port-group group-member Ethernet 0/0/3 to Ethernet 0/0/6
[LSW3-port-group]port link-type access
[LSW3-port-group]quit
[LSW3]interface Ethernet 0/0/3
[LSW3-Ethernet0/0/3]port default vlan 10
[LSW3-Ethernet0/0/3]quit
[LSW3]interface Ethernet 0/0/4
[LSW3-Ethernet0/0/4]port default vlan 20
[LSW3-Ethernet0/0/4]quit
```

```
[LSW3]interface Ethernet 0/0/5
[LSW3-Ethernet0/0/5]port default vlan 30
[LSW3-Ethernet0/0/5]quit
[LSW3]interface Ethernet 0/0/6
[LSW3-Ethernet0/0/6]port default vlan 40
[LSW3-Ethernet0/0/6]quit
[LSW3]stp mode mstp
[LSW3]stp region-configuration
[LSW3-mst-region]region-name RG1
[LSW3-mst-region]instance 1 vlan 10 30
[LSW3-mst-region]instance 2 vlan 20 40
[LSW3-mst-region]active region-configuration
[LSW3-mst-region]quit
[LSW3]
```

（5）配置交换机 LSW4，相关配置实例代码如下。

```
<Huawei>system-view
Enter system view, return user view with Ctrl+Z.
[Huawei]sysname LSW4
[LSW4]vlan batch 10 20 30 40
[LSW4]port-group 1
[LSW4]port-group group-member Ethernet 0/0/1 to Ethernet 0/0/2
[LSW4-port-group]port link-type trunk
[LSW4-port-group]port trunk allow-pass vlan all
[LSW4-port-group]quit
[LSW4]port-group group-member Ethernet 0/0/3 to Ethernet 0/0/6
[LSW4-port-group]port link-type access
[LSW4-port-group]quit
[LSW4]interface Ethernet 0/0/3
[LSW4-Ethernet0/0/3]port default vlan 10
[LSW4-Ethernet0/0/3]quit
[LSW4]interface Ethernet 0/0/4
[LSW4-Ethernet0/0/4]port default vlan 20
[LSW4-Ethernet0/0/4]quit
[LSW4]interface Ethernet 0/0/5
[LSW4-Ethernet0/0/5]port default vlan 30
[LSW4-Ethernet0/0/5]quit
[LSW4]interface Ethernet 0/0/6
[LSW4-Ethernet0/0/6]port default vlan 40
[LSW4-Ethernet0/0/6]quit
[LSW4]stp mode mstp
[LSW4]stp region-configuration
[LSW4-mst-region]region-name RG1
[LSW4-mst-region]instance 1 vlan 10 30
[LSW4-mst-region]instance 2 vlan 20 40
[LSW4-mst-region]active region-configuration
[LSW4-mst-region]quit
[LSW4]
```

（6）显示交换机 LSW3、LSW4 的配置信息，以交换机 LSW3 为例，其主要配置实例代码如下。

```
<LSW3>display current-configuration
#
sysname LSW3
#
vlan batch 10 20 30 40
#
stp region-configuration
 region-name RG1
 instance 1 vlan 10 30
 instance 2 vlan 20 40
 active region-configuration
#
interface Ethernet0/0/1
 port link-type trunk
 port trunk allow-pass vlan 2 to 4094
#
interface Ethernet0/0/2
 port link-type trunk
 port trunk allow-pass vlan 2 to 4094
#
interface Ethernet0/0/3
 port link-type access
 port default vlan 10
#
interface Ethernet0/0/4
 port link-type access
 port default vlan 20
#
interface Ethernet0/0/5
 port link-type access
 port default vlan 30
#
interface Ethernet0/0/6
 port link-type access
 port default vlan 40
#
port-group 1
#
return
<LSW3>
```

（7）设置主机 PC1 与主机 PC6 的 IP 地址，如图 3.10 所示。其他主机 IP 地址的设置这里不再赘述。

（8）测试主机 PC1 的验证结果，如图 3.11 所示。

（9）测试主机 PC6 的验证结果，如图 3.12 所示。

图 3.10 设置主机 PC1 与主机 PC6 的 IP 地址

图 3.11 测试主机 PC1 的验证结果

图 3.12 测试主机 PC6 的验证结果

（10）显示交换机 LSW1 的配置信息，使用 display eth-trunk 1 命令，显示链路聚合状态，如图 3.13 所示。

（11）显示交换机 LSW2 的配置信息，使用 display eth-trunk 1 命令，显示链路聚合状态，如图 3.14 所示。

图 3.13 交换机 LSW1 的链路聚合状态　　　　图 3.14 交换机 LSW2 的链路聚合状态

（12）显示交换机 LSW1 的配置信息，使用 display stp instance 1 brief 命令，显示 MSTP 实例 1 和实例 2 的运行状态，如图 3.15 所示。

（13）显示交换机 LSW2 的配置信息，使用 display stp instance 1 brief 命令，显示 MSTP 实例 1 和实例 2 的运行状态，如图 3.16 所示。

图 3.15　交换机 LSW1 显示 MSTP 实例 1 和实例 2 的运行状态

图 3.16　交换机 LSW2 显示 MSTP 实例 1 和实例 2 的运行状态

（14）显示交换机 LSW1 的配置信息，使用 display vrrp brief 命令，显示 VRRP 的运行状态，如图 3.17 所示。

（15）显示交换机 LSW2 的配置信息，使用 display vrrp brief 命令，显示 VRRP 的运行状态，如图 3.18 所示。

图 3.17　交换机 LSW1 显示 VRRP 的运行状态

图 3.18　交换机 LSW2 显示 VRRP 的运行状态

课后习题

1．选择题

（1）一个 VRRP 虚拟路由器配置 VRID 为 5，虚拟 IP 地址为 200.1.1.10，那么其虚拟 MAC 地址是（　　）。

　　A．00-00-5E-00-01-10　　　　B．00-00-5E-00-01-05

　　C．10-00-5E-00-01-10　　　　D．10-00-5E-00-01-05

（2）默认情况下，VRRP 路由器的优先级为（　　）。

　　A．0　　　　B．1　　　　C．100　　　　D．255

（3）若网络环境不稳定，启动 VRRP 功能后，则（　　）能防止主备路由器频繁切换。

　　A．配置 VRRP 实体间的认证　　　　B．修改 VRRP 发送通告报文时间间隔

　　C．修改 VRRP 的 track 值　　　　　D．启用备份组抢占模式

2．简答题

（1）简述 VRRP 的主要作用及应用场景。

（2）如何配置 MSTP 与 VRRP 多备份组以实现负载均衡？

项目4
RIP与高级配置

【学习目标】

- 掌握RIP路由聚合功能及配置方法。
- 掌握RIP验证功能及配置方法。
- 掌握RIP与BFD联动配置方法。

4.1 项目陈述

由于企业规模较小,某企业网络采用 RIP 进行配置。随着企业规模的不断扩大,企业网络的子网数量也增加,导致网络运行状态不够稳定,对企业业务造成了一定的影响。小张是一名刚入职的网络工程师,领导安排小张对企业网络进行优化,考虑到企业网络的安全性与稳定性,需要对企业网络进行认证使用,同时要动态检测网络的运行状况,并对企业未来的网络扩展做好准备。小张根据企业的要求制定了一份合理的网络优化方案,本项目将对小张的优化方案进行介绍。

4.2 必备知识

1. RIP 概述

路由信息协议(Routing Information Protocol,RIP)是一种内部网关协议(Interior Gateway Protocol,IGP),也是一种动态路由选择协议,用于自治系统(Autonomous System,AS)内的路由信息的传递。RIP 基于距离矢量算法(Distance Vector Algorithms),使用"跳数"(即 Metric)来衡量到达目的地址的路由距离。运行这种协议的路由器只与自己相邻的路由器交换信息,范围限制在 15 跳之内,即 16 跳就认为网络不可达。

RIP 应用于 OSI 参考模型的应用层,各厂家定义的管理距离(Administrative Distance,AD,即优先级)有所不同。例如,华为设备定义的优先级是 100,而思科设备定义的优先级是 120,在带宽、配置和管理方面要求较低,主要适用于规模较小的网络,如图 4.1 所示。RIP 中定义的相关参数比较少,它不支持可变长子网掩码(Variable Length Subnet Mask,VLSM)和无类别域间路由(Classless Inter-Domain Routing,CIDR),也不支持认证功能。

图 4.1　运行 RIP 的网络

2. RIP 的工作原理

路由器启动时，路由表中只会包含直连路由。运行 RIP 之后，路由器会发送 Request 报文，请求邻居路由器的 RIP 路由。运行 RIP 的邻居路由器收到该 Request 报文后，会根据自己的路由表，生成 Response 报文进行回复。路由器在收到 Response 报文后，会将相应的路由添加到自己的路由表中。

RIP 网络稳定以后，每个路由器会周期性地向邻居路由器发送 Response 报文，通告自己的整张路由表中的路由信息，默认周期为 30s，邻居路由器根据收到的路由信息更新自己的路由表，如图 4.2 所示。针对某一条路由信息，如果 180s 以后仍然没有接收到关于它的更新信息，那么将其标记为失效，即将 Metric 值标记为 16；在另外的 120s 以后，如果仍然没有更新信息，则该失效信息被删除。

图 4.2　更新 RIP 路由表

3. RIP 的版本

RIP 分为 3 个版本：RIPv1、RIPv2 和 RIPng，前两者用于 IPv4，RIPng 用于 IPv6。

（1）RIPv1 为有类别路由协议，不支持 VLSM 和 CIDR；RIPv1 以广播形式发送路由信息，目的 IP 地址为广播地址 255.255.255.255；不支持认证；RIPv1 通过用户数据报协议（User Datagram Protocol，UDP）交换路由信息，端口号为 520。

一个 RIPv1 路由更新消息中最多可包含 25 条路由表项，每个路由表项都携带了目的网络的地址和度量值。一个 RIP 报文体积限制为不超过 504 字节，如果整张路由表的更新消息超过该大小，则需要发送多个 RIPv1 报文。

（2）RIPv2 为无类别路由协议，支持 VLSM，支持路由聚合与 CIDR；支持以广播或组播（224.0.0.9）方式发送报文；支持明文认证和 MD5 密文认证。RIPv2 在 RIPv1 的基础上进行了扩展，但 RIPv2 的报文格式仍然同 RIPv1 类似。

RIPv1 的提出年代较早，其中有许多缺陷。为了改善 RIPv1 的不足，IETF 在 RFC1388 中提出了改进的 RIPv2，并在 RFC1723 和 RFC2453 中进行了修订。

随着 OSPF 协议和 IS-IS 协议（在项目 5 和项目 6 中进行介绍）的出现，许多人认为 RIP 已经过时了。但 RIP 仍有难以替代的优势，对于小型网络而言，RIP 所占带宽开销小，易于配置、管理和实现,仍有大量网络在运行 RIP。但 RIP 也有明显的不足,即当有多个网络时会出现环路问题。为了解决环路问题,IETF 提出了分割范围的方法,即路由器不可以通过自身端口获得路由信息,再通过本端口去通告已获得的路由信息。分割范围解决了两个路由器之间的路由环路问题,但不能阻止多个路由器形成路由环路。触发更新是解决环路问题的另一方法,它要求路由器在链路发生变化时立即广播其路由表。这加速了网络的聚合,但容易产生广播泛滥。总之,解决环路问题需要消耗一定的时间和带宽。若采用 RIP,其网络内部所经过的链路数不能超过 15,这使得 RIP 不适用于大型网络。

(3)下一代 RIP(RIP next generation,RIPng)是为 IPv6 网络设计的下一代距离矢量路由协议。与 IPv4 版本的 RIP 类似,RIPng 同样基于距离矢量算法。RIPng 保留了 RIP 的多个主要特性,例如,RIPng 规定每一跳的开销度量值为 1,最大跳数为 15,RIPng 通过 UDP 的 521 端口发送和接收路由信息。

RIPng 与 RIP 的最主要的区别在于,RIPng 使用 IPv6 组播地址 ff02::9 作为目的地址来传送路由更新报文,而 RIPv2 使用的是组播地址 224.0.0.9;IPv4 路由协议一般采用公网地址或私网地址作为路由条目的下一跳地址,而 IPv6 路由协议通常采用链路本地地址作为路由条目的下一跳地址。

4. RIP 的局限性

(1)只适用于小型网络。

协议中规定,一条有效的路由信息的 Metric 不能超过 15,使得 RIP 不能应用于大型的网络。不过正是由于设计者考虑到该协议只适用于小型网络,所以才进行了这一限制。

(2)收敛速度慢。

RIP 在实际应用中很容易出现"计数到无穷大"的现象,这使得路由收敛很慢,在网络拓扑结构变化以后,路由信息需要很长时间才能稳定下来。

(3)根据跳数选择的路由不一定是最优路由。

RIP 以跳数,即报文经过的路由器个数为衡量标准,并以此来选择路由。这一规定没有考虑网络时延、可靠性、线路负载等因素对传输质量和速度的影响,欠缺合理性。

5. RIPv1 与 RIPv2 的区别

RIPv1 使用广播方式更新路由。RIPv2 使用组播的方式向其他路由器发送更新报文,它使用的组播地址是保留的 D 类 IP 地址 224.0.0.9,使用组播方式的优点在于,对于本地网络上相连的 RIP 路由器,不需花费时间解析它们的更新广播报文。RIPv2 不是一个新的协议,它只是在 RIPv1 的基础上增加了一些扩展特性,以适用于现代网络的路由选择环境。RIPv2 的扩展特性有:每个路由条目都携带自己的子网掩码;路由选择更新更具有认证功能;每个路由条目都携带下一跳地址;具有外部路由标志;组播路由更新。其中最重要的一项是路由更新条目增加了子网掩码的字段,因而 RIPv2 可以支持 VLSM,从而使 RIPv2 协议变成了一个无类别的路由选择协议。

(1)RIPv1 是有类路由协议;RIPv2 是无类路由协议。

(2)RIPv1 不支持 VLSM;RIPv2 支持 VLSM。

(3)RIPv1 没有认证的功能;RIPv2 支持认证功能,并有明文和 MD5 两种认证方式。

(4)RIPv1 没有手动聚合的功能;RIPv2 可以在关闭自动聚合功能的前提下,进行手动聚合。

(5)RIPv1 是广播更新;RIPv2 是组播更新。

(6)RIPv1 路由没有标记的功能;RIPv2 可以对路由封装标记,用于过滤和做策略。

(7)RIPv1 的更新报文中最多可以携带 25 条路由条目;RIPv2 的更新报文在有认证的情况下最多只能携带 24 条路由条目。

（8）RIPv1 发送的更新报文中不携带下一跳地址；RIPv2 发送的更新报文中携带下一跳地址，可以用于路由更新的重定。

6. RIP 的基本配置

（1）配置 RIP 路由进程，发布路由进程关联的网络，配置命令如下。

```
[Huawei]rip ?
    INTEGER<1-65535>    Process ID
    mib-binding         Mib-Binding a process
    vpn-instance        VPN instance
    <cr>                Please press ENTER to execute command
[Huawei-rip-1]network 192.168.10.0
[Huawei-rip-1]network 192.168.20.0
```

RIP 进程号的值为 1~65535，默认情况下为 1。

（2）配置 RIP 接收和发送指定版本的数据包，配置命令如下。

```
[Huawei-rip-1]version ?
    INTEGER<1-2>    Version of RIP process
[Huawei-rip-1]version 2
```

（3）查看当前模式下的相关配置，命令如下。

```
[Huawei-rip-1]display this
[V200R003C00]
#
rip 1
 version 2
 network 192.168.10.0
 network 192.168.20.0
#
return
[Huawei-rip-1]
```

4.3 项目实施

4.3.1 RIP 路由聚合配置

当 RIP 网络规模很大时，RIP 路由表会变得十分庞大，存储路由表会占用大量的设备内存资源，传输和处理路由信息需要占用大量的网络资源，对网络传输速度和质量造成影响。使用路由聚合可以大大减小路由表的规模，此外，通过对路由进行聚合，隐藏一些具体的路由，也可以减少路由振荡对网络带来的影响。

RIP 支持两种聚合方式：自动聚合和手动聚合。自动聚合的路由优先级低于手动聚合的路由优先级，当需要将所有子网路由都发布出去时，可关闭 RIPv2 的自动聚合功能。

1. 配置 RIPv2 自动聚合的步骤

（1）使用 system-view 命令，进入系统视图。
（2）使用 rip [process-id]命令，进入 RIP 视图。
（3）使用 version 2 命令，设置 RIP 版本为 RIPv2。
（4）使用 summary 命令，使能 RIPv2 自动聚合功能。

（5）（可选）使用 summary always 命令，不论是否使能水平分割和毒性反转，都可以使能 RIPv2 自动聚合。

在 RIP 视图下使用 summary 命令进行路由聚合，即使能 RIPv2 基于有类别网络的路由聚合。

2. 配置 RIPv2 手动聚合的步骤

（1）使用 system-view 命令，进入系统视图。

（2）使用 interface interface-type interface-number 命令，进入端口视图。

（3）使用 rip summary-address ip-address mask [avoid-feedback]命令，配置 RIPv2 发布聚合的本地 IP 地址。

在端口视图下使用 rip summary-address ip-address mask [avoid-feedback]命令进行路由聚合，即使能 RIPv2 基于无类别网络的路由聚合。

3. 配置 RIP 路由聚合

（1）配置 RIP 手动聚合，如图 4.3 所示，在路由器 AR1 的 GE 0/0/0 端口上进行手动聚合，将 200.16.10.0/24 与 200.16.20.0/24 聚合成一条 200.16.0.0/16 路由并通告给路由器 AR2；在路由器 AR2 的 GE 0/0/0 端口上进行手动聚合，将 192.168.10.0/24 与 192.168.20.0/24 聚合成一条 192.168.0.0/16 路由并通告给路由器 AR1。

图 4.3 配置 RIP 手动路由聚合

V4-1 配置 RIP 手动聚合

（2）配置路由器 AR1，实例代码如下。

```
<Huawei>system-view
Enter system view, return user view with Ctrl+Z.
[Huawei]sysname AR1
[AR1]interface GigabitEthernet 0/0/0
[AR1-GigabitEthernet0/0/0]ip address 192.168.100.1 30
[AR1-GigabitEthernet0/0/0]rip version 2
[AR1-GigabitEthernet0/0/0]rip summary-address 200.16.0.0 255.255.0.0
                                //聚合成一条 200.16.0.0/16 路由并通告给路由器 AR2
[AR1-GigabitEthernet0/0/0]quit
[AR1]interface LoopBack 0
[AR1-LoopBack0]ip address 200.16.10.1 24
[AR1-LoopBack0]quit
[AR1]interface LoopBack 1
[AR1-LoopBack1]ip address 200.16.20.1 24
```

V4-2 配置 RIP 手动聚合——测试结果

```
[AR1-LoopBack1]quit
[AR1]rip
[AR1-rip-1]version 2
[AR1-rip-1]network 192.168.100.0
[AR1-rip-1]network 200.16.10.0
[AR1-rip-1]network 200.16.20.0
[AR1-rip-1]undo summary              //取消自动聚合功能
[AR1-rip-1]quit
[AR1]
```

（3）配置路由器 AR2，实例代码如下。

```
<Huawei>system-view
Enter system view, return user view with Ctrl+Z.
[Huawei]sysname AR2
[AR2]interface GigabitEthernet 0/0/0
[AR2-GigabitEthernet0/0/0]ip address 192.168.100.2 30
[AR2-GigabitEthernet0/0/0]rip version 2
[AR2-GigabitEthernet0/0/0]rip summary-address 192.168.0.0 255.255.0.0
                                //聚合成一条 192.168.0.0/16 路由并通告给路由器 AR1
[AR2-GigabitEthernet0/0/0]quit
[AR2]interface LoopBack 0
[AR2-LoopBack0]ip address 192.168.10.1 24
[AR2-LoopBack0]quit
[AR2]interface LoopBack 1
[AR2-LoopBack1]ip address 192.168.20.1 24
[AR2-LoopBack1]quit
[AR2]rip
[AR2-rip-1]version 2
[AR2-rip-1]network 192.168.100.0
[AR2-rip-1]network 192.168.10.0
[AR2-rip-1]network 192.168.20.0
[AR2-rip-1]undo summary              //取消自动聚合功能
[AR2-rip-1]quit
[AR2]
```

（4）显示路由器 AR1 的配置信息，主要配置实例代码如下。

```
<AR1>display current-configuration
#
 sysname AR1
#
interface GigabitEthernet0/0/0
 ip address 192.168.100.1 255.255.255.252
 rip version 2 multicast
 rip summary-address 200.16.0.0 255.255.0.0
#
interface LoopBack0
 ip address 200.16.10.1 255.255.255.0
#
interface LoopBack1
```

```
  ip address 200.16.20.1 255.255.255.0
#
rip 1
 undo summary
 version 2
 network 192.168.100.0
 network 200.16.10.0
 network 200.16.20.0
#
return
<AR1>
```

（5）显示路由器 AR2 的配置信息，主要配置实例代码如下。

```
<AR2>display current-configuration
#
 sysname AR2
#
interface GigabitEthernet0/0/0
 ip address 192.168.100.2 255.255.255.252
 rip version 2 multicast
 rip summary-address 192.168.0.0 255.255.0.0
#
interface LoopBack0
 ip address 192.168.10.1 255.255.255.0
#
interface LoopBack1
 ip address 192.168.20.1 255.255.255.0
#
rip 1
 undo summary
 version 2
 network 192.168.100.0
 network 192.168.10.0
 network 192.168.20.0
#
return
<AR2>
```

（6）显示路由器 AR1 的路由表信息，代码如下。

```
<AR1>display ip routing-table
Route Flags: R - relay, D - download to fib
------------------------------------------------------------
Routing Tables: Public
         Destinations : 14       Routes : 14

Destination/Mask        Proto   Pre  Cost      Flags  NextHop         Interface
        127.0.0.0/8     Direct  0    0         D      127.0.0.1       InLoopBack0
        127.0.0.1/32    Direct  0    0         D      127.0.0.1       InLoopBack0
```

127.255.255.255/32	Direct	0	0	D	127.0.0.1	InLoopBack0
192.168.0.0/16	RIP	100	1	D	192.168.100.2	GigabitEthernet 0/0/0
192.168.100.0/30	Direct	0	0	D	192.168.100.1	GigabitEthernet 0/0/0
192.168.100.1/32	Direct	0	0	D	127.0.0.1	GigabitEthernet 0/0/0
192.168.100.3/32	Direct	0	0	D	127.0.0.1	GigabitEthernet 0/0/0
200.16.10.0/24	Direct	0	0	D	200.16.10.1	LoopBack0
200.16.10.1/32	Direct	0	0	D	127.0.0.1	LoopBack0
200.16.10.255/32	Direct	0	0	D	127.0.0.1	LoopBack0
200.16.20.0/24	Direct	0	0	D	200.16.20.1	LoopBack1
200.16.20.1/32	Direct	0	0	D	127.0.0.1	LoopBack1
200.16.20.255/32	Direct	0	0	D	127.0.0.1	LoopBack1
255.255.255.255/32	Direct	0	0	D	127.0.0.1	InLoopBack0

\<AR1\>

（7）显示路由器 AR2 的路由表信息，代码如下。

```
<AR2>display ip routing-table
Route Flags: R - relay, D - download to fib
-----------------------------------------------------------------------
Routing Tables: Public
         Destinations : 14      Routes : 14
```

Destination/Mask	Proto	Pre	Cost	Flags	NextHop	Interface
127.0.0.0/8	Direct	0	0	D	127.0.0.1	InLoopBack0
127.0.0.1/32	Direct	0	0	D	127.0.0.1	InLoopBack0
127.255.255.255/32	Direct	0	0	D	127.0.0.1	InLoopBack0
192.168.10.0/24	Direct	0	0	D	192.168.10.1	LoopBack0
192.168.10.1/32	Direct	0	0	D	127.0.0.1	LoopBack0
192.168.10.255/32	Direct	0	0	D	127.0.0.1	LoopBack0
192.168.20.0/24	Direct	0	0	D	192.168.20.1	LoopBack1
192.168.20.1/32	Direct	0	0	D	127.0.0.1	LoopBack1
192.168.20.255/32	Direct	0	0	D	127.0.0.1	LoopBack1
192.168.100.0/30	Direct	0	0	D	192.168.100.2	GigabitEthernet 0/0/0
192.168.100.2/32	Direct	0	0	D	127.0.0.1	GigabitEthernet 0/0/0
192.168.100.3/32	Direct	0	0	D	127.0.0.1	GigabitEthernet 0/0/0
200.16.0.0/16	RIP	100	1	D	192.168.100.1	GigabitEthernet 0/0/0
255.255.255.255/32	Direct	0	0	D	127.0.0.1	InLoopBack0

\<AR2\>

（8）RIP 手动聚合连通性测试，如图 4.4 所示，在路由器 AR1 上访问路由器 AR2 的 192.168.0.0/16 网段地址。

```
<AR1>ping 192.168.10.1
  PING 192.168.10.1: 56  data bytes, press CTRL_C to break
    Reply from 192.168.10.1: bytes=56 Sequence=1 ttl=255 time=70 ms
    Reply from 192.168.10.1: bytes=56 Sequence=2 ttl=255 time=10 ms
    Reply from 192.168.10.1: bytes=56 Sequence=3 ttl=255 time=20 ms
    Reply from 192.168.10.1: bytes=56 Sequence=4 ttl=255 time=20 ms
    Reply from 192.168.10.1: bytes=56 Sequence=5 ttl=255 time=30 ms

  --- 192.168.10.1 ping statistics ---
    5 packet(s) transmitted
    5 packet(s) received
    0.00% packet loss
    round-trip min/avg/max = 10/30/70 ms

<AR1>ping 192.168.20.1
  PING 192.168.20.1: 56  data bytes, press CTRL_C to break
    Reply from 192.168.20.1: bytes=56 Sequence=1 ttl=255 time=20 ms
    Reply from 192.168.20.1: bytes=56 Sequence=2 ttl=255 time=10 ms
    Reply from 192.168.20.1: bytes=56 Sequence=3 ttl=255 time=10 ms
    Reply from 192.168.20.1: bytes=56 Sequence=4 ttl=255 time=20 ms
    Reply from 192.168.20.1: bytes=56 Sequence=5 ttl=255 time=10 ms

  --- 192.168.20.1 ping statistics ---
    5 packet(s) transmitted
    5 packet(s) received
    0.00% packet loss
    round-trip min/avg/max = 10/16/20 ms

<AR1>
```

图 4.4　RIP 手动聚合连通性测试

4.3.2　RIPv2 报文的认证方式配置

在安全性要求较高的网络中,可以通过配置 RIPv2 报文的认证来提高 RIP 网络的安全性。RIPv2 支持对协议报文进行认证,并提供了多种认证方式,增强了安全性。其中,简单认证使用未加密的认证字段随报文一同传送,其安全性比 MD5 认证低。

在配置 RIPv2 报文的认证方式时,如果使用 plain 选项,则密码将以明文形式保存在配置文件中,存在安全隐患;建议使用 cipher 选项,将密码加密保存。

1. 配置 RIPv2 报文的认证方式的步骤

（1）使用 system-view 命令,进入系统视图。

（2）使用 interface interface-type interface-number 命令,进入端口视图。

（3）配置 RIPv2 报文的认证方式。

① 使用 rip authentication-mode simple { plain plain-text | [cipher] password-key}命令,配置 RIPv2 报文为简单认证方式。

② 使用以下命令,配置 RIPv2 报文为 MD5 密文认证方式。

rip authentication-mode md5 usual { plain plain-text | [cipher] password-key }
rip authentication-mode md5 nonstandard { keychain keychain-name | { plain plain-text | [cipher] password-key } key-id }

如果配置 MD5 认证,则必须配置 MD5 的类型。usual 类型表示 MD5 密文认证报文使用通用报文格式（私有标准）,nonstandard 类型表示 MD5 密文认证报文使用非标准报文格式（IETF 标准）。

简单认证和 MD5 认证存在安全风险,推荐配置 HMAC-SHA256 密文认证方式。

使用 rip authentication-mode hmac-sha256 { plain plain-text | [cipher]password-key } key-id 命令,可配置 RIPv2 报文为 HMAC-SHA256 密文认证方式。

2. 配置 RIPv2 报文的认证方式

（1）配置 RIPv2 报文的认证方式,配置相关端口与 IP 地址,如图 4.5 所示,进行网络拓扑连接。

图 4.5 配置 RIPv2 报文的认证方式

（2）配置主机 PC1 和主机 PC3 的 IP 地址，如图 4.6 所示。

图 4.6 配置主机 PC1 和主机 PC3 的 IP 地址

（3）配置路由器 AR1，相关实例代码如下。

```
<Huawei>system-view
Enter system view, return user view with Ctrl+Z.
[Huawei]sysname AR1
[AR1]interface GigabitEthernet 0/0/0
[AR1-GigabitEthernet0/0/0]ip address 192.168.100.1 30
[AR1-GigabitEthernet0/0/0]rip version 2
[AR1-GigabitEthernet0/0/0]rip authentication-mode hmac-sha256 cipher 123456 1
[AR1-GigabitEthernet0/0/0]quit
[AR1]interface GigabitEthernet 0/0/1
[AR1-GigabitEthernet0/0/1]ip address 200.16.10.254 24
[AR1]interface GigabitEthernet 0/0/2
[AR1-GigabitEthernet0/0/2]ip address 200.16.20.254 24
[AR1]rip
[AR1-rip-1]version 2
[AR1-rip-1]network 192.168.100.0
[AR1-rip-1]network 200.16.10.0
[AR1-rip-1]network 200.16.20.0
```

[AR1-rip-1]undo summary   //取消自动聚合功能
[AR1-rip-1]quit
[AR1]

（4）配置路由器 AR2，相关实例代码如下。

```
<Huawei>system-view
Enter system view, return user view with Ctrl+Z.
[Huawei]sysname AR2
[AR2]interface GigabitEthernet 0/0/0
[AR2-GigabitEthernet0/0/0]ip address 192.168.100.2 30
[AR2-GigabitEthernet0/0/0]rip version 2
[AR2-GigabitEthernet0/0/0]rip authentication-mode hmac-sha256 cipher 123456 1
[AR2-GigabitEthernet0/0/0]quit
[AR2]interface GigabitEthernet 0/0/1
[AR2-GigabitEthernet0/0/1]ip address 192.168.10.254 24
[AR2]interface GigabitEthernet 0/0/2
[AR2-GigabitEthernet0/0/2]ip address 192.168.20.254 24
[AR2]rip
[AR2-rip-1]version 2
[AR2-rip-1]network 192.168.100.0
[AR2-rip-1]network 192.168.10.0
[AR2-rip-1]network 192.168.20.0
[AR2-rip-1]undo summary          //取消自动聚合功能
[AR2-rip-1]quit
[AR2]
```

（5）显示路由器 AR1、路由器 AR2 的配置信息，以路由器 AR1 为例，主要配置实例代码如下。

```
<AR1>display current-configuration
#
 sysname AR1
#
interface GigabitEthernet0/0/0
 ip address 192.168.100.1 255.255.255.252
 rip authentication-mode hmac-sha256 cipher %$%$g`:%W9%n*&DYV>>_"Q4/8=2Y%$%$ 1
 rip version 2 multicast
#
interface GigabitEthernet0/0/1
 ip address 200.16.10.254 255.255.255.0
#
interface GigabitEthernet0/0/2
 ip address 200.16.20.254 255.255.255.0
#
rip 1
 undo summary
 version 2
 network 192.168.100.0
```

```
    network 200.16.10.0
    network 200.16.20.0
#
return
<AR1>
```

（6）显示路由器 AR1、路由器 AR2 的路由表信息，这里以路由器 AR1 为例进行介绍。

```
<AR1>display ip routing-table
Route Flags: R - relay, D - download to fib
------------------------------------------------------------------------
Routing Tables: Public
         Destinations : 15      Routes : 15

Destination/Mask        Proto    Pre  Cost    Flags   NextHop          Interface

       127.0.0.0/8      Direct   0    0       D       127.0.0.1        InLoopBack0
       127.0.0.1/32     Direct   0    0       D       127.0.0.1        InLoopBack0
 127.255.255.255/32     Direct   0    0       D       127.0.0.1        InLoopBack0
    192.168.10.0/24     RIP      100  1       D       192.168.100.2    GigabitEthernet
0/0/0
    192.168.20.0/24     RIP      100  1       D       192.168.100.2    GigabitEthernet
0/0/0
    192.168.100.0/30    Direct   0    0       D       192.168.100.1    GigabitEthernet
0/0/0
    192.168.100.1/32    Direct   0    0       D       127.0.0.1        GigabitEthernet
0/0/0
    192.168.100.3/32    Direct   0    0       D       127.0.0.1        GigabitEthernet
0/0/0
     200.16.10.0/24     Direct   0    0       D       200.16.10.254    GigabitEthernet
0/0/1
    200.16.10.254/32    Direct   0    0       D       127.0.0.1        GigabitEthernet
0/0/1
    200.16.10.255/32    Direct   0    0       D       127.0.0.1        GigabitEthernet
0/0/1
     200.16.20.0/24     Direct   0    0       D       200.16.20.254    GigabitEthernet
0/0/2
    200.16.20.254/32    Direct   0    0       D       127.0.0.1        GigabitEthernet
0/0/2
    200.16.20.255/32    Direct   0    0       D       127.0.0.1        GigabitEthernet
0/0/2
 255.255.255.255/32     Direct   0    0       D       127.0.0.1        InLoopBack0
<AR1>
```

（7）测试主机 PC1 的连通性，主机 PC1 分别访问主机 PC3 和主机 PC4，测试结果如图 4.7 所示。

图 4.7 主机 PC1 分别访问 PC3 和 PC4 的测试结果

4.3.3 RIP 与 BFD 联动配置

网络上的链路故障会导致路由器重新计算路由，造成流量丢失，因此缩短路由协议的收敛时间对于提高网络性能是非常重要的。加快故障感知速度并快速通告给路由协议是一种可行的方案。

双向转发检测（Bidirectional Forwarding Detection，BFD）是一种用于检测邻居路由器之间链路故障的机制，它通常与路由协议联动，通过快速感知并通告链路故障使得路由协议能够快速地重新收敛，从而减少由于拓扑变化导致的流量丢失。在 RIP 与 BFD 的联动中，BFD 可以快速检测到链路故障并通知 RIP，从而加快 RIP 对于网络拓扑变化的响应速度。在配置 BFD 前后，RIP 的链路故障检测机制及收敛速度如表 4.1 所示。

表 4.1 RIP 的链路故障检测机制及收敛速度

是否配置 BFD	链路故障检测机制	收敛速度
否	RIP 老化定时器超时（默认配置为 180s）	秒级（>180s）
是	BFD 会话状态为 Down	秒级（<30s）

1. 工作原理

路由协议与 BFD 的联动分为静态 BFD 和动态 BFD 两种模式。

（1）静态 BFD

静态 BFD 是指通过命令行手工配置 BFD 会话参数，包括配置本地标识符和远端标识符等，并手工下发 BFD 会话建立请求。

（2）动态 BFD

动态 BFD 是指由路由协议动态触发 BFD 会话，建立 BFD 联动。在动态 BFD 中，本地标识符是动态分配的，远端标识符从对端的 BFD 报文中获取。路由协议在建立了新的邻居关系时，将对应的参数及检测参数（包括目的地址、源地址等）通告给 BFD，BFD 根据收到的参数建立会话。

当发生链路故障时，联动了 BFD 的路由协议可以快速感知到 BFD 会话状态变为 Down，从而将流量快速切换到备份路径，避免了数据的大量丢失。

静态 BFD 和动态 BFD 在不同的应用场景下各有优势。静态 BFD 可以不受对端设备的限制，在对端设备不支持 BFD 功能的情况下，本端通过静态 BFD 实现单臂 BFD 功能；而动态 BFD 比静态 BFD 的灵活性更好。

2. 应用环境

RIP 与 BFD 联动后，一旦链路发生故障，BFD 会在毫秒级时间内感知该故障并通知 RIP，则路由器在路由表中删除故障链路的路由并快速启用备份路径，提高了路由协议的收敛速度，如图 4.8 所示。

图 4.8 RIP 与 BFD 联动应用环境

（1）路由器 AR1、路由器 AR2、路由器 AR3 及路由器 AR4 建立 RIP 邻接。经过路由计算，路由器 AR1 到达路由器 AR4 的路由下一跳为路由器 AR2。在路由器 AR1 及路由器 AR2 上使能 RIP 与动态 BFD 联动。

（2）当路由器 AR1 和路由器 AR2 之间的链路出现故障时，BFD 快速感知并通告给路由器 AR1，路由器 AR1 删除下一跳为路由器 AR2 的路由。路由器 AR1 重新计算路由并选取新的路径，新的路由经过路由器 AR3、路由器 AR2 到达路由器 AR4。

（3）当路由器 AR1 与路由器 AR2 之间的链路恢复之后，二者之间的会话重新建立，路由器 AR1 收到路由器 AR2 的路由信息，重新计算路由并进行报文转发。

3. 配置 RIP 与静态 BFD 联动

BFD 能够提供轻负载、快速的链路故障检测，配置 RIP 与静态 BFD 联动是实现 BFD 功能的一种方式。在 RIP 邻居间建立 BFD 会话可以快速检测链路故障，加快 RIP 进程对网络拓扑变化的响应速度。静态 BFD 可以实现以下两种功能。

（1）单臂 BFD：当网络中存在大量设备不支持 BFD 功能，而支持 BFD 的设备需要与不支持 BFD 的设备对接时，可以通过配置静态 BFD 来实现单臂 BFD 功能。

（2）普通 BFD：在某些对故障响应速度要求高且两端设备都支持 BFD 的链路上，可以通过在两端配置静态 BFD 来实现普通 BFD 功能。

建立静态 BFD 会话需要通过命令行手工配置。

4. 配置静态 BFD 的步骤

（1）使能全局 BFD。

① 使用 system-view 命令，进入系统视图。

② 使用 bfd 命令，使能全局 BFD 功能。

③ 使用 quit 命令，返回系统视图。

注意 使能单臂 BFD 功能，请执行（2）；使能普通 BFD 功能，请执行（3）。

（2）配置单臂 BFD。

① 使用 bfd session-name bind peer-ip peer-ip interface interface-type interfacenumber [source-ip source-ip] one-arm-echo 命令，创建 BFD 会话。

这里指定了对端 IP 地址和本端端口，表示检测单跳链路，即检测以该端口为出端口、以 peer-ip 为下一跳地址的一条固定路由。

注意 在配置单臂 ECHO 功能时，必须配置源端口 IP 地址，源端口 IP 地址为合法的 IP 地址，建议配置的源端口 IP 地址与出端口 IP 地址不同。

② 使用 discriminator local discr-value 命令，配置本地标识符。

③（可选）使用 min-echo-rx-interval interval 命令，配置单臂 BFD 的最小接收间隔。

④ 使用 commit 命令，提交配置。

⑤ 使用 quit 命令，返回系统视图。

（3）配置普通 BFD。

① 使用 bfd session-name bind peer-ip ip-address [interface interface-type interface-number] [source-ip ip-address]命令，创建 BFD 会话。

这里指定了对端 IP 地址和本端端口，表示检测单跳链路，即检测以该端口为出端口、以 peer-ip 为下一跳地址的一条固定路由。

② 配置标识符。

使用 discriminator local discr-value 命令，配置本地标识符。

使用 discriminator remote discr-value 命令，配置远端标识符。

BFD 会话两端设备的本地标识符和远端标识符需要分别对应，即本端的本地标识符和对端的远端标识符相同，本端的远端标识符与对端的本地标识符相同，否则会话无法正确建立。此外，本地标识符和远端标识符配置成功后不可修改。

注意 本地设备的本地标识符 local discr-value 对应对端设备的远端标识符 remote discr-value，本地设备的远端标识符 remote discr-value 对应对端设备的本地标识符 local discr-value。

③ 使用 commit 命令，提交配置。

④ 使用 quit 命令，返回系统视图。

（4）使能端口静态 BFD。

① 使用 interface interface-type interface-number 命令，进入指定端口的端口视图。

② 使用 rip bfd static 命令，使能端口的静态 BFD 特性。

③ 使用 quit 命令，返回系统视图。

配置完成 RIP 与静态 BFD 联动之后，使用 display rip process-id interface [interface-type interface-number] verbose 命令，查看指定端口上 RIP 与 BFD 联动的配置信息。

5. 配置 RIP 与单臂静态 BFD 联动特性实例

如图 4.9 所示,在小型网络中有 4 台路由器通过 RIP 实现网络互通。其中,业务流量经过主链路路由器 AR1→路由器 AR2→路由器 AR4 进行传输。要求提高从路由器 AR1 到路由器 AR2 数据转发的可靠性,当主链路发生故障时,业务流量会快速切换到备份路径路由器 AR1→路由器 AR3→路由器 AR2→路由器 AR4 进行传输,配置相关端口与 IP 地址,进行网络拓扑连接。

图 4.9 配置 RIP 与单臂静态 BFD 联动特性实例

(1)配置路由器 AR1,相关实例代码如下。

```
<Huawei>system-view
Enter system view, return user view with Ctrl+Z.
[Huawei]sysname AR1
[AR1]interface GigabitEthernet 0/0/0
[AR1-GigabitEthernet0/0/0]ip address 192.168.2.1 24
[AR1-GigabitEthernet0/0/0]quit
[AR1]interface GigabitEthernet 0/0/1
[AR1-GigabitEthernet0/0/1]ip address 192.168.1.1 24
[AR1-GigabitEthernet0/0/1]quit
[AR1]rip
[AR1-rip-1]version 2
[AR1-rip-1]network 192.168.1.0
[AR1-rip-1]network 192.168.2.0
[AR1-rip-1]quit
[AR1]
```

(2)配置路由器 AR2,相关实例代码如下。

```
<Huawei>system-view
Enter system view, return user view with Ctrl+Z.
[Huawei]sysname AR2
[AR2]interface GigabitEthernet 0/0/0
[AR2-GigabitEthernet0/0/0]ip address 192.168.2.2 24
[AR2-GigabitEthernet0/0/0]quit
[AR2]interface GigabitEthernet 0/0/1
[AR2-GigabitEthernet0/0/1]ip address 192.168.4.1 24
[AR2-GigabitEthernet0/0/1]quit
[AR2]interface GigabitEthernet 0/0/2
[AR2-GigabitEthernet0/0/2]ip address 192.168.3.2 24
```

```
[AR2-GigabitEthernet0/0/2]quit
[AR2]rip
[AR2-rip-1]version 2
[AR2-rip-1]network 192.168.2.0
[AR2-rip-1]network 192.168.3.0
[AR2-rip-1]network 192.168.4.0
[AR2-rip-1]quit
[AR2]
```

（3）配置路由器 AR3，相关实例代码如下。

```
<Huawei>system-view
Enter system view, return user view with Ctrl+Z.
[Huawei]sysname AR3
[AR3]interface GigabitEthernet 0/0/1
[AR3-GigabitEthernet0/0/1]ip address 192.168.1.2 24
[AR3-GigabitEthernet0/0/1]quit
[AR3]interface GigabitEthernet 0/0/2
[AR3-GigabitEthernet0/0/2]ip address 192.168.3.1 24
[AR3-GigabitEthernet0/0/2]quit
[AR3]rip
[AR3-rip-1]version 2
[AR3-rip-1]network 192.168.1.0
[AR3-rip-1]network 192.168.3.0
[AR3-rip-1]quit
[AR3]
```

（4）配置路由器 AR4，相关实例代码如下。

```
<Huawei>system-view
Enter system view, return user view with Ctrl+Z.
[Huawei]sysname AR4
[AR4]interface GigabitEthernet 0/0/1
[AR4-GigabitEthernet0/0/1]ip address 192.168.4.2 24
[AR4-GigabitEthernet0/0/1]quit
[AR4]rip
[AR4-rip-1]version 2
[AR4-rip-1]network 192.168.4.0
[AR4-rip-1]quit
[AR4]
```

（5）查看路由器 AR1、路由器 AR2、路由器 AR3 及路由器 AR4 之间已经建立的邻居关系，这里以路由器 AR1 为例进行介绍。

```
<AR1>display rip 1 neighbor
----------------------------------------------------------------
  IP Address      Interface              Type   Last-Heard-Time
----------------------------------------------------------------
  192.168.1.2     GigabitEthernet0/0/1   RIP    0:0:15
  Number of RIP routes   : 1
  192.168.2.2     GigabitEthernet0/0/0   RIP    0:0:16
  Number of RIP routes   : 2
<AR1>
```

（6）查看完成配置的路由器之间互相引入的路由信息，这里以路由器 AR1 为例进行介绍。

```
<AR1>display ip routing-table
Route Flags: R - relay, D - download to fib
------------------------------------------------------------------
Routing Tables: Public
         Destinations : 12      Routes : 13

Destination/Mask        Proto    Pre  Cost    Flags   NextHop         Interface

        127.0.0.0/8     Direct   0    0       D       127.0.0.1       InLoopBack0
        127.0.0.1/32    Direct   0    0       D       127.0.0.1       InLoopBack0
  127.255.255.255/32    Direct   0    0       D       27.0.0.1        InLoopBack0
      192.168.1.0/24    Direct   0    0       D       192.168.1.1     GigabitEthernet
0/0/1
      192.168.1.1/32    Direct   0    0       D       127.0.0.1       GigabitEthernet
0/0/1
    192.168.1.255/32    Direct   0    0       D       127.0.0.1       GigabitEthernet
0/0/1
      192.168.2.0/24    Direct   0    0       D       192.168.2.1     GigabitEthernet
0/0/0
      192.168.2.1/32    Direct   0    0       D       127.0.0.1       GigabitEthernet
0/0/0
    192.168.2.255/32    Direct   0    0       D       127.0.0.1       GigabitEthernet
0/0/0
      192.168.3.0/24    RIP      100  1       D       192.168.1.2     GigabitEthernet
0/0/1
                        RIP      100  1       D       192.168.2.2     GigabitEthernet
0/0/0
      192.168.4.0/24    RIP      100  1       D       192.168.2.2     GigabitEthernet
0/0/0
  255.255.255.255/32    Direct   0    0       D       127.0.0.1       InLoopBack0
<AR1>
```

（7）配置路由器 AR1 的单臂静态 BFD 特性，相关实例代码如下。

```
[AR1]bfd           //使能 BFD 特性检测功能
[AR1-bfd]quit
[AR1] bfd 1 bind peer-ip 192.168.2.2 interface GigabitEthernet0/0/0 one-arm-echo
//配置单臂回声功能的 BFD 会话
[AR1-bfd-session-1]discriminator local ?
    INTEGER<1-8191>   Discriminator value      //取值为 1～8191
[AR1-bfd-session-1]discriminator local 1       //配置本地标识符
[AR1-bfd-session-1]min-echo-rx-interval ?
    INTEGER<10-2000>   Minimum receive interval in milliseconds, default is 1000ms
    //单臂 BFD 的最小接收间隔取值为 10～2000，默认值为 1000ms
[AR1-bfd-session-1]min-echo-rx-interval 300    //配置单臂 BFD 的最小接收间隔为 300ms
[AR1-bfd-session-1]commit
[AR1-bfd-session-1]quit
[AR1]interface GigabitEthernet 0/0/0           //使能 GE 0/0/0 端口的静态 BFD 功能
```

```
[AR1-GigabitEthernet0/0/0]rip bfd static
[AR1-GigabitEthernet0/0/0]quit
[AR1]
```

（8）显示路由器 AR1 的配置信息，主要配置实例代码如下。

```
<AR1>display current-configuration
#
 sysname AR1
#
 bfd
#
interface GigabitEthernet0/0/0
 ip address 192.168.2.1 255.255.255.0
 rip bfd static
#
interface GigabitEthernet0/0/1
 ip address 192.168.1.1 255.255.255.0
#
 bfd 1 bind peer-ip 192.168.2.2 interface GigabitEthernet0/0/0 one-arm-echo
  discriminator local 1
  min-echo-rx-interval 300
  commit
#
rip 1
 version 2
 network 192.168.1.0
 network 192.168.2.0
#
return
<AR1>
```

（9）完成上述配置之后，在路由器 AR1 上使用 display bfd session all 命令，可查看到静态 BFD 会话已经建立。

```
<AR1>display bfd session all
--------------------------------------------------------------------------------
Local Remote    PeerIpAddr      State    Type      InterfaceName
--------------------------------------------------------------------------------
1      -        192.168.2.2     Up       S_IP_IF   GigabitEthernet0/0/0
--------------------------------------------------------------------------------
     Total UP/DOWN Session Number : 1/0
<AR1>
```

（10）验证配置结果，在路由器 AR2 的 GE 0/0/0 端口上使用 shutdown 命令，模拟主链路出现故障。模拟链路出现故障为验证需要，在实际应用中不需要执行此操作。

```
[AR2]interface GigabitEthernet 0/0/0
[AR2-GigabitEthernet0/0/0]shutdown
[AR2-GigabitEthernet0/0/0]quit
[AR2]
```

（11）查看路由器 AR1 的路由表。

```
<AR1>display ip routing-table
```

```
Route Flags: R - relay, D - download to fib
------------------------------------------------------------------------
Routing Tables: Public
         Destinations : 9       Routes : 9

Destination/Mask        Proto     Pre    Cost     Flags     NextHop       Interface

      127.0.0.0/8       Direct    0      0        D         127.0.0.1     InLoopBack0
      127.0.0.1/32      Direct    0      0        D         127.0.0.1     InLoopBack0
127.255.255.255/32      Direct    0      0        D         127.0.0.1     InLoopBack0
    192.168.1.0/24      Direct    0      0        D         192.168.1.1   GigabitEthernet0/0/1
    192.168.1.1/32      Direct    0      0        D         127.0.0.1     GigabitEthernet0/0/1
  192.168.1.255/32      Direct    0      0        D         127.0.0.1     GigabitEthernet0/0/1
    192.168.3.0/24      RIP       100    1        D         192.168.1.2   GigabitEthernet0/0/1
    192.168.4.0/24      RIP       100    2        D         192.168.1.2   GigabitEthernet0/0/1
255.255.255.255/32      Direct    0      0        D         127.0.0.1     InLoopBack0
<AR1>
```

由路由器 AR1 的路由表可以看出，在主链路发生故障之后，备份链路路由器 AR1→路由器 AR3→路由器 AR2 被启用，去往 192.168.4.0/24 的路由下一跳地址是 192.168.1.2，出端口为 GE 0/0/1。

6. 配置 RIP 与动态 BFD 联动

通常情况下，RIP 通过定时接收和发送更新报文来保持邻居关系，若在老化定时器时间内没有收到邻居发送的更新报文，则宣告邻居状态变为 Down。老化定时器的默认值为 180s，如果出现链路故障，则 RIP 要经过至少 180s 才会检测到。如果网络中部署了高速数据业务，则在此期间将造成数据大量丢失。

BFD 能够提供毫秒级的故障检测机制，及时检测到被保护的链路或节点故障并通知 RIP，提高 RIP 进程对网络拓扑变化的响应速度，从而实现 RIP 的快速收敛。

配置 RIP 与动态 BFD 联动有两种方式。

（1）在 RIP 进程下使能 BFD 功能，当网络中大部分 RIP 端口需要使能 RIP 与动态 BFD 联动时，建议选择此方式。

（2）在 RIP 端口下使能 BFD 功能，当网络中只有小部分 RIP 端口需要使能 RIP 与动态 BFD 联动时，建议选择此方式。

7. 配置动态 BFD 的步骤

在 RIP 进程下使能 BFD 的步骤如下。

（1）使用 system-view 命令，进入系统视图。

（2）使用 bfd 命令，使能全局 BFD 功能。

（3）使用 quit 命令，返回系统视图。

（4）使用 rip [process-id] 命令，进入 RIP 视图。

（5）使用 bfd all-interfaces enable 命令，使能 RIP 进程的 BFD 特性，建立 BFD 会话。

当配置了全局 BFD 特性，且邻居状态为 Up 时，RIP 为该进程下所有满足上述条件的端口使用默认的 BFD 参数值建立 BFD 会话。

（6）（可选）使用 bfd all-interfaces { min-rx-interval min-receive-value | mintx-interval min-transmit-value | detect-multiplier detect-multiplier-value } *命令，配置 BFD 参数，指定用于建立 BFD 会话的各个参数值。使用该命令后，所有 RIP 端口建立 BFD 会话的参数都会改变。参数的具体配置取决于网络状况及对网络可靠性的要求。

① 对于网络可靠性要求较高的链路，可以减小 BFD 报文实际发送时间间隔。

② 对于网络可靠性要求较低的链路，可以增大 BFD 报文实际发送时间间隔。

（7）（可选）执行以下步骤阻塞 RIP 进程下某些端口创建 BFD 会话的功能。

① 使用 quit 命令，返回系统视图。

② 使用 interface interface-type interface-number 命令，进入指定端口的端口视图。

③ 使用 rip bfd block 命令，阻塞端口创建 BFD 会话的功能。

在 RIP 端口上使能 BFD 的步骤如下。

（1）使用 system-view 命令，进入系统视图。

（2）使用 bfd 命令，使能全局 BFD 功能。

（3）使用 quit 命令，返回系统视图。

（4）使用 interface interface-type interface-number 命令，进入指定端口的端口视图。

（5）使用 rip bfd enable 命令，使能端口的 BFD 特性，建立 BFD 会话。

（6）（可选）使用 rip bfd { min-rx-interval min-receive-value | min-tx-interval min-transmit-value | detect-multiplier detect-multiplier-value } *命令，配置 BFD 参数，指定用于建立 BFD 会话的各个参数值。

当链路两端均配置 RIP 与 BFD 联动后，使用 display rip process-id bfd session { interface interface-type interface-number| neighbor-id | all }命令，可以看到本地路由器上 BFDState 字段显示为 Up。

8. 配置 RIP 与动态 BFD 联动特性实例

如图 4.10 所示，在小型网络中有 4 台路由器通过 RIP 实现网络互通。采用配置 RIP 与动态 BFD 联动的方式，在各端口上配置 IP 地址，使网络可达，在各路由器上使能 RIP，基本实现网络互联，在路由器 AR1 和路由器 AR2 上配置 RIP 与动态 BFD 联动，通过 BFD 快速检测链路的状态，从而提高 RIP 的收敛速度，实现链路的快速切换。其中，业务流量经过主链路路由器 AR1→路由器 AR2→路由器 AR4 进行传输。要求提高从路由器 AR1 到路由器 AR2 数据转发的可靠

图 4.10　配置 RIP 与动态 BFD 联动特性实例

性，当主链路发生故障时，业务流量会快速切换到备份路径路由器 AR1→路由器 AR3→ 路由器 AR2→路由器 AR4 进行传输，配置相关端口与 IP 地址，进行网络拓扑连接。

（1）配置路由器 AR1，相关实例代码如下。

```
<Huawei>system-view
Enter system view, return user view with Ctrl+Z.
[Huawei]sysname AR1
[AR1]interface GigabitEthernet 0/0/0
[AR1-GigabitEthernet0/0/0]ip address 192.168.2.1 24
[AR1-GigabitEthernet0/0/0]quit
[AR1]interface GigabitEthernet 0/0/1
[AR1-GigabitEthernet0/0/1]ip address 192.168.1.1 24
[AR1-GigabitEthernet0/0/1]quit
[AR1]rip
[AR1-rip-1]version 2
[AR1-rip-1]network 192.168.1.0
[AR1-rip-1]network 192.168.2.0
[AR1-rip-1]quit
[AR1]
```

V4-8 配置 RIP 与动态 BFD 联动——测试结果

（2）配置路由器 AR2，相关实例代码如下。

```
<Huawei>system-view
Enter system view, return user view with Ctrl+Z.
[Huawei]sysname AR2
[AR2]interface GigabitEthernet 0/0/0
[AR2-GigabitEthernet0/0/0]ip address 192.168.2.2 24
[AR2-GigabitEthernet0/0/0]quit
[AR2]interface GigabitEthernet 0/0/1
[AR2-GigabitEthernet0/0/1]ip address 192.168.4.1 24
[AR2-GigabitEthernet0/0/1]quit
[AR2]interface GigabitEthernet 0/0/2
[AR2-GigabitEthernet0/0/2]ip address 192.168.3.2 24
[AR2-GigabitEthernet0/0/2]quit
[AR2]rip
[AR2-rip-1]version 2
[AR2-rip-1]network 192.168.2.0
[AR2-rip-1]network 192.168.3.0
[AR2-rip-1]network 192.168.4.0
[AR2-rip-1]quit
[AR2]
```

（3）配置路由器 AR3，相关实例代码如下。

```
<Huawei>system-view
Enter system view, return user view with Ctrl+Z.
[Huawei]sysname AR3
[AR3]interface GigabitEthernet 0/0/1
[AR3-GigabitEthernet0/0/1]ip address 192.168.1.2 24
[AR3-GigabitEthernet0/0/1]quit
[AR3]interface GigabitEthernet 0/0/2
[AR3-GigabitEthernet0/0/2]ip address 192.168.3.1 24
```

```
[AR3-GigabitEthernet0/0/2]quit
[AR3]rip
[AR3-rip-1]version 2
[AR3-rip-1]network 192.168.1.0
[AR3-rip-1]network 192.168.3.0
[AR3-rip-1]quit
[AR3]
```

（4）配置路由器 AR4，相关实例代码如下。

```
<Huawei>system-view
Enter system view, return user view with Ctrl+Z.
[Huawei]sysname AR4
[AR4]interface GigabitEthernet 0/0/1
[AR4-GigabitEthernet0/0/1]ip address 192.168.4.2 24
[AR4-GigabitEthernet0/0/1]quit
[AR4]rip
[AR4-rip-1]version 2
[AR4-rip-1]network 192.168.4.0
[AR4-rip-1]quit
[AR4]
```

（5）查看路由器 AR1、路由器 AR2、路由器 AR3 及路由器 AR4 之间已经建立的邻居关系，这里以路由器 AR1 为例进行介绍。

```
<AR1>display rip 1 neighbor
----------------------------------------------------------------
 IP Address        Interface              Type    Last-Heard-Time
----------------------------------------------------------------
 192.168.1.2       GigabitEthernet0/0/1   RIP     0:0:10
 Number of RIP routes   : 1
 192.168.2.2       GigabitEthernet0/0/0   RIP     0:0:26
 Number of RIP routes   : 2
<AR1>
```

（6）查看完成配置的路由器之间互相引入的路由信息，这里以路由器 AR1 为例进行介绍。

```
<AR1>display ip routing-table
Route Flags: R - relay, D - download to fib
----------------------------------------------------------------
Routing Tables: Public
         Destinations : 12       Routes : 13
Destination/Mask        Proto    Pre  Cost     Flags   NextHop         Interface
       127.0.0.0/8      Direct   0    0        D       127.0.0.1       InLoopBack0
       127.0.0.1/32     Direct   0    0        D       127.0.0.1       InLoopBack0
 127.255.255.255/32     Direct   0    0        D       127.0.0.1       InLoopBack0
     192.168.1.0/24     Direct   0    0        D       192.168.1.1     GigabitEthernet
0/0/1
     192.168.1.1/32     Direct   0    0        D       127.0.0.1       GigabitEthernet
0/0/1
   192.168.1.255/32     Direct   0    0        D       127.0.0.1       GigabitEthernet
0/0/1
     192.168.2.0/24     Direct   0    0        D       192.168.2.1     GigabitEthernet
```

	0/0/0						
	192.168.2.1/32	Direct	0	0	D	127.0.0.1	GigabitEthernet
	0/0/0						
	192.168.2.255/32	Direct	0	0	D	127.0.0.1	GigabitEthernet
	0/0/0						
	192.168.3.0/24	RIP	100	1	D	192.168.1.2	GigabitEthernet
	0/0/1						
		RIP	100	1	D	192.168.2.2	GigabitEthernet
	0/0/0						
	192.168.4.0/24	RIP	100	1	D	192.168.2.2	GigabitEthernet
	0/0/0						
	255.255.255.255/32	Direct	0	0	D	127.0.0.1	InLoopBack0

<AR1>

（7）配置路由器 AR1 的动态 BFD 特性，相关实例代码如下。

```
[AR1]bfd           //使能 BFD 特性检测功能
[AR1-bfd]quit
[AR1]rip 1
[AR1-rip-1]bfd all-interfaces enable
[AR1-rip-1]bfd all-interfaces min-rx-interval ?
   INTEGER<10-2000>   The minimum receive interval (milliseconds)
   //配置接收单跳 BFD 控制报文的最小时间间隔，其取值为 10～2000
[AR1-rip-1]bfd all-interfaces min-rx-interval 200 min-tx-interval ?
   INTEGER<10-2000>   The minimum transmit interval (milliseconds)
   //配置发送单跳 BFD 控制报文的最小时间间隔，其取值为 10～2000
[AR1-rip-1]bfd all-interfaces min-rx-interval 200 min-tx-interval 200 detect-multiplier ?
   INTEGER<3-50>   The detect multiplier value
   //配置单跳 BFD 检测时间倍数，其取值为 3～50
[AR1-rip-1]bfd all-interfaces min-rx-interval 200 min-tx-interval 200 detect-multiplier 10
[AR1-rip-1]quit
[AR1]
```

路由器 AR2 的配置与路由器 AR1 相同，这里不再赘述。

（8）完成上述配置之后，在路由器上使用 display rip <1-65535> bfd session all 命令可以看到路由器 AR1 与路由器 AR2 之间已经建立 BFD 会话，BFDState 字段显示为 Up。

```
<AR1>display rip 1 bfd session all
 LocalIp         :192.168.1.1     RemoteIp :192.168.1.2     BFDState  :Down
 TX              :10000           RX       :10000           Multiplier:0
 BFD Local Dis :8192              Interface :GigabitEthernet0/0/1
 Diagnostic Info:No diagnostic information

 LocalIp         :192.168.2.1     RemoteIp :192.168.2.2     BFDState  :Up
 TX              :200             RX       :200             Multiplier:10
 BFD Local Dis :8193              Interface :GigabitEthernet0/0/0
 Diagnostic Info:No diagnostic information
<AR1>
```

（9）验证配置结果，在路由器 AR2 的 GE 0/0/0 端口上使用 shutdown 命令，模拟主链路出现故障。模拟链路出现故障为验证需要，在实际应用中不需要执行此操作。

```
[AR2]interface GigabitEthernet 0/0/0
[AR2-GigabitEthernet0/0/0]shutdown
[AR2-GigabitEthernet0/0/0]quit
[AR2]
```

（10）查看路由器 AR1 的路由表。

```
<AR1>display ip routing-table
Route Flags: R - relay, D - download to fib
-----------------------------------------------------------------------------
Routing Tables: Public
         Destinations : 9        Routes : 9
Destination/Mask         Proto    Pre  Cost     Flags   NextHop         Interface
      127.0.0.0/8        Direct   0    0        D       127.0.0.1       InLoopBack0
      127.0.0.1/32       Direct   0    0        D       127.0.0.1       InLoopBack0
127.255.255.255/32       Direct   0    0        D       127.0.0.1       InLoopBack0
     192.168.1.0/24      Direct   0    0        D       192.168.1.1     GigabitEthernet
0/0/1
     192.168.1.1/32      Direct   0    0        D       127.0.0.1       GigabitEthernet
0/0/1
   192.168.1.255/32      Direct   0    0        D       127.0.0.1       GigabitEthernet
0/0/1
     192.168.3.0/24      RIP      100  1        D       192.168.1.2     GigabitEthernet
0/0/1
     192.168.4.0/24      RIP      100  2        D       192.168.1.2     GigabitEthernet
0/0/1
 255.255.255.255/32      Direct   0    0        D       127.0.0.1       InLoopBack0
<AR1>
```

由路由器 AR1 的路由表可以看出，在主链路发生故障之后，备份链路路由器 AR1→路由器 AR3→路由器 AR2 被启用，去往 192.168.4.0/24 的路由下一跳地址是 192.168.1.2，出端口为 GE 0/0/1。

（11）显示路由器 AR1 的配置信息，主要配置实例代码如下。

```
<AR1>display current-configuration
#
 sysname AR1
#
bfd
#
interface GigabitEthernet0/0/0
 ip address 192.168.2.1 255.255.255.0
#
interface GigabitEthernet0/0/1
 ip address 192.168.1.1 255.255.255.0
#
rip 1
 version 2
 network 192.168.1.0
 network 192.168.2.0
 bfd all-interfaces enable
 bfd all-interfaces min-tx-interval 200 min-rx-interval 200 detect-multiplier 10
```

#
return
<AR1>

4.3.4 RIP 引入外部路由配置

从配置和管理的角度来看，通常使用一种路由协议更容易操作。但是在实际应用中，很多情况下，一台路由器上会同时运行一种以上的路由协议。

在路由器上运行除 RIP 之外的路由协议的情况下，为了使不同路由协议之间良好地协同工作，需要在区域边界路由器（Area Border Router，ABR）上引入外部路由，RIP 可以引入其他进程或其他协议学习到的路由信息，从而丰富路由表项。

1. 配置 RIP 引入外部路由的步骤

（1）使用 system-view 命令，进入系统视图。

（2）使用 rip [process-id] 命令，进入 RIP 视图。

（3）（可选）使用 default-cost cost 命令，设定引入路由的度量值。如果在引入路由时没有指定度量值，则使用默认度量值。

（4）使用 import-route bgp [permit-ibgp] [cost { cost | transparent } | route-policyroute-policy-name] *命令或 import-route { { static | direct | unr } | { { rip | ospf | isis }[process-id] } } [cost cost | route-policy route-policy-name] *命令，引入外部路由信息。

注意　RIP 进程引入 IBGP 路由时容易造成路由环路，配置该功能前应谨慎。

（5）（可选）使用 filter-policy { acl-number | acl-name acl-name | ip-prefix ip- prefixname} export [protocol [process-id] | interface-type interface-number]命令，在引入的路由信息向外发布时进行过滤。

由于 RIP 要发布的路由信息中，有可能包含引入的其他路由协议的路由信息，所以可通指定 protocol 参数来对这些特定的路由信息进行过滤。如果没有指定 protocol 参数，则要对所要发布的路由信息进行过滤，包括引入的路由和本地 RIP 路由（相当于直连路由）。

注意　RIPv2 规定的 Tag 字段长度为 16bit，其他路由协议的 Tag 字段长度为 32bit。如果在引入其他路由协议时，应用的路由策略中使用了 Tag，则应确保 Tag 值不超过 65535，否则将导致路由策略失效或者产生错误的匹配结果。

2. 配置 RIP 引入外部路由实例

如图 4.11 所示，路由器 AR2 上运行着两个 RIP 进程：RIP 10 和 RIP 20。要求路由器 AR1 与其他网段能够互通，配置相关端口与 IP 地址，进行网络拓扑连接。

配置 RIP 引入外部路由，在各路由器上使能 RIP，实现各进程内的网络互联；在路由器 AR2 上配置 RIP 10 和 RIP 20 之间的路由相互引入，其中将引入的 RIP 20 路由的度量值设为 2，实现两个进程路由的互通；在路由器 AR2 上配置访问控制列表（Access Control List，ACL），对引入的 RIP 20 的一条路由（192.168.6.0/24）进行过滤，实现路由器 AR1 仅与网段 192.168.5.0/24 互通。

V4-9 配置 RIP 引入外部路由

图 4.11　配置 RIP 引入外部路由

（1）配置路由器 AR1，相关实例代码如下。

```
<Huawei>system-view
Enter system view, return user view with Ctrl+Z.
[Huawei]sysname AR1
[AR1]interface GigabitEthernet 0/0/0
[AR1-GigabitEthernet0/0/0]ip address 192.168.1.1 24
[AR1-GigabitEthernet0/0/0]quit
[AR1]interface GigabitEthernet 0/0/1
[AR1-GigabitEthernet0/0/1]ip address 192.168.3.1 24
[AR1-GigabitEthernet0/0/1]quit
[AR1]interface GigabitEthernet 0/0/2
[AR1-GigabitEthernet0/0/2]ip address 192.168.4.1 24
[AR1-GigabitEthernet0/0/2]quit
[AR1]rip 10
[AR1-rip-10]network 192.168.1.0
[AR1-rip-10]network 192.168.3.0
[AR1-rip-10]network 192.168.4.0
[AR1-rip-10]quit
[AR1]
```

（2）配置路由器 AR2，相关实例代码如下。

```
<Huawei>system-view
Enter system view, return user view with Ctrl+Z.
[Huawei]sysname AR2
[AR2]interface GigabitEthernet 0/0/0
[AR2-GigabitEthernet0/0/0]ip address 192.168.1.2 24
[AR2-GigabitEthernet0/0/0]quit
[AR2]interface GigabitEthernet 0/0/1
[AR2-GigabitEthernet0/0/1]ip address 192.168.2.1 24
[AR2-GigabitEthernet0/0/1]quit
[AR2]rip 10
```

[AR2-rip-10]network 192.168.1.0
[AR2-rip-10]quit
[AR2]rip 20
[AR2-rip-20]network 192.168.2.0
[AR2-rip-20]quit
[AR2]

（3）配置路由器 AR3，相关实例代码如下。

<Huawei>system-view
Enter system view, return user view with Ctrl+Z.
[Huawei]sysname AR3
[AR3]interface GigabitEthernet 0/0/0
[AR3-GigabitEthernet0/0/0]ip address 192.168.5.1 24
[AR3-GigabitEthernet0/0/0]quit
[AR3]interface GigabitEthernet 0/0/1
[AR3-GigabitEthernet0/0/1]ip address 192.168.2.2 24
[AR3-GigabitEthernet0/0/1]quit
[AR3]interface GigabitEthernet 0/0/2
[AR3-GigabitEthernet0/0/2]ip address 192.168.6.1 24
[AR3-GigabitEthernet0/0/2]quit
[AR3]rip 20
[AR3-rip-20]network 192.168.2.0
[AR3-rip-20]network 192.168.5.0
[AR3-rip-20]network 192.168.6.0
[AR3-rip-20]quit
[AR3]

路由器 AR4、路由器 AR5、路由器 AR6 及路由器 AR7 的配置与路由器 AR1、路由器 AR2、路由器 AR3 大致相同，这里不再赘述。

（4）查看路由器 AR1 的路由表。

<AR1>display ip routing-table
Route Flags: R - relay, D - download to fib

Routing Tables: Public
         Destinations : 13   Routes : 13
Destination/Mask   Proto   Pre Cost   Flags   NextHop   Interface
       127.0.0.0/8   Direct   0   0   D   127.0.0.1   InLoopBack0
       127.0.0.1/32   Direct   0   0   D   127.0.0.1   InLoopBack0
127.255.255.255/32   Direct   0   0   D   127.0.0.1   InLoopBack0
     192.168.1.0/24   Direct   0   0   D   192.168.1.1   GigabitEthernet 0/0/0
     192.168.1.1/32   Direct   0   0   D   127.0.0.1   GigabitEthernet 0/0/0
   192.168.1.255/32   Direct   0   0   D   127.0.0.1   GigabitEthernet 0/0/0
     192.168.3.0/24   Direct   0   0   D   192.168.3.1   GigabitEthernet 0/0/1
     192.168.3.1/32   Direct   0   0   D   127.0.0.1   GigabitEthernet 0/0/1

192.168.3.255/32	Direct	0	0	D	127.0.0.1	GigabitEthernet 0/0/1
192.168.4.0/24	Direct	0	0	D	192.168.4.1	GigabitEthernet 0/0/2
192.168.4.1/32	Direct	0	0	D	127.0.0.1	GigabitEthernet 0/0/2
192.168.4.255/32	Direct	0	0	D	127.0.0.1	GigabitEthernet 0/0/2
255.255.255.255/32	Direct	0	0	D	127.0.0.1	InLoopBack0

\<AR1\>

其他路由器的路由信息大致与路由器 AR1 的路由信息相同，从以上路由表中可以看出，只有本身路由器的直连路由信息，没有其他路由信息。

（5）配置 RIP 引入外部路由，在路由器 AR2 上设置默认路由值为 2，并将两个不同 RIP 进程的路由相互引入到对方的路由表中。

```
[AR2]rip 10
[AR2-rip-10]default-cost ?
  INTEGER<0-15>   Value of metric
//配置默认路由值为 2，其取值为 0～15
[AR2-rip-10]default-cost 2
[AR2-rip-10]import-route ?
  bgp       Border Gateway Protocol (BGP) routes
  direct    Direct routes
  isis      Intermediate System to Intermediate System (ISIS) routes
  ospf      Open Shortest Path First (OSPF) routes
  rip       Routing Information Protocol (RIP) routes
  static    Static routes
  unr       User Network Route
          //可以配置不同路由引入协议，如 BGP、OSPF 协议、IS-IS 协议、RIP 等
[AR2-rip-10]import-route rip 20
          //将 RIP 进程为 20 的路由引入到 RIP 进程为 10 的路由表中
[AR2-rip-10]quit
[AR2]rip 20
[AR2-rip-20]default-cost 2
[AR2-rip-20]import-route rip 10
          //将 RIP 进程为 10 的路由引入到 RIP 进程为 20 的路由表中
[AR2-rip-20]quit
[AR2]
```

（6）配置完成后，查看路由器 AR1 的路由表。

```
<AR1>display ip routing-table
Route Flags: R - relay, D - download to fib
------------------------------------------------------------------
Routing Tables: Public
         Destinations : 16        Routes : 16
```

Destination/Mask	Proto	Pre	Cost	Flags	NextHop	Interface
127.0.0.0/8	Direct	0	0	D	127.0.0.1	InLoopBack0
127.0.0.1/32	Direct	0	0	D	127.0.0.1	InLoopBack0
127.255.255.255/32	Direct	0	0	D	127.0.0.1	InLoopBack0

Destination/Mask	Proto	Pre	Cost	Flags	NextHop	Interface
192.168.1.0/24	Direct	0	0	D	192.168.1.1	GigabitEthernet 0/0/0
192.168.1.1/32	Direct	0	0	D	127.0.0.1	GigabitEthernet 0/0/0
192.168.1.255/32	Direct	0	0	D	127.0.0.1	GigabitEthernet 0/0/0
192.168.2.0/24	RIP	100	3	D	192.168.1.2	GigabitEthernet 0/0/0
192.168.3.0/24	Direct	0	0	D	192.168.3.1	GigabitEthernet 0/0/1
192.168.3.1/32	Direct	0	0	D	127.0.0.1	GigabitEthernet 0/0/1
192.168.3.255/32	Direct	0	0	D	127.0.0.1	GigabitEthernet 0/0/1
192.168.4.0/24	Direct	0	0	D	192.168.4.1	GigabitEthernet 0/0/2
192.168.4.1/32	Direct	0	0	D	127.0.0.1	GigabitEthernet 0/0/2
192.168.4.255/32	Direct	0	0	D	127.0.0.1	GigabitEthernet 0/0/2
192.168.5.0/24	RIP	100	3	D	192.168.1.2	GigabitEthernet 0/0/0
192.168.6.0/24	RIP	100	3	D	192.168.1.2	GigabitEthernet 0/0/0
255.255.255.255/32	Direct	0	0	D	127.0.0.1	InLoopBack0

<AR1>

可以看出，路由器 AR1 的路由表学习到了网段 192.168.5.0/24 与 192.168.6.0/24 的路由信息。

（7）配置 RIP 对引入路由的过滤，在路由器 AR2 上配置 ACL，并增加一条规则：拒绝源地址为 192.168.6.0/24 的数据包。

```
[AR2]acl 2000
[AR2-acl-basic-2000]rule deny source 192.168.6.0 0.0.0.255
[AR2-acl-basic-2000]rule permit
[AR2-acl-basic-2000]quit
[AR2]rip 10
[AR2-rip-10]filter-policy 2000 export
[AR2-rip-10]quit
[AR2]
```

（8）验证配置结果，查看过滤后路由器 AR1 的路由表。

```
<AR1>display ip routing-table
Route Flags: R - relay, D - download to fib
-----------------------------------------------------------------------
Routing Tables: Public
         Destinations : 15       Routes : 15
```

Destination/Mask	Proto	Pre	Cost	Flags	NextHop	Interface
127.0.0.0/8	Direct	0	0	D	127.0.0.1	InLoopBack0
127.0.0.1/32	Direct	0	0	D	127.0.0.1	InLoopBack0

127.255.255.255/32	Direct	0	0	D	127.0.0.1	InLoopBack0
192.168.1.0/24	Direct	0	0	D	192.168.1.1	GigabitEthernet 0/0/0
192.168.1.1/32	Direct	0	0	D	127.0.0.1	GigabitEthernet 0/0/0
192.168.1.255/32	Direct	0	0	D	127.0.0.1	GigabitEthernet 0/0/0
192.168.2.0/24	RIP	100	3	D	192.168.1.2	GigabitEthernet 0/0/0
192.168.3.0/24	Direct	0	0	D	192.168.3.1	GigabitEthernet 0/0/1
192.168.3.1/32	Direct	0	0	D	127.0.0.1	GigabitEthernet 0/0/1
192.168.3.255/32	Direct	0	0	D	127.0.0.1	GigabitEthernet 0/0/1
192.168.4.0/24	Direct	0	0	D	192.168.4.1	GigabitEthernet 0/0/2
192.168.4.1/32	Direct	0	0	D	127.0.0.1	GigabitEthernet 0/0/2
192.168.4.255/32	Direct	0	0	D	127.0.0.1	GigabitEthernet 0/0/2
192.168.5.0/24	RIP	100	3	D	192.168.1.2	GigabitEthernet 0/0/0
255.255.255.255/32	Direct	0	0	D	127.0.0.1	InLoopBack0

<AR1>

可以看出，路由器 AR1 的路由表中只学习到了网段 192.168.5.0/24 的路由信息，网段 192.168.6.0/24 的路由信息被过滤掉了。

4.3.5 RIPng 配置

（1）如图 4.12 所示，配置 RIPng，配置相关端口与 IP 地址，进行网络拓扑连接。路由器 AR1 和路由器 AR2 的 loopback 1 端口使用的是全球单播地址。路由器 AR1 和路由器 AR2 的物理端口在使用 RIPng 传送路由信息时，路由条目的下一跳地址只能是链路本地地址。例如，如果路由器 AR1 收到的路由条目的下一跳地址为 2005::2/64，则路由器 AR1 认为目的地址为 2004::1 的网络地址可达。

V4-10 RIPng 配置

图 4.12 配置 RIPng

（2）配置主机 PC1 和主机 PC2 的 IPv6 地址，如图 4.13 所示。

图 4.13　配置主机 PC1 和主机 PC2 的 IPv6 地址

（3）配置路由器 AR1，相关实例代码如下。

```
<Huawei>system-view
Enter system view, return user view with Ctrl+Z.
[Huawei]sysname AR1
[AR1]ipv6                    //使能 IPv6 功能，不开启此功能无法互通
[AR1]interface GigabitEthernet 0/0/1
[AR1-GigabitEthernet0/0/1]ipv6　enable
[AR1-GigabitEthernet0/0/1]ipv6 address 2001::254 64    //配置 IPv6 地址
[AR1-GigabitEthernet0/0/1]ripng 1 enable               //配置 RIPng
[AR1-GigabitEthernet0/0/1]quit
[AR1]interface GigabitEthernet 0/0/0
[AR1-GigabitEthernet0/0/0]ipv6 enable
[AR1-GigabitEthernet0/0/0]ipv6 address 2005::1 64
[AR1-GigabitEthernet0/0/0]ripng 1 enable
[AR1-GigabitEthernet0/0/0]quit
[AR1]interface LoopBack1
[AR1-LoopBack1]ipv6 address 2003::1 64
[AR1-LoopBack1]ripng 1 enable
[AR1-LoopBack1]quit
[AR1]
```

使用 ipv6 enable 命令，在路由器端口上使能 IPv6，使得端口能够接收和转发 IPv6 报文。端口的 IPv6 功能默认是未启用的。

使用 ipv6 address auto link-local 命令，为端口配置自动生成的链路本地地址。

使用 ripng process-id enable 命令，使能一个端口的 RIPng。进程 ID 可以是 1 到 65535 之间的任意值。默认情况下，端口上未使能 RIPng。

（4）配置路由器 AR2，相关实例代码如下。

```
<Huawei>system-view
Enter system view, return user view with Ctrl+Z.
[Huawei]sysname AR2
[AR2]ipv6
[AR2]interface GigabitEthernet 0/0/1
[AR2-GigabitEthernet0/0/1]ipv6   enable
[AR2-GigabitEthernet0/0/1]ipv6 address 2002::254 64
[AR2-GigabitEthernet0/0/1]ripng 1 enable
[AR2-GigabitEthernet0/0/1]quit
[AR2]interface GigabitEthernet 0/0/0
[AR2-GigabitEthernet0/0/0]ipv6 enable
[AR2-GigabitEthernet0/0/0]ipv6 address 2005::2 64
[AR2-GigabitEthernet0/0/0]ripng 1 enable
[AR2-GigabitEthernet0/0/0]quit
[AR2]interface LoopBack1
[AR2-LoopBack1]ipv6 address 2004::1 64
[AR2-LoopBack1]ripng 1 enable
[AR2-LoopBack1]quit
[AR2]
```

（5）显示路由器 AR1、路由器 AR2 的配置信息，这里以路由器 AR1 为例，其主要配置实例代码如下。

```
<AR1>display current-configuration
#
 sysname AR1
#
ipv6
#
interface GigabitEthernet0/0/0
ipv6 enable
 ipv6 address 2005::1/64
 ripng 1 enable
#
interface GigabitEthernet0/0/1
 ipv6 enable
 ipv6 address 2001::254/64
 ripng 1 enable
#
interface LoopBack1
 ipv6 enable
 ipv6 address 2003::1/64
 ripng 1 enable
#
ripng 1
#
return
<AR1>
```

（6）显示路由器 AR1 的 RIPng 路由信息，如图 4.14 所示。

图 4.14　显示路由器 AR1 的 RIPng 路由信息

使用 display ripng 命令，可以查看 RIPng 进程实例以及该实例的相关参数和统计信息。从显示信息中可以看出，RIPng 的协议优先级是 100；路由信息的更新周期是 30s；Number of routes in database 字段显示为 5，表明 RIPng 数据库中路由的条数为 5；Total number of routes in ADV DB is 字段显示为 5，表明 RIPng 正常工作并发送了 5 条路由更新信息。

（7）测试主机 PCI 的连通性，主机 PC1 分别访问路由器 AR1 和主机 PC2，测试结果如图 4.15 所示。

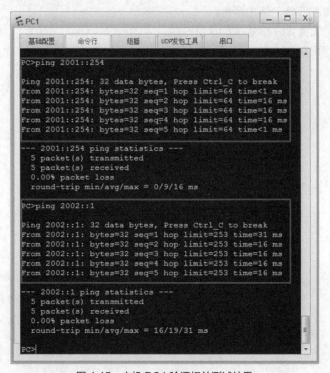

图 4.15　主机 PC1 验证相关测试结果

课后习题

1. 选择题

（1）RIP 网络中允许的最大跳数为（　　）。
    A. 13　　　　B. 14　　　　C. 15　　　　D. 16

（2）华为设备中，定义 RIP 网络的默认管理距离为（　　）。
    A. 60　　　　B. 100　　　　C. 120　　　　D. 150

（3）RIP 网络中，每个路由器会周期性地向邻居路由器通告自己的整张路由表中的路由信息，默认周期为（　　）。
    A. 30s　　　　B. 60s　　　　C. 120s　　　　D. 180s

（4）RIP 网络中，为防止路由环路，路由器不会把从邻居路由器学到的路由反向通告回去，这被称为（　　）。
    A. 定义最大值　　B. 水平分割　　C. 控制更新时间　　D. 触发更新

（5）关于 RIPv1 与 RIPv2，下列说法正确的是（　　）。
    A. RIPv1 是有类路由协议，RIPv2 是无类路由协议
    B. RIPv1 与 RIPv2 都可以支持 VLSM
    C. RIPv1 是组播更新，RIPv2 是广播更新
    D. RIPv1 与 RIPv2 都可以支持认证

2. 简答题

（1）简述 RIP 的工作原理及其局限性。
（2）简述如何进行 RIP 路由聚合。
（3）简述如何配置 RIPv2 报文的认证方式。
（4）简述如何配置 RIP 与 BFD 联动。

项目5
OSPF协议与高级配置

【学习目标】

- 掌握OSPF协议路由汇总功能及配置方法。
- 掌握OSPF协议验证功能及配置方法。
- 掌握OSPF协议特殊区域的配置方法。
- 掌握OSPF协议与BFD联动的配置方法。
- 掌握OSPF协议引入外部路由的配置方法及OSPFv3协议配置方法。

5.1 项目陈述

由于企业规模较大,某企业网络采用了开放最短路径优先(Open Shortest Path First,OSPF)协议进行配置。随着企业规模的不断扩大,企业网络的子网数量增加,导致网络运行状态不够稳定,对企业业务造成了一定的影响。企业领导安排网络工程师小张对企业网络进行优化,考虑到企业网络的安全性与稳定性,需要对企业网络进行认证使用,同时动态检测网络的运行状况,并对企业未来的网络扩展做好准备。小张根据企业的要求制定了一份合理的网络优化方案,本项目将对小张的优化方案进行介绍。

5.2 必备知识

1. OSPF协议

OSPF协议是目前广泛使用的一种动态路由协议,它属于链路状态路由协议,具有路由变化收敛速度快、无路由环路、支持VLSM和汇总、层次区域划分等优点。在网络中使用OSPF协议后,大部分路由将由OSPF协议自行计算和生成,无须网络管理员手工配置,当网络拓扑发生变化时,OSPF协议可以自动计算、更正路由,极大地方便了网络管理。项目4中介绍的RIP是一种基于距离矢量算法的路由协议,存在着收敛速度慢、易产生路由环路、可扩展性差等问题,目前已逐渐被OSPF协议所取代。

OSPF协议是一种链路状态协议。每个路由器负责发现、维护与邻居的关系,并将已知的邻居列表和链路状态更新(Link State Update,LSU)报文通过可靠的泛洪与AS内的其他路由器周期性地进行交互,学习到整个AS的网络拓扑结构,并通过AS边界的路由器注入其他AS的路由信息,从而得到整个Internet的路由信息。每隔一个特定时间或当链路状态发生变化时,重新生成链路状态广播（Link State Advertisement，LSA）数据包，路由器通过泛洪机制将新LSA通告出去，以便实现路由实时更新。

OSPF协议是一个内部网关协议，用于在单一AS内决策路由，它是基本链路状态的路由协议，链路状态是指路由器端口或链路的参数，包括端口状态、端口的IP地址、分配给端口的子网掩码、端口所连接的网络及链路开销，OSPF协议与其他路由器交换信息，但所交换的不是路由而是链路状态，OSPF协议路由器不是告知其他路由器可以到达哪些网络及其距离是多少，而是告知其网络链路状态、端口所连接的网络及链路开销。各个路由器都有其自身的链路状态，称为本地链路状态，这些本地链路状态在OSPF协议路由域内传播，直到所有的OSPF协议路由器都有完整而等同的链路状态数据库（Link State Database，LSDB）为止，一旦每台路由器都接收到所有的链路状态，则各路由器就可以构造一棵树，即以其自身为根，而分支表示到AS中所有网络的最短的或开销最低的路由。

对于规模较大的网络，OSPF协议通常将网络划分成多个OSPF协议区域，并只要求路由器与同一区域的路由器交换链路状态，而在区域边界路由器上交换区域内的汇总链路状态，这样可以减少传播的信息量，且能够降低最短路径计算强度。在区域划分时，必须要有一个骨干区域（即区域0），其他非0或非骨干区域与骨干区域必须有物理或者逻辑连接。当有物理连接时，必须有这样一台路由器：其一个端口在骨干区，而另一个端口在非骨干区。当非骨干区不可能物理连接到骨干区时，必须定义一个逻辑或虚拟链路，虚拟链路由两个端点和一个传输区来定义，其中，一个端点是路由器端口，是骨干区域的一部分；另一端点也是一个路由器端口，但在骨干区域与非骨干区域没有物理连接的另一个非骨干区域中；传输区是一个区域，介于骨干区域与非骨干区域之间。

OSPF协议的协议号为89，采用组播方式进行OSPF协议包交换，组播地址为224.0.0.5（全部OSPF协议路由器）和224.0.0.6（指定路由器）。

2. OSPF协议路由区域报文类型

OSPF协议报文用来保证路由器之间能够互相传播信息。OSPF协议报文共有5种报文类型，任意一种报文都需要加上OSPF协议的报文头，最后封装在IP中进行传送。一个OSPF报文的最大长度为1500字节，其结构如图5.1所示。

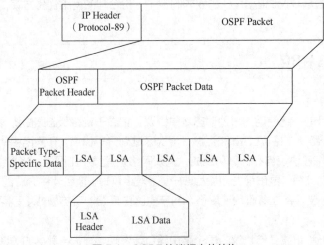

图5.1 OSPF协议报文的结构

OSPF协议直接运行在IP之上，使用IP协议号89。OSPF的5种报文都使用相同的OSPF协议报文头，如表5.1所示。

（1）Hello 报文：最常用的一种报文，用于发现、维护邻居关系，并在广播多路访问（Broadcast Multi-Access，BMA）和非广播多路访问（Non-Broadcast Multi-Access，NBMA）类型的网络中选举指定路由器（Designated Router，DR）和备份指定路由器（Backup Designated Router，BDR）。

（2）数据库描述（Database Description，DD）报文：两台路由器进行 LSDB 同步时，用 DD 报文来描述自己的 LSDB。DD 报文的内容包括 LSDB 中每一条 LSA 头（LSA 头可以唯一标识一条 LSA）。LSA 头只占一条 LSA 的整体数据量的一小部分，这样可以减少路由器之间的协议报文流量。

（3）链路状态请求（Link State Request，LSR）报文：两台路由器互相交换过 DD 报文之后，就可以知道对端的路由器有哪些 LSA 是本地 LSDB 所缺少的，此时需要发送 LSR 报文向对方请求缺少的 LSA。LSR 只包含所需 LSA 的头部，数据量很小。

（4）链路状态更新（Link State Update，LSU）报文：用来发送对端路由器所需 LSA。

（5）链路状态确认（Link State Acknowledgment，LSACK）报文：用来对接收到的 LSU 报文进行确认。

表 5.1　OSPF 协议中 5 种类型的路由协议报文

报文类型	功能描述
Hello 报文	周期性发送，发现和维护 OSPF 协议邻居关系
DD 报文	邻居间同步数据库内容
LSR 报文	向对方请求所需要的 LSA 报文
LSU 报文	向对方通告 LSA 报文
LSACK 报文	对收到的 LSU 报文进行确认

3. OSPF 协议的 LSA 类型

OSPF 协议是目前应用最广泛的内部网关协议，其 LSA 类型繁多，一共有 11 种，LSA 在 OSPF 协议环境中的作用范围如图 5.2 所示。

图 5.2　LSA 在 OSPF 协议环境中的作用范围

OSPF 协议的 LSA 一共有 11 种，分别如下。

① LSA1：路由器 LSA（Router LSA）。
② LSA2：网络 LSA（Network LSA）。
③ LSA3：网络汇总 LSA（Network Summary LSA）。
④ LSA4：ASBR 汇总 LSA（ASBR Summary LSA）。
⑤ LSA5：AS 外部 LSA（Autonomous System External LSA）。
⑥ LSA6：组成员 LSA（Group Membership LSA），目前不支持组播 OSPF 协议（MOSPF 协议）。
⑦ LSA7：NSSA 外部 LSA（NSSA External LSA）。
⑧ LSA8：BGP 的外部属性 LSA（External Attributes LSA for BGP）。
⑨ LSA9：不透明 LSA（Opaque LSA）（本地链路范围），目前主要用于多协议标签交换（Multi-Protocol Label Switching，MPLS）协议。
⑩ LSA10：不透明 LSA（本地区域范围），目前主要用于 MPLS 协议。
⑪ LSA11：不透明 LSA（AS 范围），目前主要用于 MPLS 协议。

本书主要讲解其中的 LSA1、LSA2、LSA3、LSA4、LSA5、LSA7。其余的 LSA 会在一些特殊环境中使用，暂时不对它们进行深入的探讨。

（1）LSA1：路由器 LSA。

路由器 LSA 描述了路由器的直连链路状态信息。其由每个发起路由器通告，只在本区域内传递，不会超过 ABR，这些最基本的 LSA 通告列出了路由器所有的链路和端口，并指明了它们的状态和沿每条链路方向出站的代价。

（2）LSA2：网络 LSA。

网络 LSA 描述了本区域内 BMA/NBMA（但不包括串行连接信息）的网络信息以及连接到此网络的路由器。其由本区域内的 BMA/NBMA 的 DR 或 BDR 通告，由区域内的 DR 或 BDR 路由器产生，只在本区域传递。

（3）LSA3：网络汇总 LSA。

网络汇总 LSA 描述了 OSPF 协议的区域间路由（在路由表中以 OIA 标识）。其由 ABR 产生，可以通知本区域内的路由器通往区域外的路由信息。对于位于一个区域外但仍然在 OSPF 协议 AS 内部的默认路由而言，其也可以通过 LSA3 来通告，原 LSA1 所描述的路由信息会由所在区域的 ABR 将其转换为 LSA3。LSA3 可以传播到整个 OSPF 协议的所有区域（特殊区域除外）中，其由 ABR 通告，LSA3 每穿越一个 ABR，其通告路由器（Advertisement Router，ADV Router）都会发生改变，转变为最后一次穿越的 ABR。

（4）LSA5：AS 外部 LSA。

这里先介绍 LSA5，再介绍 LSA4。AS 外部 LSA 描述的是 OSPF 协议区域以外的路由（RIP、EIGRP（增强型的内部网关路由协议）、BGP 等）。其由 ASBR 所通告，可以传播到整个 OSPF 协议的所有区域（特殊区域除外）中。注意，LSA5 的通告路由器在穿越 ABR 时不会改变。

（5）LSA4：ASBR 汇总 LSA。

ASBR 汇总 LSA 描述了 ASBR 的 Router ID。在图 5.2 中，R4（通告路由器）将 Area 3 中 R11 的 Router ID 转换为 LSA4，在整个 OSPF 协议域中泛洪传播。由于 LSA5 的通告路由器在穿越 ABR 时不会改变，图 5.2 中 Area 3 的 LSA5 在穿越 R4 到达 Area 0 时，通告路由器，仍然是 R11，因此，除了 Area 3 之外，Area 0 和其他区域都不知道 R11 的信息。此时就需要 LSA4 为 Area 0 和其他区域提供 R11 的信息，可以说，LSA4 是为 LSA5 所服务的。

此外，OSPF 协议中存在 4 类特殊区域，分别介绍如下。

（1）Stub Area（末梢区域）。

在 Stub Area 中只有域内和域间路由，只允许 LSA3 进入本区域，而不允许 LSA4/5 进入。配置 Stub Area 后，会在区域内自动生成一条默认路由（以便访问 OSPF 协议中其他区域的网络）。

（2）Total Stub Area（绝对末梢区域）。

在 Total Stub Area 中只有本区域内的路由，LSA3/4/5 均不允许进入本区域。配置 Total Stub Area 后，会在区域内自动生成一条默认路由（以便访问 OSPF 协议中其他区域的网络）。

（3）NSSA（次末梢区域）。

在 NSSA 中允许存在 ASBR，所以可以引入外部路由。该外部路由在 NSSA 内以 LSA7 的形式存在。当此 LSA7 路由离开 NSSA 进入其他区域时，NSSA 的 ABR 会进行 LSA7 向 LSA5 的转换（见图 5.2 中的 Area 6）。本区域只允许 LSA3 进入，禁止 LSA4/5 进入。所以，此区域中有域内、域间和外部路由，配置 NSSA 时，需要在区域内手工创建一条默认路由（以便访问 OSPF 协议中其他区域的网络），配置命令如下。

```
[Huawei]ospf 1
[Huawei-ospf-1]area 6
[Huawei-ospf-1-area-0.0.0.6]nssa default-route-advertise
```

（4）Total NSSA（绝对次末梢区域）。

在 Total NSSA 中允许存在 ASBR，所以可以引入外部路由。该外部路由在 Total NSSA 内以 LSA7 的形式存在。当此 LSA7 路由离开 Total NSSA 进入其他区域时，NSSA 的 ABR 会进行 LSA7 向 LSA5 的转换（见图 5.2 中的 Area 7）。本区域禁止 LSA3/4/5 进入，只有本区域内的路由和外部路由。配置 Total NSSA 后，会在区域内自动生成一条默认路由（以便访问 OSPF 协议中其他区域的网络）。

OSPF 协议中 4 类特殊区域的特点汇总如表 5.2 所示。

表 5.2　OSPF 协议中的特殊区域

区域类型	是否接收区域间路由	ABR 是否自动发送默认路由	是否可以重发布外部路由
Stub Area	是	是	否
Total Stub Area	否	是	否
NSSA	是	否	是
Total NSSA	否	是	是

Stub Area 过滤 LSA4/5，ABR 会产生默认的 LSA3，区域内不能引入外部路由。

Total Stub Area 过滤 LSA3/4/5，ABR 会产生默认的 LSA3，区域内不能引入外部路由。

NSSA 过滤 LSA4/5，ABR 会产生默认的 LSA7，区域内可以引入外部路由。

Total NSSA 过滤 LSA3/4/5，ABR 会产生默认的 LSA3，区域内可以引入外部路由。

4. DR 与 BDR 选举

每一个含有至少两个路由器的 BMA 和 NBMA 网络中都有一个 DR 和 BDR，DR 和 BDR 可以减少邻接关系的数量，从而减少链路状态信息及路由信息的交换次数，这样可以节省带宽，降低路由器的处理压力，如图 5.3 所示。

一个不是 DR/BDR 的路由器只与 DR 和 BDR 形成邻接关系，交换链路状态信息及路由信息，这样可以大大减少大型广播型网络和 NBMA 网络中的邻接关系数量。在没有 DR 的广播网络中，邻接关系的数量可以根据公式 $n(n-1)/2$ 计算得到，n 代表参与 OSPF 协议的路由器端口的数量。

如图 5.3 所示，所有路由器之间有 10 个邻接关系。当指定了 DR 后，所有的路由器都与 DR 建立邻接关系，DR 成为该广播网络中的中心点。BDR 在 DR 发生故障时接管业务，该广播网络中的所有路由器都必须同 BDR 建立邻接关系。

图 5.3　DR 与 BDR 选举

在邻居发现完成之后，路由器会根据网段类型进行 DR 选举。在 BMA 和 NBMA 网络中，路由器会根据参与选举的每个端口的优先级进行 DR 选举。优先级取值为 0~255，值越大，优先级越高。默认情况下，端口优先级为 1。如果一个端口优先级为 0，那么该端口将不会参与 DR 或者 BDR 的选举。如果优先级相同，则比较 Router ID，值越大，越优先被选举为 DR。为了给 DR 做备份，每个 BMA 和 NBMA 网络中还要选举一个 BDR，BDR 也会与网络中的所有路由器建立邻接关系。为了维护网络中邻接关系的稳定性，如果网络中已经存在 DR 和 BDR，则新添加进该网络中的路由器（即使其优先级最大）不会成为 DR 和 BDR。如果当前 DR 发生故障，则当前 BDR 自动成为新的 DR，网络中重新选举 BDR；如果当前 BDR 发生故障，则 DR 不变，重新选举 BDR。这种选举机制的目的是保持邻接关系的稳定，使拓扑结构的改变对邻接关系的影响尽量小。

5.3　项目实施

5.3.1　OSPF 协议路由聚合配置

当 OSPF 协议网络规模较大时，配置路由聚合，可以有效减少路由表中的条目，减少对系统资源的占用，不影响系统的性能。此外，如果被聚合的 IP 地址范围内的某条链路频繁改变状态，则该变化并不会通告到被聚合的 IP 地址范围外的设备，可以避免网络中的路由振荡，在一定程度上提高了网络的稳定性。

ABR 向其他区域发送路由信息时，以网段为单位生成 LSA3。当区域中存在连续的网段（具有相同前缀的路由信息）时，可以通过 abr-summary 命令将这些网段聚合成一个网段，ABR 向其他区域只发送一条聚合后的 LSA，所有指定的聚合网段范围的 LSA 将不会再被单独发送，从而减小路由表的规模，提高路由器的性能。

1. 配置 ABR 路由聚合的步骤

（1）使用 system-view 命令，进入系统视图。

（2）使用 ospf [process-id]命令，进入 OSPF 协议进程视图。

（3）使用 area area-id 命令，进入 OSPF 协议区域视图。

（4）使用 abr-summary ip-address mask [[cost { cost | inherit-minimum }] | [advertise [generate-null0-route] | not-advertise | generate-null0-route[advertise]] *] 命令，配置 OSPF 协议的 ABR 路由聚合。

2. 配置 ASBR 路由聚合的步骤

（1）使用 system-view 命令，进入系统视图。

（2）使用 ospf [process-id] 命令，进入 OSPF 协议进程视图。

（3）（可选）使用 asbr-summary type nssa-trans-type-reference [cost] nssatrans-cost-reference] 命令，配置 OSPF 协议设置聚合路由类型和开销值时考虑将 LSA7 转换为 LSA5。未配置此命令时，OSPF 协议在设置聚合路由类型和开销时不考虑将 LSA7 转换为 LSA5。

（4）使用 asbr-summary ip-address mask [not-advertise | tag tag |cost cost |distribute-delay interval] *命令，配置 OSPF 协议的 ASBR 路由聚合。

> **注意** 在配置路由聚合后，本地 OSPF 协议设备的路由表保持不变。但是其他 OSPF 协议设备的路由表中将只有一条聚合路由，没有具体路由，直到网络中被聚合的路由都因出现故障而消失后，该聚合路由才会消失。

3. 配置 OSPF 协议路由聚合

（1）配置 OSPF 协议手动聚合。在一个 OSPF 协议 AS 中，配置 ABR 与 ASBR，路由器 AR1、路由器 AR2 运行在骨干 Area 0 中，路由器 AR2、路由器 AR3 运行在 Area 1 中，路由器 AR2 为 ABR 路由器，路由器 AR3 为 ASBR 路由器。在路由器 AR3 的 GE 0/0/0 与 GE 0/0/2 端口上进行手动聚合，将 150.1.1.0/24 与 150.1.2.0/24 聚合为一条 150.1.0.0/16 路由并引入到 OSPF 协议中，配置相关端口与 IP 地址，如图 5.4 所示，进行网络拓扑连接。

V5-1 配置 OSPF 协议路由聚合

图 5.4　配置 OSPF 协议路由聚合

（2）配置路由器 AR1，相关实例代码如下。
```
<Huawei>system-view
[Huawei]sysname AR1
[AR1]router id  1.1.1.1
[AR1]interface GigabitEthernet 0/0/0
[AR1-GigabitEthernet0/0/0]ip address 192.168.1.1 24
[AR1-GigabitEthernet0/0/0]quit
[AR1]interface LoopBack 0
[AR1-LoopBack0]ip address 1.1.1.2 255.255.255.255
[AR1-LoopBack0]quit
[AR1]ospf 1
[AR1-ospf-1]area 0
[AR1-ospf-1-area-0.0.0.0] network 1.1.1.1 0.0.0.0
[AR1-ospf-1-area-0.0.0.0]network 192.168.1.0 0.0.0.255
[AR1-ospf-1-area-0.0.0.0]quit
[AR1-ospf-1]quit
[AR1]
```

（3）配置路由器 AR2，相关实例代码如下。
```
<Huawei>system-view
[Huawei]sysname AR2
[AR2]router id  2.2.2.2
[AR2]interface GigabitEthernet 0/0/0
[AR2-GigabitEthernet0/0/0]ip address 192.168.1.2 24
[AR2-GigabitEthernet0/0/0]quit
[AR2]interface GigabitEthernet 0/0/1
[AR2-GigabitEthernet0/0/1]ip address 192.168.2.1 24
[AR2-GigabitEthernet0/0/1]quit
[AR2]ospf 1
[AR2-ospf-1]area 0
[AR2-ospf-1-area-0.0.0.0]network 2.2.2.2 0.0.0.0
[AR2-ospf-1-area-0.0.0.0]network 192.168.1.0 0.0.0.255
[AR2-ospf-1-area-0.0.0.0]quit
[AR2-ospf-1]quit
[AR2]ospf 1
[AR2-ospf-1]area 1
[AR2-ospf-1-area-0.0.0.1]network 192.168.2.0 0.0.0.255
[AR2-ospf-1-area-0.0.0.1]quit
[AR2-ospf-1]quit
[AR2]
```

（4）配置路由器 AR3，相关实例代码如下。
```
<Huawei>system-view
[Huawei]sysname AR3
[AR3]router id 3.3.3.3
[AR3]interface GigabitEthernet 0/0/0
[AR3-GigabitEthernet0/0/0]ip address 150.1.1.254 24
[AR3-GigabitEthernet0/0/0]quit
[AR3]interface GigabitEthernet 0/0/1
```

```
[AR3-GigabitEthernet0/0/1]ip address 192.168.2.2 24
[AR3-GigabitEthernet0/0/1]quit
[AR3]interface GigabitEthernet 0/0/2
[AR3-GigabitEthernet0/0/2]ip address 150.1.2.254 24
[AR3-GigabitEthernet0/0/2]quit
[AR3]ospf 1
[AR3-ospf-1]asbr-summary 150.1.0.0 255.255.0.0        //对引入的路由进行聚合
[AR3-ospf-1]import-route direct                        //引入端口的直连路由
[AR3-ospf-1]area 1
[AR3-ospf-1-area-0.0.0.1]network 3.3.3.3 0.0.0.0
[AR3-ospf-1-area-0.0.0.1]network 192.168.2.0 0.0.0.255
[AR3-ospf-1-area-0.0.0.1]quit
[AR3-ospf-1]quit
[AR3]
```

（5）查看路由器 AR1、路由器 AR2、路由器 AR3 的路由信息，这里以路由器 AR1 为例，主要配置实例代码如下。

```
<AR1>display ip routing-table
Route Flags: R - relay, D - download to fib
-----------------------------------------------------------------------------
Routing Tables: Public
        Destinations : 10       Routes : 10

Destination/Mask        Proto   Pre  Cost   Flags   NextHop         Interface

        1.1.1.2/32      Direct  0    0      D       127.0.0.1       LoopBack0
        127.0.0.0/8     Direct  0    0      D       127.0.0.1       InLoopBack0
        127.0.0.1/32    Direct  0    0      D       127.0.0.1       InLoopBack0
  127.255.255.255/32    Direct  0    0      D       127.0.0.1       InLoopBack0
        150.1.0.0/16    O_ASE   150  2      D       192.168.1.2     GigabitEthernet
0/0/0
        192.168.1.0/24  Direct  0    0      D       192.168.1.1     GigabitEthernet
0/0/0
        192.168.1.1/32  Direct  0    0      D       127.0.0.1       GigabitEthernet
0/0/0
      192.168.1.255/32  Direct  0    0      D       127.0.0.1       GigabitEthernet
0/0/0
        192.168.2.0/24  OSPF    10   2      D       192.168.1.2     GigabitEthernet
0/0/0
  255.255.255.255/32    Direct  0    0      D       127.0.0.1       InLoopBack0
<AR1>
```

（6）显示路由器 AR1、路由器 AR2、路由器 AR3 的配置信息，这里以路由器 AR3 为例，主要配置实例代码如下。

```
<AR3>display current-configuration
#
 sysname AR3
#
router id 3.3.3.3
```

```
#
interface GigabitEthernet0/0/0
 ip address 150.1.1.254 255.255.255.0
#
interface GigabitEthernet0/0/1
 ip address 192.168.2.2 255.255.255.0
#
interface GigabitEthernet0/0/2
 ip address 150.1.2.254 255.255.255.0
#
interface NULL0
#
ospf 1
 asbr-summary 150.1.0.0 255.255.0.0
 import-route direct
 area 0.0.0.1
  network 3.3.3.3 0.0.0.0
  network 192.168.2.0 0.0.0.255
#
return
<AR3>
```

5.3.2 OSPF 协议认证方式配置

为了提高网络的安全性，需要配置 OSPF 协议认证。例如，若网络中的所有路由器都启动了 OSPF 协议，在该网络拓扑环境中，某人想学习 OSPF 协议网络中的路由信息，其在 OSPF 协议网络中的一个节点上又连接了一台路由器，并且启动了 OSPF 协议，如果 OSPF 协议网络中没有配置加密，这台路由就会轻易地学习到其他所有路由器的信息。这是极不安全的，所以通常需要启动加密认证来管理 OSPF 协议网络。

1. 区域认证方式

使用区域认证方式时，一个区域中的所有路由器在该区域下的验证模式和口令必须一致。例如，在 Area 0 中的所有路由器上配置认证模式为简单认证，口令为 123。

注意 在配置区域认证模式时，如果使用 plain 选项，则密码将以明文形式保存在配置文件中，存在安全隐患。建议使用 cipher 选项，即将密码加密保存。
简单认证、MD5 和 HMAC-MD5 密文认证存在安全风险，推荐使用 HMAC-SHA256 密文认证模式。

配置区域认证方式的步骤如下。
（1）使用 system-view 命令，进入系统视图。
（2）使用 ospf [process-id]命令，进入 OSPF 协议进程视图。
（3）使用 area area-id 命令，进入 OSPF 协议区域视图。
（4）根据需求，配置 OSPF 协议区域的认证模式。

使用 authentication-mode simple [plain plain-text | [cipher] cipher-text]命令，配置 OSPF 协议区域的认证模式。

① simple 表示使用简单认证模式。
② plain 表示明文口令类型。
③ cipher 表示密文口令类型。对于 MD5/HMAC-MD5 认证模式，默认为 cipher 类型。

使用 authentication-mode { md5 | hmac-md5 | hmac-sha256 } [key-id{ plain plain-text | [cipher] cipher-text }]命令，配置 OSPF 区域的认证模式。

① md5 表示使用 MD5 密文认证模式。
② hmac-md5 表示使用 HMAC-MD5 密文认证模式。
③ hmac-sha256 表示使用 HMAC-SHA256 密文认证模式。
④ key-id 表示密文认证的认证字标识符。

使用 authentication-mode keychain keychain-name 命令，表示配置 OSPF 协议区域的 Keychain 认证模式。

2．端口认证方式

端口认证方式用于在相邻的路由器之间设置认证模式和口令，其优先级高于区域认证方式。

注意　在配置端口认证模式时，如果使用 plain 选项，则密码将以明文形式保存在配置文件中，存在安全隐患。建议使用 cipher 选项，即将密码加密保存。
简单认证模式、MD5 认证模式和 HMAC-MD5 认证模式存在安全风险，推荐使用 HMAC-SHA256 认证模式。

配置端口认证方式的步骤如下。
（1）使用 system-view 命令，进入系统视图。
（2）使用 interface interface-type interface-number 命令，进入运行 OSPF 协议的端口视图。
（3）根据需求，配置端口认证方式。

使用 ospf authentication-mode simple [plain plain-text | [cipher] cipher-text]命令，配置 OSPF 协议端口的认证模式。

① simple 表示使用简单认证模式。
② plain 表示明文口令类型。对于简单认证模式，默认为 plain 类型。
③ cipher 表示密文口令类型。对于 MD5 或者 HMAC-MD5 认证模式，默认为 cipher 类型。

使用 ospf authentication-mode { md5 | hmac-md5 | hmac-sha256 } [key-id{ plain plain-text | [cipher] cipher-text }]命令，配置 OSPF 协议端口的认证模式。

① md5 表示使用 MD5 密文认证模式。
② hmac-md5 表示使用 HMAC-MD5 密文认证模式。
③ hmac-sha256 表示使用 HMAC-SHA256 密文认证模式。

使用 ospf authentication-mode null 命令，表示不对 OSPF 协议端口进行认证。
使用 ospf authentication-mode keychain keychain-name 命令，表示配置 OSPF 协议区域的 Keychain 认证模式。

注意　使用 Keychain 认证模式时，需要在系统视图下配置 Keychain 信息。必须保证本端 ActiveSendKey 和对端 ActiveRecvKey 的 key-id、algorithm、key-string 相同，才能建立 OSPF 协议邻居。

5.3.3　OSPF 协议认证方式配置实例

如图 5.5 所示，配置 OSPF 协议认证方式，配置相关端口与 IP 地址，进行网络拓扑连接。

图 5.5　配置 OSPF 协议认证方式

V5-2　配置 OSPF 协议认证方式

1. 区域认证方式的配置

配置路由器 AR1，相关实例代码如下。

```
<Huawei>system-view
[Huawei]sysname AR1
[AR1]router id  1.1.1.1
[AR1]interface GigabitEthernet 0/0/0
[AR1-GigabitEthernet0/0/0]ip address 192.168.1.1 24
[AR1-GigabitEthernet0/0/0]quit
[AR1]interface LoopBack 0
[AR1-LoopBack0]ip address 1.1.1.2 255.255.255.255
[AR1-LoopBack0]quit
[AR1]ospf 1
[AR1-ospf-1]area 0
[AR1-ospf-1-area-0.0.0.0] network 1.1.1.1 0.0.0.0
[AR1-ospf-1-area-0.0.0.0]network 192.168.1.0 0.0.0.255
[AR1-ospf-1-area-0.0.0.0]quit
[AR1-ospf-1]quit
[AR1]ospf 1
[AR1-ospf-1]area 0
[AR1-ospf-1-area-0.0.0.0]authentication-mode？              //4 种认证方式
  hmac-md5    Use HMAC-MD5 algorithm
  keychain    Keychain authentication mode
  md5         Use MD5 algorithm
  simple      Simple authentication mode
[AR1-ospf-1-area-0.0.0.0]authentication-mode keychain？
  STRING<1-47>    The keychain name
[AR1-ospf-1-area-0.0.0.0]authentication-mode keychain admin
[AR1-ospf-1-area-0.0.0.0]authentication-mode hmac-md5 1？
  STRING<1-255>/<20-392>   The password (key)
  cipher                   Encryption type (Cryptogram)     //加密
  plain                    Encryption type (Plain text)     //明文
[AR1-ospf-1-area-0.0.0.0]authentication-mode hmac-md5 1 cipher 123456
[AR1-ospf-1-area-0.0.0.0]quit
[AR1-ospf-1]quit
[AR1]
```

路由器 AR2 与路由器 AR1 的认证方式相同，这里不再赘述。

2．端口认证方式的配置

配置路由器 AR1，相关实例代码如下。

```
<Huawei>system-view
[Huawei]sysname AR1
[AR1]router id  1.1.1.1
[AR1]interface GigabitEthernet 0/0/0
[AR1-GigabitEthernet0/0/0]ip address 192.168.1.1 24
[AR1-GigabitEthernet0/0/0]quit
[AR1]interface LoopBack 0
[AR1-LoopBack0]ip address 1.1.1.2 255.255.255.255
[AR1-LoopBack0]quit
[AR1]ospf 1
[AR1-ospf-1]area 0
[AR1-ospf-1-area-0.0.0.0] network 1.1.1.1 0.0.0.0
[AR1-ospf-1-area-0.0.0.0]network 192.168.1.0 0.0.0.255
[AR1-ospf-1-area-0.0.0.0]quit
[AR1-ospf-1]quit
[AR1]interface GigabitEthernet 0/0/0
[AR1-GigabitEthernet0/0/0]ospf authentication-mode？  //4 种认证方式
  hmac-md5    Use HMAC-MD5 algorithm
  keychain    Keychain authentication mode
  md5         Use MD5 algorithm
  null        Use null authentication
  simple      Simple authentication mode
[AR1-GigabitEthernet0/0/0]ospf authentication-mode hmac-md5？
  INTEGER<1-255>   Key ID
  <cr>             Please press ENTER to execute command
[AR1-GigabitEthernet0/0/0]ospf authentication-mode hmac-md5 1？
  STRING<1-255>/<20-392>   The password (key)
  cipher                   Encryption type (Cryptogram)
  plain                    Encryption type (Plain text)
[AR1-GigabitEthernet0/0/0]ospf authentication-mode hmac-md5 1 cipher 123456
[AR1-GigabitEthernet0/0/0]quit
[AR1]
```

路由器 AR2 与路由器 AR1 的认证方式相同，这里不再赘述。

5.3.4 OSPF 协议的特殊区域虚连接配置

OSPF 协议要求每个区域与骨干区域（Area 0）必须直接相连，但是在实际组网中，网络情况非常复杂，在划分区域时，有时无法保证每个区域都满足此要求。此时就需要使用虚链接（Virtual Link）技术来解决这个问题。

虚连接是指在两台 ABR 之间，穿过一个非骨干区域（也称为转换区域，Transit Area），建立的一条逻辑上的连接通道（必须在两端的 ABR 上同时配置）。

（1）如图 5.6 所示，配置 OSPF 协议虚连接，配置相关端口与 IP 地址，进行网络拓扑连接，Area 2 没有与骨干区域直接相连，Area 1 被作为转换区域来连接 Area 2 和 Area 0，路由器 AR1 和路由器 AR2 之间配置了一条虚连接。

V5-3 配置 OSPF 协议虚连接

图 5.6 配置 OSPF 协议虚连接

（2）配置路由器 AR1，相关实例代码如下。

```
<Huawei>system-view
[Huawei]sysname AR1
[AR1]interface GigabitEthernet 0/0/0
[AR1-GigabitEthernet0/0/0]ip address 192.168.2.1 24
[AR1]interface GigabitEthernet 0/0/1
[AR1-GigabitEthernet0/0/1]ip address 192.168.1.2 24
[AR1-GigabitEthernet0/0/1]quit
[AR1]ospf 1
[AR1-ospf-1]area 0
[AR1-ospf-1-area-0.0.0.0]network 192.168.1.0 0.0.0.255
[AR1-ospf-1-area-0.0.0.0]quit
[AR1-ospf-1]area 1
[AR1-ospf-1-area-0.0.0.1]network 192.168.2.0 0.0.0.255
[AR1-ospf-1-area-0.0.0.1]quit
[AR1-ospf-1]quit
[AR1]
```

（3）配置路由器 AR2，相关实例代码如下。

```
<Huawei>system-view
[Huawei]sysname AR2
[AR2]interface GigabitEthernet 0/0/0
[AR2-GigabitEthernet0/0/0]ip address 192.168.2.2 24
[AR2]interface GigabitEthernet 0/0/1
[AR2-GigabitEthernet0/0/1]ip address 192.168.3.2 24
[AR2-GigabitEthernet0/0/1]quit
[AR2]ospf 1
```

```
[AR2-ospf-1]area 1
[AR2-ospf-1-area-0.0.0.1]network 192.168.2.0 0.0.0.255
[AR2-ospf-1-area-0.0.0.1]quit
[AR2-ospf-1]area 2
[AR2-ospf-1-area-0.0.0.2]network 192.168.3.0 0.0.0.255
[AR2-ospf-1-area-0.0.0.2]quit
[AR2-ospf-1]quit
[AR2]
```

（4）配置路由器 AR3，相关实例代码如下。

```
<Huawei>system-view
[Huawei]sysname AR3
[AR3]interface GigabitEthernet 0/0/1
[AR3-GigabitEthernet0/0/1]ip address 192.168.1.1 24
[AR3-GigabitEthernet0/0/1]quit
[AR3]ospf 1
[AR3-ospf-1]area 0
[AR3-ospf-1-area-0.0.0.0]network 192.168.1.0 0.0.0.255
[AR3-ospf-1-area-0.0.0.0]quit
[AR3-ospf-1]quit
[AR3]
```

（5）配置路由器 AR4，相关实例代码如下。

```
<Huawei>system-view
[Huawei]sysname AR4
[AR4]interface GigabitEthernet 0/0/1
[AR4-GigabitEthernet0/0/1]ip address 192.168.3.1 24
[AR4-GigabitEthernet0/0/1]quit
[AR4]ospf 1
[AR4-ospf-1]area 2
[AR4-ospf-1-area-0.0.0.2]network 192.168.3.0 0.0.0.255
[AR4-ospf-1-area-0.0.0.2]quit
[AR4-ospf-1]quit
[AR4]
```

（6）查看路由器 AR1 的路由表，由于 Area 2 没有与 Area 0 直接相连，所以路由器 AR1 的路由表中没有 Area 2 中的路由。

```
<AR1>display ospf routing
         OSPF Process 1 with Router ID 192.168.2.1
                 Routing Tables
Routing for Network
Destination          Cost    Type      NextHop        AdvRouter      Area
192.168.1.0/24       1       Transit   192.168.1.2    192.168.2.1    0.0.0.0
192.168.2.0/24       1       Transit   192.168.2.1    192.168.2.1    0.0.0.1
Total Nets: 2
Intra Area: 2  Inter Area: 0   ASE: 0   NSSA: 0
<AR1>
```

（7）在路由器 AR1 与路由器 AR2 上配置虚连接。

在路由器 AR1 上配置虚连接。

```
[AR1]router id 1.1.1.1
[AR1]ospf 1
[AR1-ospf-1]area 1
[AR1-ospf-1-area-0.0.0.1]vlink-peer 2.2.2.2
[AR1-ospf-1-area-0.0.0.1]vlink-peer 192.168.2.2
[AR1-ospf-1-area-0.0.0.1]quit
[AR1-ospf-1]quit
[AR1]
```

在路由器 AR2 上配置虚连接。

```
[AR2]router id 2.2.2.2
[AR2]ospf 1
[AR2-ospf-1]area 1
[AR2-ospf-1-area-0.0.0.1]vlink-peer 1.1.1.1
[AR2-ospf-1-area-0.0.0.1]vlink-peer 192.168.2.1
[AR2-ospf-1-area-0.0.0.1]quit
[AR2-ospf-1]quit
[AR2]
```

（8）验证配置结果，查看路由器 AR2 的路由表。

```
<AR2>display ospf routing
        OSPF Process 1 with Router ID 192.168.2.2
                Routing Tables
Routing for Network
Destination       Cost   Type       NextHop         AdvRouter       Area
192.168.2.0/24    1      Transit    192.168.2.2     192.168.2.2     0.0.0.1
192.168.3.0/24    1      Transit    192.168.3.2     192.168.2.2     0.0.0.2
192.168.1.0/24    2      Inter-area 192.168.2.1     192.168.2.1     0.0.0.1
Total Nets: 3
Intra Area: 2  Inter Area: 1   ASE: 0   NSSA: 0
```

5.3.5　Stub Area 配置

OSPF 协议路由器需要用 LSA5 来了解 OSPF 协议 AS 以外的路径。随着 LSA5 记录数量的变多，OSPF 协议 Database（数据库）会变得非常庞大，路由表的外部地址也会增大，占用路由器大量的资源。这个问题的解决方案为：这就像计算机主机只专注于应用程序一样，它们把所有路由的工作都交给网关。Stub Area 中的路由器不保存任何 OSPF 协议 AS 外部的路径，它们把 ABR 当作默认网关。

Stub Area 就是一个对区域概念的最典型的应用。Stub Area 的设计思想为：在划分了区域之后，非骨干区域中的路由器对于区域外的路由而言，一定要通过 ABR 来转发；对于区域内的路由器来说，ABR 是一个通往外部的必经之路。既然如此，对于区域内的路由器来说，就没有必要知道通往外部的详细的路由了，以由 ABR 向该区域发布一条默认路由来指导报文的发送即可。这样，区域内的路由器中只有为数不多的区域内路由和一条指向 ABR 的默认路由，无论区域外的路由如何变化，都不会影响到区域内路由器的路由表。由于区域内的路由器通常是由一些处理能力有限的低端路由器组成的，所以有了 Stub 属性之后，网络的规划更符合实际设备的特点。处于 Stub Area 内的低端设备既不需要保存庞大的路由表，又不需要经常性地进行路由计算。

1. 配置当前区域为 Stub Area 的步骤

（1）使用 system-view 命令，进入系统视图。

（2）使用 ospf [process-id]命令，进入 OSPF 协议进程视图。

（3）使用 area area-id 命令，进入 OSPF 协议区域视图。

（4）使用 stub [no-summary | default-route-advertise backbone-peer-ignore] *命令，配置当前区域为 Stub Area。

① no-summary 用来禁止 ABR 向 Stub Area 内发送 LSA3。

② default-route-advertise 用来在 ABR 上配置产生默认的 LSA3 到 Stub Area 中。

③ backbone-peer-ignore 用来忽略检查骨干区域的邻居状态，即骨干区域中只要存在 Up 状态的端口，无论其是否存在 Full 状态的邻居，ABR 都会产生默认的 LSA3 到 Stub Area 中。

2.（可选）配置发送到 Stub Area 的默认路由的开销

（1）使用 stub [no-summary]命令，配置当前区域为 Stub Area（可选）。

no-summary 用来禁止 ABR 向 Stub Area 内发送 LSA3。

（2）使用 default-cost cost 命令，配置发送到 Stub Area 的默认路由的开销。

cost 为发送到 Stub Area 的 LSA3 默认路由的开销，默认值为 1。

当区域配置为 Stub Area 后，为保证到 AS 外的路由可达，Stub Area 的 ABR 将生成一条默认路由，并发布给 Stub Area 中的其他路由器。

> **注意** 所有连接到 Stub Area 的路由器必须使用 stub 命令将该区域配置为 Stub Area 属性。配置或取消 Stub Area 属性时，可能会触发区域更新。只有在上一次区域更新完成后，才能再次进行配置或取消配置操作。

3. 检查 OSPF 协议的 Stub Area 的配置结果

（1）使用以下命令查看 OSPF 协议的 LSDB 信息。

display ospf [process-id] lsdb [brief]
display ospf [process-id] lsdb [{ router | network | summary | asbr | ase | nssa |opaque-link | opaque-area | opaque-as } [link-state-id]] [originate-router[advertising-router-id] | self-originate] [age { min-value min-age-value | max-valuemax-age-value } *]

（2）使用以下命令查看 OSPF 协议路由表的信息。

display ospf [process-id] routing [ip-address [mask | mask-length]] [interface interface-type interface-number] [nexthop nexthop-address]
display ospf [process-id] routing router-id [router-id]

（3）使用 display ospf [process-id] abr-asbr [router-id]命令，可以查看 OSPF 协议 ABR 及 ASBR 的信息。

4. 配置 OSPF 协议的 Stub Area 实例

如图 5.7 所示，所有的路由器都运行 OSPF 协议，整个 AS 划分为 3 个区域。其中，路由器 AR1 和路由器 AR2 作为 ABR 来转发区域之间的路由，路由器 AR4 作为 ASBR 引入了外部路由（静态路由）。要求将 Area 1 配置为 Stub Area，减少通告到此区域内的 LSA 数量，但不影响路由的可达性，配置相关端口与 IP 地址，进行网络拓扑连接。

V5-4 配置 OSPF 协议的 Stub Area

图 5.7 配置 OSPF 协议的 Stub Area 实例

（1）配置路由器 AR1，相关实例代码如下。

V5-5 配置 OSPF 协议的 Stub Area ——测试结果

```
<Huawei>system-view
[Huawei]sysname AR1
[AR1]router id 1.1.1.1
[AR1]interface GigabitEthernet 0/0/0
[AR1-GigabitEthernet0/0/0]ip address 192.168.1.1 24
[AR1]interface GigabitEthernet 0/0/1
[AR1-GigabitEthernet0/0/1]ip address 192.168.2.1 24
[AR1-GigabitEthernet0/0/1]quit
[AR1]ospf 1
[AR1-ospf-1]area 0
[AR1-ospf-1-area-0.0.0.0]network 192.168.1.0 0.0.0.255
[AR1-ospf-1-area-0.0.0.0]quit
[AR1-ospf-1]area 1
[AR1-ospf-1-area-0.0.0.1]network 192.168.2.0 0.0.0.255
[AR1-ospf-1-area-0.0.0.1]quit
[AR1-ospf-1]quit
[AR1]
```

（2）配置路由器 AR2，相关实例代码如下。

```
<Huawei>system-view
[Huawei]sysname AR2
[AR2]router id 2.2.2.2
[AR2]interface GigabitEthernet 0/0/0
[AR2-GigabitEthernet0/0/0]ip address 192.168.1.2 24
[AR2]interface GigabitEthernet 0/0/1
[AR2-GigabitEthernet0/0/1]ip address 192.168.3.1 24
[AR2-GigabitEthernet0/0/1]quit
```

```
[AR2]ospf 1
[AR2-ospf-1]area 0
[AR2-ospf-1-area-0.0.0.0]network 192.168.1.0 0.0.0.255
[AR2-ospf-1-area-0.0.0.0]quit
[AR2-ospf-1]area 2
[AR2-ospf-1-area-0.0.0.2]network 192.168.3.0 0.0.0.255
[AR2-ospf-1-area-0.0.0.2]quit
[AR2-ospf-1]quit
[AR2]
```

（3）配置路由器 AR3，相关实例代码如下。

```
<Huawei>system-view
[Huawei]sysname AR3
[AR3]router id 3.3.3.3
[AR3]interface GigabitEthernet 0/0/1
[AR3-GigabitEthernet0/0/1]ip address 192.168.2.2 24
[AR3]interface GigabitEthernet 0/0/2
[AR3-GigabitEthernet0/0/2]ip address 172.16.1.1 24
[AR3-GigabitEthernet0/0/2]quit
[AR3]ospf 1
[AR3-ospf-1]area 1
[AR3-ospf-1-area-0.0.0.1]network 192.168.2.0 0.0.0.255
[AR3-ospf-1-area-0.0.0.1]network 172.16.1.0 0.0.0.255
[AR3-ospf-1-area-0.0.0.1]quit
[AR3-ospf-1]quit
[AR3]
```

（4）配置路由器 AR4，相关实例代码如下。

```
<Huawei>system-view
[Huawei]sysname AR4
[AR4]router id 4.4.4.4
[AR4]interface GigabitEthernet 0/0/1
[AR4-GigabitEthernet0/0/1]ip address 192.168.3.2 24
[AR4]interface GigabitEthernet 0/0/2
[AR4-GigabitEthernet0/0/2]ip address 172.16.2.1 24
[AR4-GigabitEthernet0/0/2]quit
[AR4]ospf 1
[AR4-ospf-1]area 2
[AR4-ospf-1-area-0.0.0.2]network 192.168.3.0 0.0.0.255
[AR4-ospf-1-area-0.0.0.2]network 172.16.2.0 0.0.0.255
[AR4-ospf-1-area-0.0.0.2]quit
[AR4-ospf-1]quit
[AR4]
```

（5）配置路由器 AR5，相关实例代码如下。

```
<Huawei>system-view
[Huawei]sysname AR5
[AR5]router id 5.5.5.5
[AR5]interface GigabitEthernet 0/0/2
```

```
[AR5-GigabitEthernet0/0/2]ip address 172.16.1.2 24
[AR5-GigabitEthernet0/0/2]quit
[AR5]ospf 1
[AR5-ospf-1]area 1
[AR5-ospf-1-area-0.0.0.1]network 172.16.1.0 0.0.0.255
[AR5-ospf-1-area-0.0.0.1]quit
[AR5-ospf-1]quit
[AR5]
```

（6）配置路由器 AR6，相关实例代码如下。

```
<Huawei>system-view
[Huawei]sysname AR6
[AR6]router id 6.6.6.6
[AR6]interface GigabitEthernet 0/0/2
[AR6-GigabitEthernet0/0/2]ip address 172.16.2.1 24
[AR6-GigabitEthernet0/0/2]quit
[AR6]ospf 1
[AR6-ospf-1]area 2
[AR6-ospf-1-area-0.0.0.2]network 172.16.2.0 0.0.0.255
[AR6-ospf-1-area-0.0.0.2]quit
[AR6-ospf-1]quit
[AR6]
```

（7）配置路由器 AR4，引入静态路由。

```
[AR4]ip route-static 100.0.0.0 8 NULL 0
[AR4]ospf 1
[AR4-ospf-1]import-route static type 1
[AR4-ospf-1]quit
[AR4]
```

（8）查看路由器 AR3 的 ABR 及 ASBR 的信息。

```
<AR3>display ospf abr-asbr
        OSPF Process 1 with Router ID 3.3.3.3
            Routing Table to ABR and ASBR
```

RtType	Destination	Area	Cost	Nexthop	Type
Intra-area	1.1.1.1	0.0.0.1	1	192.168.2.1	ABR
Inter-area	4.4.4.4	0.0.0.1	3	192.168.2.1	ASBR

`<AR3>`

（9）查看路由器 AR3 的 OSPF 路由表信息。

```
<AR3>display ospf routing
        OSPF Process 1 with Router ID 3.3.3.3
            Routing Tables
```

Routing for Network

Destination	Cost	Type	NextHop	AdvRouter	Area
172.16.1.0/24	1	Transit	172.16.1.1	3.3.3.3	0.0.0.1
192.168.2.0/24	1	Transit	192.168.2.2	3.3.3.3	0.0.0.1
172.16.2.0/24	4	Inter-area	192.168.2.1	1.1.1.1	0.0.0.1
192.168.1.0/24	2	Inter-area	192.168.2.1	1.1.1.1	0.0.0.1
192.168.3.0/24	3	Inter-area	192.168.2.1	1.1.1.1	0.0.0.1

Routing for ASEs

Destination	Cost	Type	Tag	NextHop	AdvRouter
100.0.0.0/8	4	Type1	1	192.168.2.1	4.4.4.4

Total Nets: 6
  Intra Area: 2   Inter Area: 3   ASE: 1   NSSA: 0
<AR3>

可以看出，当路由器 AR3 所在区域为普通区域时，可以看到路由表中存在 AS 外部的路由。

（10）配置 Area 1 为 Stub Area。

配置路由器 AR1。

```
[AR1]ospf 1
[AR1-ospf-1]area 1
[AR1-ospf-1-area-0.0.0.1]stub
[AR1-ospf-1-area-0.0.0.1]quit
[AR1-ospf-1]quit
[AR1]
```

配置路由器 AR3。

```
[AR3]ospf 1
[AR3-ospf-1]area 1
[AR3-ospf-1-area-0.0.0.1]stub
[AR3-ospf-1-area-0.0.0.1]quit
[AR3-ospf-1]quit
[AR3]
```

配置路由器 AR5。

```
[AR5]ospf 1
[AR5-ospf-1]area 1
[AR5-ospf-1-area-0.0.0.1]stub
[AR5-ospf-1-area-0.0.0.1]quit
[AR5-ospf-1]quit
[AR5]
```

（11）配置完成后，再次查看路由器 AR3 的 OSPF 协议路由表信息。

```
<AR3>display ospf routing
        OSPF Process 1 with Router ID 3.3.3.3
          Routing Tables
Routing for Network
```

Destination	Cost	Type	NextHop	AdvRouter	Area
172.16.1.0/24	1	Transit	172.16.1.1	3.3.3.3	0.0.0.1
192.168.2.0/24	1	Transit	192.168.2.2	3.3.3.3	0.0.0.1
0.0.0.0/0	2	Inter-area	192.168.2.1	1.1.1.1	0.0.0.1
172.16.2.0/24	4	Inter-area	192.168.2.1	1.1.1.1	0.0.0.1
192.168.1.0/24	2	Inter-area	192.168.2.1	1.1.1.1	0.0.0.1
192.168.3.0/24	3	Inter-area	192.168.2.1	1.1.1.1	0.0.0.1

Total Nets: 6
  Intra Area: 2   Inter Area: 4   ASE: 0   NSSA: 0
<AR3>

当把路由器 AR3 所在区域配置为 Stub Area 时，已经看不到 AS 外部的路由，取而代之的是一条默认路由。

（12）配置禁止向 Stub Area 通告 LSA3，配置路由器 AR1。

```
[AR1]ospf 1
[AR1-ospf-1]area 1
[AR1-ospf-1-area-0.0.0.1]stub no-summary
[AR1-ospf-1-area-0.0.0.1]quit
[AR1-ospf-1]quit
[AR1]
```

（13）验证配置结果，查看路由器 AR3 的 OSPF 协议路由表信息，主要配置实例代码如下。

```
<AR3>display ospf routing
         OSPF Process 1 with Router ID 3.3.3.3
                  Routing Tables
Routing for Network
Destination          Cost      Type       NextHop        AdvRouter     Area
172.16.1.0/24        1         Transit    172.16.1.1     3.3.3.3       0.0.0.1
192.168.2.0/24       1         Transit    192.168.2.2    3.3.3.3       0.0.0.1
0.0.0.0/0            2         Inter-area 192.168.2.1    1.1.1.1       0.0.0.1
Total Nets: 3
Intra Area: 2   Inter Area: 1   ASE: 0   NSSA: 0
<AR3>
```

从以上信息可以看出，在禁止向 Stub Area 通告 LSA3 后，Stub 路由器的路由表项进一步减少，只保留了一条通往区域外部的默认路由。

（14）查看路由器 AR1 的 OSPF 路由表信息。

```
<AR1>display ospf routing
         OSPF Process 1 with Router ID 1.1.1.1
                  Routing Tables
Routing for Network
Destination          Cost      Type       NextHop        AdvRouter     Area
192.168.1.0/24       1         Transit    192.168.1.1    1.1.1.1       0.0.0.0
192.168.2.0/24       1         Transit    192.168.2.1    1.1.1.1       0.0.0.1
172.16.1.0/24        2         Transit    192.168.2.2    5.5.5.5       0.0.0.1
172.16.2.0/24        3         Inter-area 192.168.1.2    2.2.2.2       0.0.0.0
192.168.3.0/24       2         Inter-area 192.168.1.2    2.2.2.2       0.0.0.0
Routing for ASEs
Destination          Cost      Type       Tag            NextHop       AdvRouter
100.0.0.0/8          3         Type1      1              192.168.1.2   4.4.4.4
Total Nets: 6
Intra Area: 3   Inter Area: 2   ASE: 1   NSSA: 0
<AR1>
```

从以上信息可以看出，路由器 AR1 的 OSPF 协议路由表信息是完整的。

5.3.6　NSSA 配置

OSPF 协议是目前互联网中应用最为广泛的一种 IGP，而 NSSA 则是在该协议发展过程中产生的一种新的属性，它的英文全称是 Not-so-Stubby Area。在 NSSA 内，所有路由器都必须支持该属性（包括 NSSA 的 ABR），而 AS 中的其他路由器则不需要。

Stub Area 虽然为合理地规划网络描绘了美好的前景，但其在实际的组网中不具备可操作性，但此时的 OSPF 协议已经基本成形，不可能再做大的修改。为了弥补缺陷，协议设计者提出了一种新的概念——NSSA，并被作为 OSPF 协议的一种扩展属性在 RFC1587 中单独进行了描述。

AS 外的 ASE 路由不可以进入到 NSSA 中，但是 NSSA 内的路由器引入的 ASE 路由可以在 NSSA 中传播并发送到区域之外，即取消了 Stub 关于 ASE 的双向传播的限制（区域外的进不来，区域内的出不去），改为单向限制（区域外的进不来，区域内的能出去）。

由于 NSSA 是 OSPF 协议的一种扩展属性，因此应尽量减少与不支持该属性的路由器协调工作时的冲突和兼容性问题。

为了解决 ASE 单向传递的问题，NSSA 中重新定义了一种 LSA，即 LSA7，作为区域内的路由器引入外部路由时使用，该类型的 LSA 除了类型标识与 LSA5 不同之外，其他内容基本一样。这样区域内的路由器就可以通过 LSA 的类型来判断该路由是否来自本区域内。但由于 LSA7 是新定义的，对于不支持 NSSA 属性的路由器无法识别，因此协议规定：在 NSSA 的 ABR 上，将 NSSA 内部产生的 LSA7 转换为 LSA5 后再发布出去，并更改 LSA 的发布者为 ABR 自己。这样 NSSA 外的路由器就无须支持该属性。

NSSA 是由 Stub Area 的概念改进得来的，其对 LSA 的影响如下。

（1）LSA7 在一个 NSSA 内携带外部信息。

（2）LSA7 在 NSSA 的 ABR 上被转换为 LSA5。

（3）不允许外部 LSA 进入。

（4）引入汇总 LSA。

NSSA 适用于既需要引入外部路由又要避免外部路由带来的资源消耗的场景。

NSSA 是 OSPF 协议的特殊的区域类型。NSSA 与 Stub Area 有许多相似的地方，如两者都不传播来自 OSPF 网络其他区域的外部路由，其差别在于，Stub Area 不能引入外部路由，NSSA 能够将区域外部路由引入并传播到整个 OSPF 协议区域中。

在 NSSA 中，使用 LSA7 描述引入的外部路由信息。LSA7 由 NSSA 的 ASBR 产生，其扩散范围仅限于边界路由器所在的协议 NSSA。

NSSA 的 ABR 收到 LSA7 中时，会有选择地将其转换为 LSA5，以便将外部路由信息通告到 OSPF 协议网络的其他区域中。

1. 配置 OSPF 协议的 NSSA 的步骤

（1）使用 system-view 命令，进入系统视图。

（2）使用 ospf [process-id]命令，进入 OSPF 协议进程视图。

（3）使用 area area-id 命令，进入 OSPF 协议区域视图。

（4）使用 nssa [{ default-route-advertise [backbone-peer-ignore] | ppress-defaul troute} | flush-waiting-timer interval-value | no-import-route | no-summary | set-n-bit suppress-forwarding-address | translator-always | translator-interval interval-value |zero-address-forwarding | translator-strict]*命令，配置当前区域为 NSSA。

> **注意** 所有连接到 NSSA 的设备必须使用 nssa 命令将该区域配置为 NSSA 属性。
> 配置或取消 NSSA 属性时，可能会触发区域更新及邻居中断。只有在上一次区域更新完成后，才能再次进行配置或取消配置操作。

nssa 命令参数的适用场景如下。

① default-route-advertise 用来在 ASBR 上配置产生默认的 LSA7 到 NSSA 中。

在 ABR 上，无论路由表中是否存在默认路由 0.0.0.0/0，都会产生 LSA7 默认路由；在 ASBR

上,只有当路由表中存在默认路由 0.0.0.0/0 时,才会产生 LSA7 默认路由。

② backbone-peer-ignore 用来忽略检查骨干区域的邻居状态,即骨干区域中只要存在 Up 状态的端口,无论其是否存在 Full 状态的邻居,ABR 都会自动产生默认的 LSA7 到 NSSA 中。

③ 当 ASBR 所在的区域被配置为 NSSA 时,LSA 泛洪区域中的其他路由器上仍会保留已经无用的 LSA5,这些 LSA 必须等到老化时间达到 3600s 后才会被删除。大量的 LSA 会占用路由器内存,对设备性能造成影响。此时,通过配置 flush-waiting-timer 参数产生老化时间被置为最大值(3600s)的 LSA5,及时清除其他路由器上已经无用的 LSA5。

注意　当 LSA 报文头部的 LS age(老化时间)达到 3600s 时,该 LSA 会被删除。
当 ASBR 同时是 ABR 时,flush-waiting-timer 参数不会生效,防止删除非 NSSA 的 LSA5。

④ 当 ASBR 同时是 ABR 时,配置 no-import-route 参数使 OSPF 协议通过 import-route 命令引入的外部路由不被通告到 NSSA。

⑤ 为了继续减少发送到 NSSA 的 LSA 的数量,可以配置 ABR 的 no-summary 属性,禁止 ABR 向 NSSA 内发送 LSA3。

注意　配置 nssa default-route-advertise backbone-peer-ignore no-summary 命令后,骨干区域中只要存在 Up 状态的端口,无论其是否存在 Full 状态的邻居,ABR 会同时产生默认的 LSA7 和默认的 LSA3,且默认的 LSA3 优先生效。

⑥ 设置了 set-n-bit 关键字后,当路由器与邻居路由器同步时,会在 DD 报文中设置 N-bit 位的标志。

⑦ 当 NSSA 中有多个 ABR 时,系统会根据规则自动选择一个 ABR 作为转换器(通常情况下为 Router ID 最大的设备),将 LSA7 转换为 LSA5。通过在 ABR 上配置 translator-always 参数,可以将某一个 ABR 指定为转换器。如果需要指定某两个 ABR 进行负载分担,则可以通过配置 translator-always 来指定两个转换器同时工作。如果需要使用某个固定的转换器,防止由于转换器变动引起 LSA 重新泛洪,则可以预先使用此命令进行指定。

⑧ translator-interval 参数主要用于转换器切换过程,以保障切换平滑进行。所以,interval-value 参数的默认间隔要大于泛洪的时间。

(5)(可选)使用 default-cost cost 命令,配置 ABR 发送到 NSSA 的 LSA3 的默认路由的开销。

当区域配置为 NSSA 后,为保证到 AS 外的路由可达,NSSA 的 ABR 将生成一条默认路由,并发布给 NSSA 中的其他路由器。配置 NSSA 的默认路由的开销时,需调整默认路由的选路。

默认路由也可以通过 LSA7 来表示,用于指导流量流向其他区域。

在 NSSA 中,可能同时存在多个边界路由器。为了防止路由环路产生,边界路由器之间不计算对方发布的默认路由。

默认情况下,ABR 发送到 NSSA 的默认路由的开销为 1。

2. Stub Area 和 NSSA 的区别

(1)Stub Area:不传播来自其他区域的外部路由;不传播 ASBR 引入的 AS 外部的路由,即在 Stub Area 中不会传播 LSA5。

（2）NSSA：不传播来自其他区域的外部路由；可以传播 ASBR 引入的 AS 外部的路由。NSSA 的 ASBR 产生 LSA7（其扩散范围仅限于 ASBR 所在的 NSSA），NSSA 的 ABR 收到 LSA7 后，会将其转换为 LSA5，并将外部路由信息发布到 OSPF 协议网络的其他区域；配置区域为 NSSA 后，可以避免大量外部路由对路由器带宽和存储资源的消耗。

3. 配置 OSPF 协议的 NSSA 实例

如图 5.8 所示，所有的路由器都运行 OSPF 协议，整个 AS 划分为两个区域。其中，路由器 AR1 和路由器 AR2 作为 ABR 来转发区域间的路由，路由器 AR3 和路由器 AR4 作为 ASBR 分别引入外部静态路由 10.0.0.0/8 和 20.0.0.0/8。此时，可在不影响路由可达的前提下，达到减少通告到 Area 1 内的 LSA 数量，但引入 AS 外部路由的目的。需要将 Area 1 配置为 NSSA，并配置 NSSA 中的路由器 AR1 为转换路由器，配置相关端口与 IP 地址，进行网络拓扑连接。

图 5.8 配置 OSPF 协议的 NSSA 实例

V5-6 配置 OSPF 协议的 NSSA

V5-7 配置 OSPF 协议的 NSSA ——测试结果

（1）配置路由器 AR1，相关实例代码如下。

```
<Huawei>system-view
[Huawei]sysname AR1
[AR1]router id 1.1.1.1
[AR1]interface GigabitEthernet 0/0/0
[AR1-GigabitEthernet0/0/0]ip address 192.168.0.1 24
[AR1]interface GigabitEthernet 0/0/1
[AR1-GigabitEthernet0/0/1]ip address 192.168.1.1 24
[AR1-GigabitEthernet0/0/1]quit
[AR1]interface GigabitEthernet 0/0/2
[AR1-GigabitEthernet0/0/2]ip address 192.168.2.1 24
[AR1-GigabitEthernet0/0/2]quit
[AR1]ospf 1
[AR1-ospf-1]area 0
[AR1-ospf-1-area-0.0.0.0]network 192.168.1.0 0.0.0.255
[AR1-ospf-1-area-0.0.0.0]quit
[AR1-ospf-1]area 1
```

```
[AR1-ospf-1-area-0.0.0.1]network 192.168.0.0 0.0.0.255
[AR1-ospf-1-area-0.0.0.1]network 192.168.2.0 0.0.0.255
[AR1-ospf-1-area-0.0.0.1]quit
[AR1-ospf-1]quit
[AR1]
```

（2）配置路由器 AR2，相关实例代码如下。

```
<Huawei>system-view
[Huawei]sysname AR2
[AR2]router id 2.2.2.2
[AR2]interface GigabitEthernet 0/0/0
[AR2-GigabitEthernet0/0/0]ip address 192.168.0.2 24
[AR2]interface GigabitEthernet 0/0/1
[AR2-GigabitEthernet0/0/1]ip address 192.168.4.2 24
[AR2-GigabitEthernet0/0/1]quit
[AR2]interface GigabitEthernet 0/0/2
[AR2-GigabitEthernet0/0/2]ip address 192.168.3.2 24
[AR2-GigabitEthernet0/0/2]quit
[AR2]ospf 1
[AR2-ospf-1]area 0
[AR2-ospf-1-area-0.0.0.0]network 192.168.3.0 0.0.0.255
[AR2-ospf-1-area-0.0.0.0]quit
[AR2-ospf-1]area 1
[AR2-ospf-1-area-0.0.0.1]network 192.168.0.0 0.0.0.255
[AR2-ospf-1-area-0.0.0.1]network 192.168.4.0 0.0.0.255
[AR2-ospf-1-area-0.0.0.1]quit
[AR2-ospf-1]quit
[AR2]
```

（3）配置路由器 AR3，相关实例代码如下。

```
<Huawei>system-view
[Huawei]sysname AR3
[AR3]router id 3.3.3.3
[AR3]interface GigabitEthernet 0/0/1
[AR3-GigabitEthernet0/0/1]ip address 192.168.1.2 24
[AR3-GigabitEthernet0/0/1]quit
[AR3]interface GigabitEthernet 0/0/2
[AR3-GigabitEthernet0/0/2]ip address 192.168.3.1 24
[AR3-GigabitEthernet0/0/2]quit
[AR3]ospf 1
[AR3-ospf-1]area 0
[AR3-ospf-1-area-0.0.0.0]network 192.168.1.0 0.0.0.255
[AR3-ospf-1-area-0.0.0.0]network 192.168.3.0 0.0.0.255
[AR3-ospf-1-area-0.0.0.0]quit
[AR3-ospf-1]quit
[AR3]
```

（4）配置路由器 AR4，相关实例代码如下。

```
<Huawei>system-view
[Huawei]sysname AR4
```

```
[AR4]router id 4.4.4.4
[AR4]interface GigabitEthernet 0/0/1
[AR4-GigabitEthernet0/0/1]ip address 192.168.4.1 24
[AR4-GigabitEthernet0/0/1]quit
[AR4]interface GigabitEthernet 0/0/2
[AR4-GigabitEthernet0/0/2]ip address 192.168.2.2 24
[AR4-GigabitEthernet0/0/2]quit
[AR4]ospf 1
[AR4-ospf-1]area 1
[AR4-ospf-1-area-0.0.0.1]network 192.168.2.0 0.0.0.255
[AR4-ospf-1-area-0.0.0.1]network 192.168.4.0 0.0.0.255
[AR4-ospf-1-area-0.0.0.1]quit
[AR4-ospf-1]quit
[AR4]
```

（5）配置路由器 AR3，引入静态路由 10.0.0.0/8。

```
[AR3]ip route-static 10.0.0.0 8 null 0
[AR3]ospf 1
[AR3-ospf-1]import-route static
[AR3-ospf-1]quit
[AR3]
```

（6）配置路由器 AR4，引入静态路由 20.0.0.0/8。

```
[AR4]ip route-static 20.0.0.0 8 NULL 0
[AR4]ospf 1
[AR4-ospf-1]import-route static
[AR4-ospf-1]quit
[AR4]
```

（7）查看路由器 AR3 的 OSPF 协议路由表信息。

```
<AR3>display ospf routing
        OSPF Process 1 with Router ID 3.3.3.3
              Routing Tables
 Routing for Network
 Destination        Cost      Type         NextHop        AdvRouter      Area
 192.168.1.0/24     1         Transit      192.168.1.2    3.3.3.3        0.0.0.0
 192.168.3.0/24     1         Transit      192.168.3.1    3.3.3.3        0.0.0.0
 192.168.0.0/24     2         Inter-area   192.168.3.2    2.2.2.2        0.0.0.0
 192.168.0.0/24     2         Inter-area   192.168.1.1    1.1.1.1        0.0.0.0
 192.168.2.0/24     2         Inter-area   192.168.1.1    1.1.1.1        0.0.0.0
 192.168.4.0/24     2         Inter-area   192.168.3.2    2.2.2.2        0.0.0.0
 Routing for ASEs
 Destination        Cost      Type         Tag       NextHop        AdvRouter
 20.0.0.0/8         1         Type2        1         192.168.3.2    4.4.4.4
 20.0.0.0/8         1         Type2        1         192.168.1.1    4.4.4.4
 Total Nets: 8
 Intra Area: 2   Inter Area: 4   ASE: 2   NSSA: 0
<AR3>
```

从以上信息可以看出，配置 NSSA 前，路由器 AR3 的路由表中有路由器 AR4 引入的 AS 外部的静态路由 20.0.0.0/8。

（8）查看路由器 AR4 的 OSPF 协议路由表信息。

```
<AR4>display ospf routing
         OSPF Process 1 with Router ID 4.4.4.4
                Routing Tables
 Routing for Network
 Destination        Cost    Type        NextHop         AdvRouter       Area
 192.168.2.0/24     1       Transit     192.168.2.2     4.4.4.4         0.0.0.1
 192.168.4.0/24     1       Transit     192.168.4.1     4.4.4.4         0.0.0.1
 192.168.0.0/24     2       Transit     192.168.2.1     2.2.2.2         0.0.0.1
 192.168.0.0/24     2       Transit     192.168.4.2     2.2.2.2         0.0.0.1
 192.168.1.0/24     2       Inter-area  192.168.2.1     1.1.1.1         0.0.0.1
 192.168.3.0/24     2       Inter-area  192.168.4.2     2.2.2.2         0.0.0.1
 Routing for ASEs
 Destination        Cost    Type        Tag      NextHop         AdvRouter
 10.0.0.0/8         1       Type2       1        192.168.2.1     3.3.3.3
 10.0.0.0/8         1       Type2       1        192.168.4.2     3.3.3.3
 Total Nets: 8
 Intra Area: 4   Inter Area: 2   ASE: 2   NSSA: 0
<AR4>
```

从以上信息可以看出，配置 NSSA 前，路由器 AR4 的路由表中有路由器 AR3 引入的 AS 外部的静态路由 10.0.0.0/8。

（9）配置 Area 1 为 NSSA。

配置路由器 AR1。

```
[AR1]ospf 1
[AR1-ospf-1]area 1
[AR1-ospf-1-area-0.0.0.1]nssa
[AR1-ospf-1-area-0.0.0.1]quit
[AR1-ospf-1]quit
[AR1]
```

配置路由器 AR2。

```
[AR2]ospf 1
[AR2-ospf-1]area 1
[AR2-ospf-1-area-0.0.0.1]nssa
[AR2-ospf-1-area-0.0.0.1]quit
[AR2-ospf-1]quit
[AR2]
```

配置路由器 AR4。

```
[AR4]ospf 1
[AR4-ospf-1]area 1
[AR4-ospf-1-area-0.0.0.1]nssa
[AR4-ospf-1-area-0.0.0.1]quit
[AR4-ospf-1]quit
[AR4]
```

注意 所有连接到 NSSA 的路由器必须使用 nssa 命令将该区域配置为 NSSA 属性。

（10）配置完成后，再次查看路由器 AR3 的 OSPF 协议路由表信息。

```
<AR3>display ospf routing
       OSPF Process 1 with Router ID 3.3.3.3
          Routing Tables
Routing for Network
Destination        Cost      Type        NextHop       AdvRouter      Area
192.168.1.0/24     1         Transit     192.168.1.2   3.3.3.3        0.0.0.0
192.168.3.0/24     1         Transit     192.168.3.1   3.3.3.3        0.0.0.0
192.168.0.0/24     2         Inter-area  192.168.3.2   2.2.2.2        0.0.0.0
192.168.0.0/24     2         Inter-area  192.168.1.1   1.1.1.1        0.0.0.0
192.168.2.0/24     2         Inter-area  192.168.1.1   1.1.1.1        0.0.0.0
192.168.4.0/24     2         Inter-area  192.168.3.2   2.2.2.2        0.0.0.0
Routing for ASEs
Destination        Cost      Type        Tag    NextHop        AdvRouter
20.0.0.0/8         1         Type2       1      192.168.3.2    2.2.2.2
Total Nets: 7
Intra Area: 2   Inter Area: 4   ASE: 1   NSSA: 0
<AR3>
```

从以上信息可以看出，配置 NSSA 后，路由器 AR3 的路由表中仍然有路由器 AR4 引入的 AS 外部的静态路由 20.0.0.0/8，说明 NSSA 可以传播 ASBR 引入的 AS 外部的路由。

由 AdvRouter 字段可以看出，发布路由器的 Router ID 为 2.2.2.2，即路由器 AR2 为 NSSA 中的转换路由器，这是因为默认情况下 OSPF 协议会选举 Router ID 较大的 ABR 作为转换路由器。

（11）配置完成后，查看路由器 AR4 的 OSPF 协议路由表信息。

```
<AR4>display ospf routing
       OSPF Process 1 with Router ID 4.4.4.4
          Routing Tables
Routing for Network
Destination        Cost      Type        NextHop       AdvRouter      Area
192.168.2.0/24     1         Transit     192.168.2.2   4.4.4.4        0.0.0.1
192.168.4.0/24     1         Transit     192.168.4.1   4.4.4.4        0.0.0.1
192.168.0.0/24     2         Transit     192.168.4.2   2.2.2.2        0.0.0.1
192.168.0.0/24     2         Transit     192.168.2.1   2.2.2.2        0.0.0.1
192.168.1.0/24     2         Inter-area  192.168.2.1   1.1.1.1        0.0.0.1
192.168.3.0/24     2         Inter-area  192.168.4.2   2.2.2.2        0.0.0.1
Routing for NSSAs
Destination        Cost      Type        Tag    NextHop        AdvRouter
0.0.0.0/0          1         Type2       1      192.168.4.2    2.2.2.2
0.0.0.0/0          1         Type2       1      192.168.2.1    1.1.1.1
Total Nets: 8
Intra Area: 4   Inter Area: 2   ASE: 0   NSSA: 2
<AR4>
```

从以上信息可以看出，配置 NSSA 后，路由器 AR4 的路由表中没有路由器 AR3 引入的 AS 外部的静态路由 10.0.0.0/8，说明 NSSA 不传播来其他区域的外部路由。

（12）配置 AR1 转换路由器。

```
[AR1]ospf 1
[AR1-ospf-1]area 1
[AR1-ospf-1-area-0.0.0.1]nssa default-route-advertise translator-always
```

[AR1-ospf-1-area-0.0.0.1]quit
[AR1-ospf-1]quit
[AR1]

（13）配置完成后，查看路由器 AR3 的 OSPF 协议路由表信息。

```
<AR3>display ospf routing
     OSPF Process 1 with Router ID 3.3.3.3
         Routing Tables
Routing for Network
Destination       Cost   Type        NextHop       AdvRouter     Area
192.168.1.0/24    1      Transit     192.168.1.2   3.3.3.3       0.0.0.0
192.168.3.0/24    1      Transit     192.168.3.1   3.3.3.3       0.0.0.0
192.168.0.0/24    2      Inter-area  192.168.3.2   2.2.2.2       0.0.0.0
192.168.0.0/24    2      Inter-area  192.168.1.1   1.1.1.1       0.0.0.0
192.168.2.0/24    2      Inter-area  192.168.1.1   1.1.1.1       0.0.0.0
192.168.4.0/24    2      Inter-area  192.168.3.2   2.2.2.2       0.0.0.0
Routing for ASEs
Destination       Cost   Type        Tag           NextHop       AdvRouter
20.0.0.0/8        1      Type2       1             192.168.3.2   1.1.1.1
Total Nets: 7
Intra Area: 2   Inter Area: 4   ASE: 1   NSSA: 0
<AR3>
```

从 AdvRouter 字段可以看出，发布路由器的 Router ID 变为 1.1.1.1，即路由器 AR1 成为转换路由器。

（14）配置禁止向 NSSA 通告 LSA3，缩小路由表的规模。

[AR1]ospf 1
[AR1-ospf-1]area 1
[AR1-ospf-1-area-0.0.0.1]nssa default-route-advertise no-summary
[AR1-ospf-1-area-0.0.0.1]quit
[AR1-ospf-1]quit
[AR1]

 注意 配置 no-summary 参数可能会导致路由振荡。

（15）验证配置结果，查看路由器 AR4 的 OSPF 协议路由表信息。

```
<AR4>display ospf routing
     OSPF Process 1 with Router ID 4.4.4.4
         Routing Tables
Routing for Network
Destination       Cost   Type        NextHop       AdvRouter     Area
192.168.2.0/24    1      Transit     192.168.2.2   4.4.4.4       0.0.0.1
192.168.4.0/24    1      Transit     192.168.4.1   4.4.4.4       0.0.0.1
0.0.0.0/0         2      Inter-area  192.168.2.1   1.1.1.1       0.0.0.1
192.168.0.0/24    2      Transit     192.168.4.2   2.2.2.2       0.0.0.1
192.168.0.0/24    2      Transit     192.168.2.1   2.2.2.2       0.0.0.1
192.168.1.0/24    3      Inter-area  192.168.4.2   2.2.2.2       0.0.0.1
192.168.3.0/24    2      Inter-area  192.168.4.2   2.2.2.2       0.0.0.1
```

```
    Total Nets: 7
     Intra Area: 4   Inter Area: 3   ASE: 0   NSSA: 0
<AR4>
```

从以上信息可以看出，ABR 不再向 NSSA 内发送 LSA3，路由表规模减小。

（16）显示路由器 AR1 的配置信息。

```
<AR1>display current-configuration
#
 sysname AR1
#
 #
router id 1.1.1.1
#
interface GigabitEthernet0/0/0
 ip address 192.168.0.1 255.255.255.0
#
interface GigabitEthernet0/0/1
 ip address 192.168.1.1 255.255.255.0
#
interface GigabitEthernet0/0/2
 ip address 192.168.2.1 255.255.255.0
#
ospf 1
 area 0.0.0.0
  network 192.168.1.0 0.0.0.255
 area 0.0.0.1
  network 192.168.0.0 0.0.0.255
  network 192.168.2.0 0.0.0.255
  nssa default-route-advertise no-summary
#
return
<AR1>
```

5.3.7　OSPF 协议与 BFD 联动配置

如果需要提高链路状态变化时 OSPF 协议的收敛速度，可以在运行 OSPF 协议的链路上配置 BFD 特性。当 BFD 检测到链路出现故障时，能够将故障通告给路由协议，触发路由协议的快速收敛；当邻居关系为 Down 时，将动态删除 BFD 会话。

OSPF 协议通过周期性地向邻居发送 Hello 报文来实现邻居检测，检测到故障所需时间比较长，超过 1s。随着科技的发展，语音、视频等业务应用广泛，而这些业务对于丢包和时延非常敏感，当数据达到吉比特速率级时，较长的检测时间会导致大量数据丢失，无法满足电信级网络高可靠性的需求。

为了解决上述问题，配置指定进程或指定端口的 BDF for OSPF 特性，可以快速检测链路的状态，故障检测时间可以达到毫秒级，提高链路状态变化时 OSPF 协议的收敛速度。

注意　如果因绑定的对端 IP 地址改变而引起路由切换到其他链路上，则只有当原链路转发不通时，才会重新建立 BFD 会话。

1. 配置 OSPF 协议与 BFD 联动的基本流程

在配置 OSPF 协议与 BFD 联动前，需要配置端口的网络层地址，使各相邻节点网络层可达，再配置 OSPF 协议的基本功能。OSPF 协议与 BFD 联动的基本流程如图 5.9 所示。

（1）配置全局 BFD 功能。
（2）配置 OSPF 协议的 BFD 特性。
（3）（可选）阻止端口动态创建 BFD 会话。
（4）（可选）配置指定端口的 BFD 特性。

2. 配置 OSPF 协议与 BFD 联动的步骤

（1）配置全局 BFD 功能。

① 使用 system-view 命令，进入系统视图。
② 使用 bfd 命令，配置全局 BFD 功能并进入全局 BFD 视图。

（2）配置 OSPF 协议的 BFD 特性。

图 5.9　OSPF 协议与 BFD 联动的基本流程

① 使用 system-view 命令，进入系统视图。
② 使用 ospf [process-id]命令，进入 OSPF 协议进程视图。
③ 使用 bfd all-interfaces enable 命令，使能 OSPF 协议的 BFD 特性，建立 BFD 会话。

当配置了全局 BFD 特性，且邻居状态为 Full 时，OSPF 协议会为该进程下所有具有邻接关系的邻居建立 BFD 会话。

如果需要配置 BFD 参数，则可使用 bfd all-interfaces { min-rx-interval receive-interval | min-tx-interval transmit-interval | detect-multiplier multiplier-value | frr-binding } *命令，指定需要建立 BFD 会话的各个参数值。

min-rx-interval receive-interval 表示期望从对端接收 BFD 报文的最小接收间隔。
min-tx-interval transmit-interval 表示向对端发送 BFD 报文的最小发送间隔。
detect-multiplier multiplier-value 表示本地检测倍数。
frr-binding 表示将 BFD 会话状态与 OSPF IP FRR 绑定。

> **注意**　如果只是通过 bfd all-interfaces { min-rx-interval receive-interval | min-tx-interval transmit-interval| detect-multiplier multiplier-value | frr-binding } * 命令配置了 BFD 参数，而没有使用 bfd all-interfaces enable 命令，则不会使能 BFD 特性。

BFD 报文实际发送时间间隔和检测倍数一般推荐使用默认值，即不使用该命令。

具体参数如何配置取决于网络状况以及对网络可靠性的要求，对于网络可靠性要求较高的链路，可以减小 BFD 报文实际发送时间间隔；对于网络可靠性要求较低的链路，可以增大 BFD 报文实际发送时间间隔。

（3）（可选）阻止端口动态创建 BFD 会话。

在 OSPF 协议进程中使用 bfd all-interfaces enable 命令后，该进程下所有使能 BFD 协议且邻居状态为 Full 的邻居都将创建 BFD 会话。

① 使用 system-view 命令，进入系统视图。
② 使用 interface interface-type interface-number 命令，进入运行 OSPF 协议与 BFD 联动的端口视图。

③ 使用 ospf bfd block 命令，阻止端口动态创建 BFD 会话。

（4）（可选）配置指定端口的 BFD 特性。

在某些应用场景下，需要只对某些指定的端口配置 BFD for OSPF 特性，当这些端口的链路发生故障时，路由器可以快速地感知到，并及时通知 OSPF 协议重新计算路由，从而提高 OSPF 协议的收敛速度。且当邻居关系为 Down 时，动态删除 BFD 会话。

OSPF 协议创建 BFD 会话时，需要先配置全局 BFD 功能。可在指定端口配置 BFD 会话的路由器上进行以下配置。

① 使用 system-view 命令，进入系统视图。

② 使用 interface interface-type interface-number 命令，进入运行 OSPF 协议与 BFD 联动的端口视图。

③ 使用 ospf bfd enable 命令，打开端口 BFD 特性的开关，建立 BFD 会话。

当配置了全局 BFD 特性，且邻居状态为 Full 时，OSPF 协议为指定的端口建立了使用默认参数值的 BFD 会话。

> **注意** 端口上配置的 BFD for OSPF 特性的优先级高于进程中配置的 BFD for OSPF 特性的优先级。
> 如果需要单独配置 BFD 参数，则可使用 ospf bfd { min-rx-interval receive-interval | min-tx-interval transmit- interval | detect-multiplier multiplier-value } * 命令，指定 BFD 会话的参数值。

3. 配置 OSPF 协议与 BFD 联动特性实例

如图 5.10 所示，路由器 AR1、路由器 AR2、路由器 AR3 和路由器 AR4 之间运行 OSPF 协议。使能路由器 AR1、路由器 AR2、路由器 AR3 和路由器 AR4 的 OSPF 协议进程的 BFD 特性；业务流量在主链路路由器 AR1→路由器 AR2→路由器 AR4 上传送，链路路由器 AR1→路由器 AR3→路由器 AR2→路由器 AR4 作为备份链路；在路由器 AR1 和路由器 AR2 之间的链路上配置端口的 BFD 特性，当路由器 AR1 和路由器 AR2 之间的链路出现故障时，BFD 能够快速检测到故障并通告 OSPF 协议，使业务流量使用备份链路传送，配置相关端口与 IP 地址，进行网络拓扑连接。

V5-8 配置 OSPF 协议与 BFD 联动特性实例

图 5.10 配置 OSPF 协议与 BFD 联动特性实例

（1）对路由器 AR1 进行相关配置，相关实例代码如下。

V5-9 配置 OSPF 协议与 BFD 联动特性实例
——测试结果

```
<Huawei>system-view
[Huawei]sysname AR1
[AR1]router id 1.1.1.1
[AR1]interface GigabitEthernet 0/0/0
[AR1-GigabitEthernet0/0/0]ip address 192.168.1.1 24
[AR1]interface GigabitEthernet 0/0/1
[AR1-GigabitEthernet0/0/1]ip address 192.168.2.1 24
[AR1-GigabitEthernet0/0/1]quit
[AR1]ospf 1
[AR1-ospf-1]area 0
[AR1-ospf-1-area-0.0.0.0]network 192.168.1.0 0.0.0.255
[AR1-ospf-1-area-0.0.0.0]network 192.168.2.0 0.0.0.255
[AR1-ospf-1-area-0.0.0.0]quit
[AR1-ospf-1]quit
[AR1]
```

（2）对路由器 AR2 进行相关配置，相关实例代码如下。

```
<Huawei>system-view
[Huawei]sysname AR2
[AR2]router id 2.2.2.2
[AR2]interface GigabitEthernet 0/0/0
[AR2-GigabitEthernet0/0/0]ip address 192.168.1.2 24
[AR2]interface GigabitEthernet 0/0/1
[AR2-GigabitEthernet0/0/1]ip address 192.168.4.1 24
[AR2-GigabitEthernet0/0/1]quit
[AR2]interface GigabitEthernet 0/0/2
[AR2-GigabitEthernet0/0/2]ip address 192.168.3.1 24
[AR2-GigabitEthernet0/0/2]quit
[AR2]ospf 1
[AR2-ospf-1]area 0
[AR2-ospf-1-area-0.0.0.0]network 192.168.1.0 0.0.0.255
[AR2-ospf-1-area-0.0.0.0]network 192.168.3.0 0.0.0.255
[AR2-ospf-1-area-0.0.0.0]network 192.168.4.0 0.0.0.255
[AR2-ospf-1-area-0.0.0.1]quit
[AR2-ospf-1]quit
[AR2]
```

（3）对路由器 AR3 进行相关配置，相关实例代码如下。

```
<Huawei>system-view
[Huawei]sysname AR3
[AR3]router id 3.3.3.3
[AR3]interface GigabitEthernet 0/0/1
[AR3-GigabitEthernet0/0/1]ip address 192.168.2.2 24
[AR3-GigabitEthernet0/0/1]quit
[AR3]interface GigabitEthernet 0/0/2
[AR3-GigabitEthernet0/0/2]ip address 192.168.3.2 24
[AR3-GigabitEthernet0/0/2]quit
[AR3]ospf 1
```

```
[AR3-ospf-1]area 0
[AR3-ospf-1-area-0.0.0.0]network 192.168.2.0 0.0.0.255
[AR3-ospf-1-area-0.0.0.0]network 192.168.3.0 0.0.0.255
[AR3-ospf-1-area-0.0.0.1]quit
[AR3-ospf-1]quit
[AR3]
```

（4）对路由器 AR4 进行相关配置，相关实例代码如下。

```
<Huawei>system-view
[Huawei]sysname AR4
[AR4]router id 4.4.44
[AR4]interface GigabitEthernet 0/0/1
[AR4-GigabitEthernet0/0/1]ip address 192.168.4.2 24
[AR4-GigabitEthernet0/0/1]quit
[AR4]ospf 1
[AR4-ospf-1]area 0
[AR4-ospf-1-area-0.0.0.0]network 192.168.4.0 0.0.0.255
[AR4-ospf-1-area-0.0.0.1]quit
[AR4-ospf-1]quit
[AR4]
```

（5）配置完成后，使用 display ospf peer 命令，可以看到路由器 AR1、路由器 AR2、路由器 AR3 和路由器 AR4 之间互相建立了邻居关系，这里以路由器 AR1 为例进行介绍。

```
<AR1>display ospf peer
         OSPF Process 1 with Router ID 1.1.1.1
             Neighbors
 Area 0.0.0.0 interface 192.168.1.1(GigabitEthernet0/0/0)'s neighbors
 Router ID: 2.2.2.2           Address: 192.168.1.2
    State: Full   Mode:Nbr is    Master   Priority: 1
    DR: 192.168.1.1  BDR: 192.168.1.2  MTU: 0
    Dead timer due in 33   sec
    Retrans timer interval: 5
    Neighbor is up for 00:07:13
    Authentication Sequence: [ 0 ]
             Neighbors
 Area 0.0.0.0 interface 192.168.2.1(GigabitEthernet0/0/1)'s neighbors
 Router ID: 3.3.3.3           Address: 192.168.2.2
    State: Full   Mode:Nbr is    Master   Priority: 1
    DR: 192.168.2.1  BDR: 192.168.2.2  MTU: 0
    Dead timer due in 31   sec
    Retrans timer interval: 5
    Neighbor is up for 00:03:35
    Authentication Sequence: [ 0 ]
<AR1>
```

查看路由器 AR1 的 OSPF 协议路由表的信息，应有去往路由器 AR2 和路由器 AR3 的路由表项，去往 192.168.4.0/24 的路由的下一跳地址为 192.168.1.2，流量在主链路路由器 AR1→路由器 AR21→路由器 AR4 上传送。

（6）跟踪数据流量走向，这里以路由器 AR1 为例进行介绍。

```
<AR1>tracert 192.168.4.2
```

```
traceroute to   192.168.4.2(192.168.4.2), max hops: 30 ,packet length: 40,press
CTRL_C to break
 1 192.168.1.2 30 ms    30 ms    20 ms
 2 192.168.4.2 40 ms    30 ms    10 ms
<AR1>
```

（7）查看路由器 AR1 的路由表信息。

```
<AR1>display ospf routing
         OSPF Process 1 with Router ID 1.1.1.1
                 Routing Tables
 Routing for Network
 Destination         Cost    Type      NextHop          AdvRouter        Area
 192.168.1.0/24      1       Transit   192.168.1.1      1.1.1.1          0.0.0.0
 192.168.2.0/24      1       Transit   192.168.2.1      1.1.1.1          0.0.0.0
 192.168.3.0/24      2       Transit   192.168.1.2      2.2.2.2          0.0.0.0
 192.168.3.0/24      2       Transit   192.168.2.2      2.2.2.2          0.0.0.0
 192.168.4.0/24      2       Transit   192.168.1.2      2.2.2.2          0.0.0.0
 Total Nets: 5
 Intra Area: 5  Inter Area: 0   ASE: 0   NSSA: 0
<AR1>
```

（8）配置 OSPF 协议与 BFD 联动。

在路由器 AR1、路由器 AR2、路由器 AR3 和路由器 AR4 上使能全局 BFD 特性。这里以路由器 AR1 为例进行介绍，其他路由器的设置与路由器 AR1 相同，这里不再赘述。

```
[AR1]bfd
[AR1-bfd]quit
[AR1]ospf 1
[AR1-ospf-1]bfd all-interfaces enable
[AR1-ospf-1]quit
[AR1]
```

（9）配置完成后，在路由器 AR1 或路由器 AR2、路由器 AR3、路由器 AR4 上使用 display ospf bfd session all 命令，可以看到 BFDState 为 Up，这里以路由器 AR1 为例进行介绍。

```
<AR1>display ospf bfd session all
         OSPF Process 1 with Router ID 1.1.1.1

   Area 0.0.0.0 interface 192.168.1.1(GigabitEthernet0/0/0)'s BFD Sessions
 NeighborId:2.2.2.2         AreaId:0.0.0.0         Interface:GigabitEthernet0/0/0
 BFDState:up                rx     :1000           tx            :1000
 Multiplier:3               BFD Local Dis:8192     LocalIpAdd:192.168.1.1
 RemoteIpAdd:192.168.1.2    Diagnostic Info:No diagnostic information

   Area 0.0.0.0 interface 192.168.2.1(GigabitEthernet0/0/1)'s BFD Sessions
 NeighborId:3.3.3.3         AreaId:0.0.0.0         Interface:GigabitEthernet0/0/1
 BFDState:up                rx     :1000           tx            :1000
 Multiplier:3               BFD Local Dis:8193     LocalIpAdd:192.168.2.1
 RemoteIpAdd:192.168.2.2    Diagnostic Info:No diagnostic information
<AR1>
```

（10）配置端口的 BFD 特性，在路由器 AR1 的 GE 0/0/0 端口上配置 BFD 特性，并指定最小发送和接收时间间隔为 300ms，本地检测时间倍数为 4。

```
[AR1]interface GigabitEthernet 0/0/0
[AR1-GigabitEthernet0/0/0]ospf bfd enable
[AR1-GigabitEthernet0/0/0]ospf bfd min-rx-interval 300 min-tx-interval 300 detect-multiplier 4
[AR1-GigabitEthernet0/0/0]quit
[AR1]
```

（11）配置端口的 BFD 特性，在路由器 AR2 的 GE 0/0/0 端口上配置 BFD 特性，并指定最小发送和接收时间间隔为 300ms，本地检测时间倍数为 4。

```
[AR2]interface GigabitEthernet 0/0/0
[AR2-GigabitEthernet0/0/0]ospf bfd enable
[AR2-GigabitEthernet0/0/0]ospf bfd min-rx-interval 300 min-tx-interval 300 detect-multiplier 4
[AR2-GigabitEthernet0/0/0]quit
[AR2]
```

（12）配置完成后，在路由器 AR1 或路由器 AR2 上使用 display ospf bfd session all 命令，可以看到 BFDState 为 Up，这里以路由器 AR1 为例进行介绍。

```
<AR1>display ospf bfd session all
     OSPF Process 1 with Router ID 1.1.1.1
   Area 0.0.0.0 interface 192.168.1.1(GigabitEthernet0/0/0)'s BFD Sessions
 NeighborId:2.2.2.2         AreaId:0.0.0.0          Interface:GigabitEthernet0/0/0
 BFDState:up                rx        :300          tx         :300
 Multiplier:4               BFD Local Dis:8192      LocalIpAdd:192.168.1.1
 RemoteIpAdd:192.168.1.2    Diagnostic Info:No diagnostic information
   Area 0.0.0.0 interface 192.168.2.1(GigabitEthernet0/0/1)'s BFD Sessions
 NeighborId:3.3.3.3         AreaId:0.0.0.0          Interface:GigabitEthernet0/0/1
 BFDState:up                rx        :1000         tx         :1000
 Multiplier:3               BFD Local Dis:8193      LocalIpAdd:192.168.2.1
 RemoteIpAdd:192.168.2.2    Diagnostic Info:No diagnostic information
<AR1>
```

（13）验证配置结果，对路由器 AR2 的 GE 0/0/0 端口使用 shutdown 命令，模拟主链路出现故障。

```
[AR2]interface GigabitEthernet 0/0/0
[AR2-GigabitEthernet0/0/0]shutdown
[AR2-GigabitEthernet0/0/0]quit
[AR2]
```

（14）查看路由器 AR1 的 OSPF 协议路由表信息。可以看出，在主链路失效后，备份链路路由器 AR1→路由器 AR3→路由器 AR2→路由器 AR4 生效，去往 192.168.4.0/24 的路由的下一跳地址为 192.168.2.2。

```
<AR1>display ospf routing
      OSPF Process 1 with Router ID 1.1.1.1
              Routing Tables
 Routing for Network
 Destination        Cost    Type     NextHop         AdvRouter       Area
 192.168.2.0/24     1       Transit  192.168.2.1     1.1.1.1         0.0.0.0
 192.168.3.0/24     2       Transit  192.168.2.2     2.2.2.2         0.0.0.0
 192.168.4.0/24     3       Transit  192.168.2.2     2.2.2.2         0.0.0.0
 Total Nets: 3
 Intra Area: 3   Inter Area: 0   ASE: 0   NSSA: 0
<AR1>
```

（15）跟踪数据流量走向，这里以路由器 AR1 为例进行介绍。

```
<AR1>tracert 192.168.4.2
 traceroute to    192.168.4.2(192.168.4.2), max hops: 30 ,packet length: 40,press
CTRL_C to break
 1 192.168.2.2 20 ms    20 ms    20 ms
 2 192.168.3.1 30 ms    30 ms    40 ms
 3 192.168.4.2 40 ms    40 ms    30 ms
<AR1>
```

5.3.8　OSPF 协议引入外部路由配置

当 OSPF 协议网络中的设备需要访问运行其他协议的网络中的设备时，需要将其他协议的路由引入 OSPF 协议网络中。

OSPF 协议是一种无环路的动态路由协议，但这是针对域内路由和域间路由而言的，其对引入的外部路由环路没有很好的防范机制，所以在配置 OSPF 协议引入外部路由时一定要慎重，以防止手工配置引起路由环路。

1. 配置 OSPF 协议引入其他协议的路由

（1）使用 system-view 命令，进入系统视图。

（2）使用 ospf [process-id] 命令，进入 OSPF 协议进程视图。

（3）使用 import-route { limit limit-number | { bgp [permit-ibgp] | direct | unr |rip [process-id-rip] | static | isis [process-id-isis] | ospf [process-id-ospf] }[cost cost | type type | tag tag | route-policy route-policy-name] * }命令，引入其他协议的路由信息。

2. 配置 OSPF 协议引入路由时的相关参数

（1）使用 system-view 命令，进入系统视图。

（2）使用 ospf [process-id] 命令，进入 OSPF 协议进程视图。

（3）使用 default { cost { cost-value | inherit-metric } | limit limit | tag tag | type type } * 命令，配置引入路由时的参数默认值（路由度量、标记、类型）。

当 OSPF 协议引入外部路由时，可以配置一些额外参数的默认值，如开销、路由度量、标记和类型。路由标记可以用来标识协议相关的信息，如 OSPF 协议接收 BGP 时用来区分 AS 的编号。

默认情况下，OSPF 协议引入外部路由的度量值为 1，标记值为 1，引入的外部路由类型为 LSA2。

> **注意**　可以通过以下 3 条命令设置引入路由的开销值，其优先级依次递减。
> （1）通过 apply cost 命令设置引入路由的开销值。
> （2）通过 import-route 命令设置引入路由的开销值。
> （3）通过 default 命令设置引入路由的默认开销值。

3. 配置 OSPF 协议将默认路由通告到 OSPF 协议路由区域

在实际组网中，OSPF 协议区域边界和 AS 边界通常是由多个路由器组成的多出口冗余备份或者负载均衡。此时，为了减小路由表的容量，可以配置默认路由，保证网络的高可用性。

OSPF 默认路由通常应用于以下两种情况。

（1）由 ABR 发布 LSA3，用来指导区域内路由器进行区域之间报文的转发。

（2）由 ASBR 发布 LSA5 或 LSA7，用来指导 OSPF 协议路由域内路由器进行域外报文的转发。

当路由器无精确匹配的路由时，就可以通过默认路由进行报文转发。LSA3 默认路由的优先级高于 LSA5 或 LSA7 默认路由的优先级。OSPF 协议默认路由的发布方式取决于引入默认路由的区域类型，如表 5.3 所示。

表 5.3　OSPF 默认路由的发布方式

区域类型	产生条件	发布方式	产生 LSA 的类型	泛洪范围
普通区域	通过 default-route-advertise 命令配置	ASBR 发布	LSA5	普通区域
Stub Area	自动产生	ABR 发布	LSA3	Stub Area
NSSA	通过 nssa [default-route-advertise] 命令配置	ASBR 发布	LSA7	NSSA
Total NSSA	自动产生	ASBR 发布	LSA3	NSSA

4．配置 OSPF 协议将默认路由通告到 OSPF 协议路由区域的步骤

（1）使用 system-view 命令，进入系统视图。

（2）使用 ospf [process-id] 命令，进入 OSPF 进程视图。

（3）根据实际情况进行以下配置。

如果要将默认路由通告到 OSPF 协议路由区域，则使用 default-route-advertise [[always | permit-calculate-other] | cost cost | type type | route-policy-route-policy-name [match-any]] * 命令。

① always 表示无论本机是否存在激活的非本 OSPF 协议进程的默认路由，都会产生并发布一个描述默认路由的 LSA。

② permit-calculate-other 表示在发布默认路由后，仍允许计算其他路由器发布的默认路由。

③ route-policy route-policy-name 表示当路由表中有匹配的非本 OSPF 协议进程产生的默认路由表项时，按路由策略所配置的参数发布默认路由的匹配规则。

如果要指定 LSA3 的默认开销值，则应使用 default-route-advertise summary cost cost 命令。在选用该参数时，必须先配置虚拟专用网络（Virtual Private Network，VPN），否则无法配置该命令。

注意　本机必须存在激活的非本 OSPF 协议进程的默认路由，才会产生并发布一个默认路由的 ASE LSA。
在 OSPF 协议路由域中，在通告默认路由前，会比较默认路由的优先级，如果其中某台 OSPF 协议路由器同时配置了静态默认路由，要使 OSPF 协议通告的默认路由加入到当前的路由表中，就必须保证所配置的静态默认路由的优先级比 OSPF 协议通告的默认路由的优先级低。

5．配置 OSPF 协议引入外部路由

如图 5.11 所示，配置 OSPF 协议引入外部路由，配置相关端口与 IP 地址，进行网络拓扑连接。

V5-10 配置 OSPF 协议引入外部路由

图 5.11　配置 OSPF 协议引入外部路由

（1）对路由器 AR1 进行相关配置，相关实例代码如下。

```
<Huawei>system-view
[Huawei]sysname AR1
[AR1]router id 1.1.1.1
[AR1]interface GigabitEthernet 0/0/0
[AR1-GigabitEthernet0/0/0]ip address 192.168.1.1 24
[AR1]interface GigabitEthernet 0/0/1
[AR1-GigabitEthernet0/0/1]ip address 192.168.2.1 24
[AR1-GigabitEthernet0/0/1]quit
[AR1]ospf 1
[AR1-ospf-1]area 0
[AR1-ospf-1-area-0.0.0.0]network 192.168.1.0 0.0.0.255
[AR1-ospf-1-area-0.0.0.0]quit
[AR1-ospf-1]import-route rip 1 cost 50          //引入 RIP 路由
[AR1-ospf-1]quit
[AR1]rip 1
[AR1-rip-1]version 2
[AR1-rip-1]network 192.168.2.0
[AR1-rip-1]import-route ospf 1 cost 1           //引入 OSPF 协议路由
[AR1-rip-1]quit
[AR1]
```

（2）对路由器 AR2 进行相关配置，相关实例代码如下。

```
<Huawei>system-view
[Huawei]sysname AR2
[AR2]router id 2.2.2.2
[AR2]interface GigabitEthernet 0/0/0
[AR2-GigabitEthernet0/0/0]ip address 192.168.1.2 24
[AR2]interface GigabitEthernet 0/0/1
[AR2-GigabitEthernet0/0/1]ip address 192.168.3.1 24
[AR2-GigabitEthernet0/0/1]quit
[AR2]ospf 1
[AR2-ospf-1]area 0
[AR2-ospf-1-area-0.0.0.0]network 192.168.1.0 0.0.0.255
```

[AR2-ospf-1-area-0.0.0.0]quit
[AR2-ospf-1]area 1
[AR2-ospf-1-area-0.0.0.1]network 192.168.3.0 0.0.0.255
[AR2-ospf-1]quit
[AR2]

（3）对路由器 AR3 进行相关配置，相关实例代码如下。

<Huawei>system-view
[Huawei]sysname AR3
[AR3]router id 3.3.3.3
[AR3]interface GigabitEthernet 0/0/1
[AR3-GigabitEthernet0/0/1]ip address 192.168.2.2 24
[AR3-GigabitEthernet0/0/1]quit
[AR3]rip 1
[AR3-rip-1]version 2
[AR3-rip-1]network 192.168.2.0
[AR3-rip-1]quit
[AR3]

（4）对路由器 AR4 进行相关配置，相关实例代码如下。

<Huawei>system-view
[Huawei]sysname AR4
[AR4]router id 4.4.4.4
[AR4]interface GigabitEthernet 0/0/1
[AR4-GigabitEthernet0/0/1]ip address 192.168.3.2 24
[AR4-GigabitEthernet0/0/1]quit
[AR4]ospf 1
[AR4-ospf-1]area 1
[AR4-ospf-1-area-0.0.0.1]network 192.168.3.0 0.0.0.255
[AR4-ospf-1-area-0.0.0.1]quit
[AR4-ospf-1]quit
[AR4]

（5）查看路由器 AR1 的配置信息。

<AR1>display current-configuration
#
 sysname AR1
#
 router id 1.1.1.1
#
interface GigabitEthernet0/0/0
 ip address 192.168.1.1 255.255.255.0
#
interface GigabitEthernet0/0/1
 ip address 192.168.2.1 255.255.255.0
#
ospf 1
 import-route rip 1 cost 50
 area 0.0.0.0
  network 192.168.1.0 0.0.0.255
#
rip 1
 version 2

```
   network 192.168.2.0
 import-route ospf 1 cost 1
#
return
<AR1>
```

（6）查看路由器 AR4 的路由表信息。

```
<AR4>display ospf routing
       OSPF Process 1 with Router ID 192.168.3.2
              Routing Tables
Routing for Network
Destination        Cost      Type           NextHop           AdvRouter         Area
192.168.3.0/24     1         Transit        192.168.3.2       192.168.3.2       0.0.0.1
192.168.1.0/24     2         Inter-area     192.168.3.1       2.2.2.2           0.0.0.1
Routing for ASEs
Destination        Cost      Type           Tag           NextHop         AdvRouter
192.168.2.0/24     50        Type2          1             192.168.3.1     1.1.1.1
Total Nets: 3
Intra Area: 1   Inter Area: 1   ASE: 1   NSSA: 0
<AR4>
```

（7）验证配置测试结果，在路由器 AR4 上访问路由器 AR3 的 GE 0/0/1 端口。

```
<AR4>ping 192.168.2.2
   PING 192.168.2.2: 56   data bytes, press CTRL_C to break
     Reply from 192.168.2.2: bytes=56 Sequence=1 ttl=253 time=50 ms
     Reply from 192.168.2.2: bytes=56 Sequence=2 ttl=253 time=40 ms
     Reply from 192.168.2.2: bytes=56 Sequence=3 ttl=253 time=40 ms
     Reply from 192.168.2.2: bytes=56 Sequence=4 ttl=253 time=50 ms
     Reply from 192.168.2.2: bytes=56 Sequence=5 ttl=253 time=30 ms
   --- 192.168.2.2 ping statistics ---
     5 packet(s) transmitted
     5 packet(s) received
     0.00% packet loss
     round-trip min/avg/max = 30/42/50 ms
<AR4>
```

5.3.9　OSPFv3 协议配置

OSPFv3 协议是运行在 IPv6 网络中的 OSPF 协议。运行 OSPFv3 的路由器使用物理端口的链路本地单播地址为源地址来发送 OSPF 协议报文。相同链路上的路由器互相学习与之相连的其他路由器的链路本地地址，并在报文转发的过程中将学习到的地址作为下一跳信息使用。OSPFv3 协议中使用组播地址 ff02::5 来表示 All Routers，而 OSPFv2 协议中使用的是组播地址 224.0.0.5。需要注意的是，OSPFv3 协议和 OSPFv2 协议互不兼容。

Router ID 在 OSPFv3 中同样用于标识路由器。与 OSPFv2 协议的 Router ID 不同，OSPFv3 协议的 Router ID 必须手工配置；如果没有手工配置 Router ID，则 OSPFv3 协议将无法正常运行。OSPFv3 协议在 BMA 和 NBMA 网络中选举 DR 和 BDR 的过程与 OSPFv2 协议相似。但 IPv6 中使用组播地址 FF02::6 表示 All Routers，而 OSPFv2 协议中使用的是组播地址 224.0.0.6。

OSPFv3 协议是基于链路而不是网段的。在配置 OSPFv3 协议时，不需要考虑路由器的端口是否配置在同一网段中，只要路由器的端口连接在同一链路上，就可以不配置 IPv6 全局地址而直接建立联系。这一变化影响了 OSPFv3 协议报文的接收、Hello 报文的内容及 LSA2 的内容。

OSPFv3 协议直接使用 IPv6 的扩展头部（认证头（Authentication Header，AH）和封装安全载荷（Encapsulating Security Payload，ESP））来实现认证及安全处理，不再需要 OSPFv3 协议自身来完成认证。

（1）配置 OSPFv3 协议，配置相关端口与 IP 地址，如图 5.12 所示，进行网络拓扑连接。

V5-11 配置 OSPFv3 协议

图 5.12 配置 OSPFv3 协议

（2）对路由器 AR1 进行相关配置，相关实例代码如下。

```
<Huawei>system-view
[Huawei]sysname AR1
[AR1]ipv6
[AR1]ospfv3
[AR1-ospfv3-1]router-id 10.10.10.10
[AR1-ospfv3-1]quit
[AR1]interface GigabitEthernet 0/0/0
[AR1-GigabitEthernet0/0/0]ipv6 enable
[AR1-GigabitEthernet0/0/0]ipv6 address 2005::1 64
[AR1-GigabitEthernet0/0/0]ospfv3 1 area 0
[AR1-GigabitEthernet0/0/0]quit
[AR1]interface GigabitEthernet 0/0/1
[AR1-GigabitEthernet0/0/1]ipv6 enable
[AR1-GigabitEthernet0/0/1]ipv6 address 2001::254 64
[AR1-GigabitEthernet0/0/1]ospfv3 1 area 0
[AR1-GigabitEthernet0/0/1]quit
[AR1]interface GigabitEthernet 0/0/2
[AR1-GigabitEthernet0/0/2]ipv6 enable
[AR1-GigabitEthernet0/0/2]ipv6 address 2002::254 64
[AR1-GigabitEthernet0/0/2]ospfv3 1 area 0
[AR1]interface loopback 1
[AR1-LoopBack1]ipv6 enable
[AR1-LoopBack1]ipv6 address 2007::1 64
[AR1-LoopBack1]ospfv3 1 area 0
[AR1-LoopBack1]quit
[AR1]
```

（3）对路由器 AR2 进行相关配置，相关实例代码如下。

```
<Huawei>system-view
[Huawei]sysname AR2
[AR2]ipv6
[AR2]ospfv3
[AR2-ospfv3-1]router-id 20.20.20.20
[AR2-ospfv3-1]quit
[AR2]interface GigabitEthernet 0/0/0
[AR2-GigabitEthernet0/0/0]ipv6 enable
[AR2-GigabitEthernet0/0/0]ipv6 address 2005::2 64
[AR2-GigabitEthernet0/0/0]ospfv3 1 area 0
[AR2-GigabitEthernet0/0/0]quit
[AR2]interface GigabitEthernet 0/0/1
[AR2-GigabitEthernet0/0/1]ipv6 enable
[AR2-GigabitEthernet0/0/1]ipv6 address 2003::254 64
[AR2-GigabitEthernet0/0/1]ospfv3 1 area 0
[AR2-GigabitEthernet0/0/1]quit
[AR2]interface GigabitEthernet 0/0/2
[AR2-GigabitEthernet0/0/2]ipv6 enable
[AR2-GigabitEthernet0/0/2]ipv6 address 2004::254 64
[AR2-GigabitEthernet0/0/2]ospfv3 1 area 0
[AR2]interface loopback 1
[AR2-LoopBack1]ipv6 enable
[AR2-LoopBack1]ipv6 address 2008::1 64
[AR2-LoopBack1]ospfv3 1 area 0
[AR2-LoopBack1]quit
[AR2]
```

（4）显示路由器 AR1、路由器 AR2 的配置信息，这里以路由器 AR1 为例进行介绍，其主要配置实例代码如下。

```
<AR1>display current-configuration
#
 sysname AR1
#
ipv6
ospfv3 1
 router-id 10.10.10.10
#
interface GigabitEthernet0/0/0
 ipv6 enable
 ipv6 address 2005::1/64
 ospfv3 1 area 0.0.0.0
#
interface GigabitEthernet0/0/1
 ipv6 enable
ipv6 address 2001::254/64
 ospfv3 1 area 0.0.0.0
#
interface GigabitEthernet0/0/2
 ipv6 enable
```

```
 ipv6 address 2002::254/64
 ospfv3 1 area 0.0.0.0
#
interface LoopBack1
 ipv6 enable
 ipv6 address 2007::1/64
 ospfv3 1 area 0.0.0.0
#
return
<AR1>
```

（5）显示路由器 AR1 的 OSPFv3 协议路由信息，如图 5.13 所示。

在邻居路由器上完成 OSPFv3 协议配置后，使用 display ospfv3 命令可以验证 OSPFv3 协议配置及其相关参数。从显示信息中可以看到，正在运行的 OSPFv3 协议进程为 1，Router ID 为 10.10.10.10，完全邻居数值为 1。

（6）主机 PC1 验证相关测试结果，如图 5.14 所示。

图 5.13　显示路由器 AR1 的 OSPFv3 协议路由信息

图 5.14　主机 PC1 验证相关测试结果

课后习题

1. 选择题

（1）关于 OSPF 协议路由聚合的描述错误的是（　　）。
　　A. 路由聚合是指将相同前缀的路由信息聚合一起，发布一条路由到其他区域
　　B. 通过路由聚合可以减少路由信息，从而缩小路由表的规模，提高路由器的性能
　　C. OSPF 协议有两种路由聚合方式，即 ABR 聚合和 ASBR 聚合
　　D. OSPF 协议中任意一台路由器都可以进行路由聚合的操作

（2）关于路由协议的开销值，以下描述不正确的是（　　）。
　　A. 通常，RIP 的开销基于路数，取值范围很小
　　B. 通常，IS-IS 协议和 OSPF 协议的开销值基于带宽，取值范围很大
　　C. 不同的路由协议计算路由开销的依据不同，在引入路由时一般建议自动转换
　　D. 不同的路由协议计算路由开销的依据不同，在引入路由时一般建议手动转换

（3）目前使用最广泛的 IGP 是（　　）。

　　　A．RIP　　　　　B．BGP　　　　　C．IS-IS 协议　　　　D．OSPF 协议

（4）OSPF 协议使用的组播地址为（　　）。

　　　A．224.0.0.6　　　B．224.0.0.7　　　C．224.0.0.8　　　D．224.0.0.9

（5）在 OSPF 协议中，Stub Area 与 Total Stub Area 的区别在于对（　　）的处理不同。

　　　A．LSA2　　　　B．LSA3　　　　C．LSA5　　　　D．LSA7

（6）在 OSPF 协议中，NSSA 与 Total Stub Area 的区别在于对（　　）的处理不同。

　　　A．LSA2　　　　B．LSA4　　　　C．LSA5　　　　D．LSA7

（7）关于配置 OSPF 协议中的 Stub 区域，下列说法错误的是（　　）。

　　　A．骨干区域不能配置成 Stub 区域，虚连接不能穿过 Stub 区域

　　　B．区域内的所有路由器不是必须配置该属性

　　　C．Stub 区域中不能存在 ASBR

　　　D．一个区域配置成 Stub 区域后，其他区域的 LSA3 可以在该区域中传播

（8）以下（　　）不是 OSPF 协议的特点。

　　　A．支持非区域划分　　B．支持身份验证　　C．无路由自环　　D．路由自动聚合

（9）下列关于 OSPF 和 RIPv2 的描述，正确的是（　　）。

　　　A．只能采取组播更新　　　　　　B．只传递路由状态信息

　　　C．都支持 VLSM　　　　　　　　D．都采用了水平分割的机制

（10）OSPF 协议的协议号是（　　）。

　　　A．53　　　　　B．69　　　　　C．89　　　　　D．520

（11）OSPF 协议计算开销时，依据的主要参数是（　　）。

　　　A．MTU　　　　B．带宽　　　　C．跳数　　　　D．时延

（12）下列不是 OSPF 协议中 Hello 报文的主要作用的是（　　）。

　　　A．发现邻居　　　　　　　　　　B．协商参数

　　　C．选举 DR、BDR　　　　　　　D．协商交换 DD 报文时的主从关系

（13）在 OSPF 协议中，当一个稳定的网络中加入一台优先级比原 DR 和 BDR 更高的路由器时，该路由器会（　　）。

　　　A．立刻成为 DR　　　　　　　　B．立刻成为 BDR

　　　C．等 DR 失效后，立刻成为 DR　D．成为 DROther 路由器

（14）在 OSPF 协议中，对于各种网络类型中的 DR 和 BDR 的说法中，错误的是（　　）。

　　　A．任何类型的网络都需要有 DR，但是不一定需要有 BDR

　　　B．点对点类型的网络中没有 DR

　　　C．NBMA 类型的网络中有 DR

　　　D．点对多点网络中没有 DR

（15）【多选】OSPF 协议中的骨干区域内肯定不存在（　　）。

　　　A．LSA3　　　　B．LSA4　　　　C．LSA5　　　　D．LSA7

2. 简答题

（1）简述 OSPF 协议认证方式的配置方法。

（2）简述 OSPF 协议的特殊区域的配置方法。

（3）简述 OSPF 协议与 BFD 联动的配置方法。

（4）简述 OSPF 协议引入外部路由的配置方法。

项目6
IS-IS协议

【学习目标】
- 掌握IS-IS协议的基本概念。
- 掌握IS-IS协议的工作原理。
- 掌握IS-IS协议与BFD联动的配置方法。

6.1 项目陈述

为什么要学习中间系统到中间系统（Intermediate System to Intermediate System，IS-IS）协议，这是我们需要了解的。IS-IS 协议与 OSPF 协议都是链路状态路由协议，它们不仅适用于LAN 环境，还适用于城域网环境。在 IP 城域网中，关键技术包括路由技术、端到端的 QoS 管理、接入网技术和用户/业务管理，路由技术中最常用的就是 BGP、OSPF 协议和 IS-IS 协议，想要了解城域网的知识，需要先学习 IS-IS 协议。

6.2 必备知识

6.2.1 IS-IS协议概述

IS-IS 协议属于 IGP，用于 AS 内部。IS-IS 也是一种链路状态协议，使用最短路径优先（Shortest Path First，SPF）算法进行路由计算。

IS-IS 协议最初是国际标准化组织（International Organization for Standardization，ISO）为无连接网络协议（Connection Less Network Protocol，CLNP）设计的一种动态路由协议。为了提供对 IP 路由的支持，ISD 对 IS-IS 协议进行了扩充和修改，使 IS-IS 协议能够同时应用在TCP/IP 和 OSI 环境中，形成了集成化 IS-IS（Integrated IS-IS 或 Dual IS-IS）协议。现在提到的 IS-IS 协议都是指集成化的 IS-IS 协议，主要用于城域网和承载网。

ISO 网络和 IP 网络的网络层地址的编址方式不同。IP 网络的三层地址是常见的 IPv4 地址或 IPv6 地址，IS-IS 协议将 ISO 网络层地址称为网络服务接入点（Network Service Access Point，NSAP），用来描述 ISO 模型的网络地址结构。

运行 IS-IS 协议的网络包含了终端系统（End System，ES）、中间系统（Intermediate

System，IS)、区域和路由域(Routing Domain)。路由器是 IS，主机是 ES。主机和路由器之间运行的协议称为 ES-IS，路由器与路由器之间运行的协议称为 IS-IS。

1. IS-IS 协议的报文类型

在 IS-IS 协议中，协议报文总共有 9 种，所有的协议报文都可根据层次划分为 Level1(简称 L1)和 Level2(简称 L2)报文。

(1) Hello 报文，用于建立和维护毗邻关系。

Hello 报文可以具体细分为 L1IIH、L2IIH 和 P-2-PIIH。

L1IIH 的组播地址为 0180-C200-0014。

L2IIH 的组播地址为 0180-C200-0015。

P-2-PIIH 采用单播地址进行通信。

Hello 报文的作用为发现、建立和维系邻居关系，其在功能上类似于 OSPF 协议中的 Hello 报文。

(2) 链路状态包(Link-State Packet，LSP)报文，用于发布链路状态信息。

LSP 分为两种：L1LinkStatePDU 和 L2LinkStatePDU。L2LinkStatePDU 包含 IS-IS 路由区域中所有可到达前缀的信息，L1LinkStatePDU 只用于本地区域。

LSP 报文描述了本路由器中所有的链路状态信息，其在功能上类似于 OSPF 协议中的 LSA。

(3) 完整序列号协议数据单元(Complete Sequence Number Protocol Data Unit，CSNP)报文，用于发布完整链路状态数据库。

CSNP 分为两种：L1CSNP 和 L2CSNP。

CSNP 用于数据库同步，以范围来描述 LSDB 中的所有的 LSPs，包含地址范围及各 LSP 的简要信息，如标识符、序列号、校验和、剩余时间等重要信息。在广播网络上，CSNP 报文由 DIS 定期发送(默认周期为 10s)；在点到点串行线路上，只在第一次形成邻接时发送 CSNP 报文。如果路由器的 LSDB 非常大，则将分为多个 CSNP 报文发出。其在功能上类似于 OSPF 协议中的 DD 报文。

(4) 部分序列号协议数据单元(Partial Sequence Number Protocol Data Unit，PSNP)报文，用于确认和请求链路状态信息。

PSNP 分为两种：L1PSNP 和 L2PSNP。

PSNP 报文用于数据库同步，主要有以下功能：在点对点链路上，其用来相互交换作为 Ack 应答以确认收到某个 LSP；在广播网络中，其用来请求发送最新的 LSP，当路由器从邻居收到 CSNP 报文时，若注意到某些本地数据库中没有 LSP(或自己的 LSP 比较旧)，则路由器会发送 PSNP 报文请求新的 LSP。PSNP 报文在头中利用标识符、序列号、校验和、剩余时间来描述 LSPs，仅仅包含 LSP 报文的头部。在点对点网络类型中，其在功能上类似于 OSPF 协议中的 LSAck 报文；在广播网络中，其在功能上类似于 OSPF 协议中的 LSRequest 报文。

2. IS-IS 协议的拓扑结构

为了支持大规模的路由网络，IS-IS 协议在 AS 内采用骨干区域与非骨干区域两级的分层结构。一般来说，将 L1 路由器部署在非骨干区域，L2 路由器和 L1/2 路由器部署在骨干区域。每一个非骨干区域都通过 L1/2 路由器与骨干区域相连。

图 6.1 所示为一个运行 IS-IS 协议的网络拓扑结构，它与 OSPF 协议的多区域网络拓扑结构非常相似，整个骨干区域不仅包括 Area 1 中的所有路由器，还包括其他区域的 L1/2 路由器。

图 6.1 IS-IS 协议拓扑结构 1

图 6.2 所示为 IS-IS 协议的另一种拓扑结构。在这个拓扑结构中，L2 的路由器没有在同一个区域中，而是分别属于不同的区域。此时，所有物理连续的 L1/2 和 L2 路由器就构成了 IS-IS 协议的骨干区域。

图 6.2 IS-IS 拓扑结构 2

通过以上两种拓扑结构可以体现 IS-IS 协议与 OSPF 协议的不同点。

（1）在 IS-IS 协议中，每台路由器都只属于一个区域；而在 OSPF 协议中，一台路由器的不同端口可以属于不同的区域。

（2）在 IS-IS 协议中，单个区域没有骨干区域与非骨干区域的概念；而在 OSPF 协议中，Area 0 被定义为骨干区域。

（3）在 IS-IS 协议中，L1 和 L2 级别的路由都采用了 SPF 算法，分别生成最短路径树（Shortest Path Tree，SPT）；而在 OSPF 协议中，只有在同一个区域内才使用 SPF 算法，区域之间的路由需要通过骨干区域来转发。

3. IS-IS 路由器的分类

（1）L1 路由器。

L1 路由器负责区域内的路由，它只与属于同一区域的 L1 和 L1/2 路由器形成邻居关系，属于不同区域的 L1 路由器不能形成邻居关系。L1 路由器只负责维护 L1 的 LSDB，该 LSDB 中包含本区域的路由信息，发送到本区域外的报文转发给最近的 L1/2 路由器。

（2）L2 路由器。

L2 路由器负责区域间的路由，它可以与相同或者不同区域的 L2 路由器或者其他区域的 L1/2 路由器形成邻居关系。L2 路由器维护一个 L2 的 LSDB，该 LSDB 中包含区域间的路由信息。

所有 L2 级别（即形成 L2 邻居关系）的路由器组成路由域的骨干网，负责不同区域间通信。路由域中 L2 级别的路由器必须是物理连续的，以保证骨干网的连续性。只有 L2 级别的路由器才能直接与区域外的路由器交换数据报文或路由信息。

（3）L1/2 路由器。

同时属于 L1 和 L2 的路由器称为 L1/2 路由器，它可以与同一区域的 L1 和 L1/2 路由器建立 L1 邻居关系，也可以与其他区域的 L2 和 L1/2 路由器建立 L2 邻居关系。L1 路由器必须通过 L1/2 路由器才能连接至其他区域。

L1/2 路由器维护两个 LSDB，L1 的 LSDB 用于区域内路由，L2 的 LSDB 用于区域间路由。

4. IS-IS 协议的网络类型

IS-IS 协议只支持两种类型的网络，根据物理链路不同可分为以下两种。

（1）广播链路：如 Ethernet（以太网）、Token-Ring（令牌环网）等。

（2）点对点链路：如 PPP 链路等。

注意 对于 NBMA 网络，如 ATM，需对其配置子端口，并注意子端口类型应配置为 P2P。IS-IS 协议不能在点对多点链路上运行。

5. DIS 和伪节点

在广播网络中，IS-IS 协议需要在所有的路由器中选举一个路由器作为指定中间系统（Designated Intermediate System，DIS）。DIS 用来创建和更新伪节点，并负责生成伪节点的链路状态协议数据单元，用来描述这个网络中有哪些网络设备。

伪节点是用来模拟广播网络的一个虚拟节点，并非真实的路由器。在 IS-IS 协议中，伪节点用 DIS 的 System ID 和一个字节的 Circuit ID（非 0 值）标识。

使用伪节点可以简化网络拓扑，使路由器产生的 LSP 长度较小。另外，当网络发生变化时，需要产生的 LSP 数量也会较少，减少 SPF 算法的资源消耗。

L1 和 L2 的 DIS 是分别选举的，用户可以为不同级别的 DIS 选举设置不同的优先级。DIS 优先级数值最大的被选为 DIS。如果优先级数值最大的路由器有多台，则其中 MAC 地址最大的路由器会被选中。不同级别的 DIS 可以是同一台路由器，也可以是不同的路由器。

IS-IS 协议中的 DIS 与 OSPF 协议中的指定路由器（Designated Router，DR）的区别如下。

（1）在 IS-IS 协议的广播网中，优先级为 0 的路由器也参与 DIS 的选举，而在 OSPF 协议中，优先级为 0 的路由器不参与 DR 的选举。

（2）在 IS-IS 协议的广播网中，当有新的路由器加入并符合成为 DIS 的条件时，这台路由器会被选举成为新的 DIS，原有的伪节点被删除，此更改会引起一组新的 LSP 泛洪。而在 OSPF 协议中，当一台新路由器加入后，即使它的 DR 优先级值最大，也不会立即成为该网段中的 DR。

（3）在 IS-IS 协议的广播网中，同一网段上的同一级别的路由器之间都会建立邻接关系，所有的非 DIS 路由器之间也会建立邻接关系。而在 OSPF 中，路由器只与 DR 和 BDR 建立邻接关系。

注意 IS-IS 协议的广播网中的所有路由器之间都会形成邻接关系，但 LSDB 的同步仍然依靠 DIS 来保证。

6. IS-IS 协议的地址结构

NSAP 是 OSI 参考模型中用于定位资源的地址。NSAP 的地址结构如图 6.3 所示，它由初始域部分（Initial Domain Part，IDP）和域特定部分（Domain Specific Part，DSP）组成。IDP 和 DSP 的长度都是可变的，NSAP 总长最多为 20 字节，最少为 8 字节。

图 6.3 NSAP 的地址结构

（1）IDP 相当于 IP 地址中的主网络号。它是由 ISO 规定的，并由权限和格式标识符（Authority and Format Identifier，AFI）与初始域标识符（Initial Domain Identifier，IDI）两部分组成。AFI 用来表示地址分配机构和地址格式，IDI 用来标识域。

（2）DSP 相当于 IP 地址中的子网号和主机地址。它由高阶数字信号处理器（High Order DSP）、系统标识（System ID）和指示服务类型（SEL）三部分组成。High Order DSP 用来分割区域，System ID 用来区分主机，SEL（NSAP Selector）用来指示服务类型。

（3）IDP 和 DSP 中的 High Order DSP 一起，既能够标识路由域，又能够标识路由域中的区域，因此，它们一起被称为区域地址（Area Address），相当于 OSPF 协议中的区域编号。同一 L1 区域内的所有路由器必须具有相同的区域地址，L2 区域内的路由器可以具有不同的区域地址。

一般情况下，一个路由器只需要配置一个区域地址，且同一区域中所有节点的区域地址都要相同。为了支持区域的平滑合并、分割及转换，在设备的实现中，一个 IS-IS 协议进程下最多可配置 3 个区域地址。

（4）System ID 用来在区域内唯一标识主机或路由器。在设备的实现中，它的长度固定为 48bit（6 字节）。

在实际应用中，一般使用 Router ID 与 System ID 进行对应。假设一台路由器使用端口 Loopback 0 的 IP 地址 192.16.10.1 作为 Router ID，则它在 IS-IS 协议中使用的 System ID 可通过如下方法转换得到。

首先，将 IP 地址 192.16.10.1 的每个十进制数都扩展为 3 位，不足 3 位的在前面补 0，得到 192.016.010.001。

然后，将扩展后的地址分为 3 部分，每部分由 4 位数字组成，得到 1920.1601.0001。重新组合的 1920.1601.0001 就是 System ID。

实际上，System ID 的指定可以有不同的方法，但是要保证能够唯一地标识主机或路由器。

（5）SEL 的作用类似于 IP 中的"协议标识符"，不同的传输协议对应不同的 SEL。IP 的 SEL 均为 00。

网络实体名称（Network Entity Title，NET）指的是设备本身的网络层信息，可以看作一类特殊的 NSAP（SEL=00）。NET 的长度与 NSAP 的相同，最多为 20 字节，最少为 8 字节。在路由器上配置 IS-IS 时，只需要考虑 NET 即可，不用关注 NSAP。

例如，NSAP=50.0001.aaaa.bbbb.cccc.00。

IS-IS 协议：Area = 50.0001，System ID = aaaa.bbbb.cccc，SEL = 00。

ISO-IGRP：Domain = 50，Area = 0001，System ID = aaaa.bbbb.cccc，SEL = 00。

6.2.2 IS-IS 协议工作原理

IS-IS 协议是一种链路状态路由协议，每一台路由器都会生成一个 LSP，它包含了该路由器所有使能 IS-IS 协议端口的链路状态信息。通过与相邻设备建立 IS-IS 协议邻接关系，互相更新本地设备的 LSDB，可以使得 LSDB 与整个 IS-IS 协议网络的其他设备的 LSDB 实现同步，并根据 LSDB 运用 SPF 算法计算出 IS-IS 协议路由。如果此 IS-IS 协议路由是到达目的地址的最优路由，则此路由会下发到 IP 路由表中，并指导报文的转发。

1．邻居关系的建立

两台运行 IS-IS 协议的路由器在交互协议报文实现路由功能之前必须先建立邻居关系。在不同类型的网络中，IS-IS 协议的邻居建立方式并不相同。

（1）广播链路邻居关系的建立。

图 6.4 以 L2 路由器为例，描述了广播链路中邻居关系的建立过程，L1 路由器之间邻居关系的建立过程与此相同。

图 6.4　广播链路中邻居关系的建立过程

① RouterA 广播发送 L2 LAN IIH 报文，此报文中无邻居标识。

② RouterB 收到此报文后，将自己与 RouterA 的邻居状态标识为 Initial。RouterB 再向 RouterA 回复 L2 LAN IIH 报文，此报文中标识 RouterA 为 RouterB 的邻居。

③ RouterA 收到此报文后，将自己与 RouterB 的邻居状态标识为 Up。RouterA 再向 RouterB 发送一个标识 RouterB 为 RouterA 邻居的 L2 LAN IIH 报文。

④ RouterB 收到此报文后，将自己与 RouterA 的邻居状态标识为 Up。两个路由器成功建立邻居关系。

因为是广播网络，需要选举 DIS，所以在邻居关系建立后，路由器会等待两个 Hello 报文间隔，再进行 DIS 的选举。Hello 报文中包含 Priority 字段，优先级最高的路由器将被选举为该广播网的 DIS，若优先级相同，则端口 MAC 地址较大的被选举为 DIS。

（2）P2P 链路邻居关系的建立。

在 P2P 链路上，邻居关系的建立不同于广播链路，分为两次握手机制和三次握手机制。

① 两次握手机制：只要路由器收到对端发来的 Hello 报文，就单方面宣布邻居为 Up 状态，建立邻居关系。

② 三次握手机制：通过三次发送 P2P 的 IS-IS Hello PDU 最终建立邻居关系，类似于广播邻居关系的建立。

两次握手机制存在明显的缺陷。当路由器间存在两条及以上的链路时，如果某条链路上到达对端的单向状态为 Down，而另一条链路同方向的状态为 Up，则路由器之间仍能建立邻接关系。SPF 在计算时会使用状态为 Up 的链路上的参数，导致没有检测到故障的路由器在转发报文时仍然试图通过状态为 Down 的链路。三次握手机制解决了上述不可靠点对点链路中存在的问题。在这种方式下，路由器只有在知道邻居路由器也接收到它的报文时，才宣布邻居路由器处于 Up 状态，从而建立邻居关系。

2. IS-IS 协议建立邻居关系的原则

（1）只有同一层次的相邻路由器才有可能成为邻居。
（2）对于 L1 路由器来说，区域号必须一致。
（3）链路两端 IS-IS 协议端口的网络类型必须一致。
（4）链路两端 IS-IS 协议端口的地址必须处于同一网段。

由于 IS-IS 协议是直接运行在数据链路层上的协议，并且最早是为 CLNP 设计的，因此 IS-IS 协议邻居关系的形成与 IP 地址无关。但在实际的实现中，由于只在 IP 上运行 IS-IS 协议，所以要检查对方的 IP 地址。如果端口配置了从 IP 地址，那么只要双方有某个 IP 地址（主 IP 地址或者从 IP 地址）在同一网段中，就能建立邻居，不一定需要主 IP 地址相同。

注意 通过将以太网端口模拟成 P2P 端口，可以建立 P2P 链路邻居关系。
当链路两端 IS-IS 协议端口的地址不在同一网段中时，即使配置端口对接收的 Hello 报文不做 IP 地址检查，也可以建立邻居关系。对于 P2P 端口，可以配置端口忽略 IP 地址检查；对于以太网端口，需要先将以太网端口模拟成 P2P 端口，再配置端口忽略 IP 地址检查。

3. IS-IS 协议的默认配置

在使用和配置过程中，IS-IS 协议的默认配置如表 6.1 所示。

表 6.1 IS-IS 协议的默认配置

参数	默认配置
IS-IS 协议	未使能
DIS 优先级	64
DIS 默认发送 Hello 报文的时间间隔	3.3s
设备级别	L1/2
发送 Hello 报文的时间间隔	10s
发送 LSP 的最小时间间隔	50ms
发送 LSP 的最大数目	10 条
LSP 刷新时间间隔	900s
LSP 最大有效时间	1200s
带宽参考值	100Mbit/s

6.3 项目实施

6.3.1 IS-IS 协议基础配置

1. 创建 IS-IS 协议进程

（1）使用 system-view 命令，进入系统视图。

（2）使用 isis [process-id] [vpn-instance vpn-instance-name]命令，创建 IS-IS 协议进程并进入 IS-IS 视图。

其中，参数 process-id 用来指定一个 IS-IS 协议进程。如果不指定参数 process-id，则系统默认的进程为 1。IS-IS 协议进程可以与 VPN 实例相关联，此时需要指定 VPN 实例名。

（3）（可选）使用 description description 命令，为 IS-IS 协议进程设置描述信息。

（4）（可选）配置 IS-IS 协议在 Purge LSP 报文中添加 POI TLV。使用 purge-originator-identification enable [always]命令，根据配置的认证情况决定在 Purge LSP 报文中是否添加 POI TLV 以及是否使能主机名 TLV 的功能。

① 如果使用 purge-originator-identification enable 命令，同时配置任意认证，则生成 Purge LSP 时不携带 POI/Hostname TLV。

② 如果使用 purge-originator-identification enable 命令，且没有配置任何认证，则生成 Purge LSP 时携带 POI/Hostname TLV。

③ 如果使用 purge-originator-identification enable always 命令，则无论是否配置认证，生成 Purge LSP 时均携带 POI/Hostname TLV。

（5）（可选）配置去使能 IS-IS 协议自动修改冲突 System ID 的功能。使用 isis system-id auto-recover disable 命令，当检测到 System ID 冲突时，使自动修改 IS-IS System ID 的功能失效。

自动修改 IS-IS System ID 的功能默认启用，当需要取消该功能时，可以执行此操作。

2. 配置网络实体名

网络实体名是 NSAP 的特殊形式，在进入 IS-IS 视图之后，必须完成 IS-IS 协议进程的 NET 配置，IS-IS 协议才能真正启动。

通常情况下，一个 IS-IS 协议进程下配置一个 NET 即可。当区域需要重新划分，如将多个区域合并或者将一个区域划分为多个区域时，配置多个 NET 可以在重新配置时保证路由的正确性。由于一个 IS-IS 协议进程中区域地址最多可配置 3 个，所以 NET 最多也只能配置 3 个。在配置多个 NET 时，必须保证它们的 System ID 都相同。

（1）使用 system-view 命令，进入系统视图。

（2）使用 isis [process-id]命令，进入 IS-IS 视图。

（3）使用 network-entity net 命令，设置网络实体名称。

> **注意**
> 建议将 Loopback 端口的地址转换为 NET，以保证 NET 在网络中的唯一性。如果网络中的 NET 不唯一，则容易引发路由振荡，因此要做好前期网络规划。
> IS-IS 协议在建立 L2 邻居时，不检查区域地址是否相同；而在建立 L1 邻居时，区域地址必须相同，否则无法建立邻居。

3. 配置全局 Level

建议根据网络规划的需要配置设备的 Level。

（1）当 Level 为 L1 时，设备只与属于同一区域的 L1 和 L1/2 设备形成邻居关系，并且只负责维护 L1 的链路状态数据库。

（2）当 Level 为 L2 时，设备可以与同一区域或者不同区域的 L2 设备或者其他区域的 L1/2 设备形成邻居关系，并且只维护一个 L2 的 LSDB。

（3）当 Level 级别为 L1/2 时，设备会为 L1 和 L2 分别建立邻居，分别维护 L1 和 L2 两个 LSDB。

注意 在网络运行过程中，改变 IS-IS 协议设备的级别可能会导致 IS-IS 协议进程重启并可能会造成 IS-IS 协议邻居断连，建议用户在配置 IS-IS 协议时先完成设备级别的配置。

配置设备 Level 的步骤如下。

（1）使用 system-view 命令，进入系统视图。

（2）使用 isis [process-id]命令，进入 IS-IS 视图。

（3）使用 is-level { level-1 | level-1-2 | level-2 }命令，设置设备的 Level。

默认情况下，设备的 Level 为 L1/2。

4. 建立 IS-IS 协议邻居

由于 IS-IS 协议在广播网和 P2P 网络中建立邻居的方式不同，因此，针对不同类型的端口，可以配置不同的 IS-IS 协议属性。

在广播网中，IS-IS 协议需要选择 DIS，因此通过配置 IS-IS 协议端口的 DIS 优先级，可以使拥有最高端口优先级的设备被优选为 DIS。

在 P2P 网络中，IS-IS 协议不需要选择 DIS，因此无须配置端口的 DIS 优先级。但是为了保证 P2P 链路的可靠性，可以配置 IS-IS 协议在使用 P2P 端口建立邻居时采用 3-way 模式，以检测单向链路故障。

通常情况下，IS-IS 协议会对收到的 Hello 报文进行 IP 地址检查，只有当收到的 Hello 报文的源地址和本地接收报文的端口地址在同一网段中时，才会建立邻居。但当两端端口 IP 地址不在同一网段中时，如果均配置了 isis peer-ip-ignore 命令，则会忽略对对端 IP 地址的检查，此时，链路两端的 IS-IS 协议端口间可以建立正常的邻居关系。

（1）在广播链路上建立 IS-IS 协议邻居。

① 使用 system-view 命令，进入系统视图。

② 使用 interface interface-type interface-number 命令，进入端口视图。

③ 使用 isis enable [process-id]命令，使能 IS-IS 协议端口。

使用该命令后，IS-IS 协议将通过该端口建立邻居、扩散 LSP 报文。

注意 由于 Loopback 端口不需要建立邻居，因此如果在 Loopback 端口下使能 IS-IS 协议，则只会将该端口所在的网段路由通过其他 IS-IS 协议端口发布出去。

④ 使用 isis circuit-level [level-1 | level-1-2 | level-2]命令，设置端口的 Level。

默认情况下，端口的 Level 为 L1/2。

两台 L1/2 设备建立邻居关系时，默认情况下，会分别建立 L1 和 L2 邻居关系，如果只希望建立 L1 或者 L2 的邻居关系，则可以通过修改端口的 Level 来实现。

注意 只有当 IS-IS 协议设备的 Level 为 L1/2 时,改变端口的级别才有意义,否则将由 IS-IS 协议设备的级别决定所能建立的邻接关系层次。

（2）在 P2P 链路上建立 IS-IS 协议邻居。

① 使用 system-view 命令,进入系统视图。

② 使用 interface interface-type interface-number 命令,进入端口视图。

③ 使用 isis enable [process-id]命令,使能 IS-IS 协议端口。

④ 使用 isis circuit-level [level-1 | level-1-2 | level-2]命令,设置端口的 Level。默认情况下,端口的 Level 为 L1/2。

⑤ 使用 isis circuit-type p2p [strict-snpa-check]命令,设置端口的网络类型为 P2P。默认情况下,端口网络类型根据物理端口决定。

在使能 IS-IS 协议的端口上,当端口网络类型发生改变时,相关配置也会发生改变,具体如下。

使用 isis circuit-type p2p 命令将广播网络端口模拟成 P2P 端口时,端口发送 Hello 报文的间隔时间、宣告邻居失效前 IS-IS 协议没有收到的邻居 Hello 报文数目、点对点链路上 LSP 报文的重传间隔时间及 IS-IS 协议各种认证均恢复为默认配置,而 DIS 优先级、DIS 名称、广播网络中发送 CSNP 报文的间隔时间等配置均失效。

使用 undo isis circuit-type 命令恢复端口的默认网络类型时,端口发送 Hello 报文的间隔时间、宣告邻居失效前 IS-IS 协议没有收到的邻居 Hello 报文数目、点对点链路上 LSP 报文的重传间隔时间、IS-IS 协议各种认证、DIS 优先级和广播网络中发送 CSNP 报文的间隔时间均恢复为默认配置。

⑥ 使用 isis ppp-negotiation { 2-way | 3-way [only] }命令,指定端口使用的协商模型。默认情况下,使用 3-way 协商模式。

⑦ 使用 isis peer-ip-ignore 命令,配置对接收的 Hello 报文不做 IP 地址检查。默认情况下,IS-IS 协议检查对端 Hello 报文的 IP 地址。

⑧ 使用 isis ppp-osicp-check 命令,指定 PPP 链路端口进行 OSICP 状态检查。默认情况下,PPP 链路协议的 OSICP 状态不影响 IS-IS 协议端口的状态。

5. 检查 IS-IS 协议基本功能的配置结果

（1）使用 display isis peer [verbose] [process-id | vpn-instance vpn-instance-name]命令,查看 IS-IS 协议的邻居信息。

（2）使用 display isis interface [verbose] [process-id | vpn-instance vpn-instance-name]命令,查看使能 IS-IS 协议的端口信息。

（3）使用 display isis route [process-id | vpn-instance vpn-instance-name] [ipv4][verbose | [level-1 | level-2] | ip-address [mask | mask-length]] *命令,查看 IS-IS 协议的路由信息。

（4）使用 display isis purge packet process-id packet-number 命令,查看 IS-IS 协议收到携带 POI TLV 的 Purge LSP 报文的统计信息。

6.3.2 IS-IS 协议基本功能配置实例

如图 6.5 所示,网络中有 4 台路由器,用户希望在这 4 台路由器上实现网络互联。路由器 AR1 和路由器 AR2 的性能相对较低,配置路由器 AR1 和路由器 AR2 为 L1 路由器,使这两台路由器处理相对较少的数据量,配置相关端口与 IP 地址,进行网络拓扑连接。

V6-1 配置 IS-IS 协议基本功能实例

图 6.5 IS-IS 协议基本功能配置实例

（1）对路由器 AR1 进行相关配置，相关实例代码如下。

```
<Huawei>system-view
[Huawei]sysname AR1
[AR1]isis 1                                                    //使能 IS-IS 协议
[AR1-isis-1]is-level level-1                                   //配置级别
[AR1-isis-1]network-entity 10.0000.0000.0001.00                //网络实体名，相当于定义区域
[AR1]interface GigabitEthernet 0/0/0
[AR1-GigabitEthernet0/0/0]ip address 192.168.1.1 24
[AR1-GigabitEthernet0/0/0]isis enable                          //使能 IS-IS 协议
[AR1-GigabitEthernet0/0/0]quit
[AR1-isis-1]quit
[AR1]
```

（2）对路由器 AR2 进行相关配置，相关实例代码如下。

```
<Huawei>system-view
[Huawei]sysname AR2
[AR2]isis 1
[AR2-isis-1]is-level level-1
[AR2-isis-1]network-entity 10.0000.0000.0002.00
[AR2-isis-1]quit
[AR2]interface GigabitEthernet 0/0/0
[AR2-GigabitEthernet0/0/0]ip address 192.168.2.1 24
[AR2-GigabitEthernet0/0/0]isis enable
[AR2-GigabitEthernet0/0/0]quit
[AR2]
```

（3）对路由器 AR3 进行相关配置，相关实例代码如下。

```
<Huawei>system-view
[Huawei]sysname AR3
[AR3]isis 1
[AR3-isis-1]is-level level-1-2
[AR3-isis-1]network-entity 10.0000.0000.0003.00
[AR3-isis-1]quit
[AR3]interface GigabitEthernet 0/0/0
```

```
[AR3-GigabitEthernet0/0/0]ip address 192.168.1.2 24
[AR3-GigabitEthernet0/0/0]isis enable
[AR3-GigabitEthernet0/0/0]quit
[AR3]interface GigabitEthernet 0/0/1
[AR3-GigabitEthernet0/0/1]ip address 192.168.2.2 24
[AR3-GigabitEthernet0/0/1]isis enable
[AR3-GigabitEthernet0/0/1]quit
[AR3]interface GigabitEthernet 0/0/2
[AR3-GigabitEthernet0/0/2]ip address 192.168.3.2 24
[AR3-GigabitEthernet0/0/2]isis enable
[AR3-GigabitEthernet0/0/2]quit
[AR3]
```

（4）对路由器 AR4 进行相关配置，相关实例代码如下。

```
<Huawei>system-view
[Huawei]sysname AR4
[AR4]isis 1
[AR4-isis-1]is-level level-2
[AR4-isis-1]network-entity 20.0000.0000.0004.00
[AR4-isis-1]quit
[AR4]interface GigabitEthernet 0/0/2
[AR4-GigabitEthernet0/0/0]ip address 192.168.3.1 24
[AR4-GigabitEthernet0/0/0]isis enable
[AR4-GigabitEthernet0/0/0]quit
[AR4]
```

（5）显示路由器 AR3 的配置信息，主要配置实例代码如下。

```
<AR3>display current-configuration
#
 sysname AR3
#
isis 1
 network-entity 10.0000.0000.0003.00
#
interface GigabitEthernet0/0/0
 ip address 192.168.1.2 255.255.255.0
 isis enable 1
#
interface GigabitEthernet0/0/1
 ip address 192.168.2.2 255.255.255.0
 isis enable 1
#
interface GigabitEthernet0/0/2
 ip address 192.168.3.2 255.255.255.0
 isis enable 1
#
return
<AR3>
```

（6）验证配置结果，显示各路由器的 IS-IS LSDB 信息，查看 LSDB 是否同步，这里以路由器 AR1 为例进行介绍。

```
<AR1>display isis lsdb
                    Database information for ISIS(1)
                    -------------------------------
                       Level-1 Link State Database

LSPID              Seq Num      Checksum    Holdtime   Length  ATT/P/OL
------------------------------------------------------------------------
0000.0000.0001.00-00* 0x0000000c  0x81fa      937        68      0/0/0
0000.0000.0001.01-00* 0x00000006  0xc7bf      937        55      0/0/0
0000.0000.0002.00-00  0x00000009  0xd4a6      1060       68      0/0/0
0000.0000.0002.01-00  0x00000006  0xd0b4      1060       55      0/0/0
0000.0000.0003.00-00  0x0000000d  0x8c5d      383        111     1/0/0
Total LSP(s): 5
       *(In TLV)-Leaking Route, *(By LSPID)-Self LSP, +-Self LSP(Extended),
        ATT-Attached, P-Partition, OL-Overload
<AR1>
```

（7）显示各路由器的 IS-IS 协议路由信息。L1 路由器的路由表中应该有一条默认路由，且路由下一跳为 L1/2 路由器，L2 路由器应该有所有 L1 和 L2 的路由，这里以路由器 AR1 为例进行介绍。

```
<AR1>display isis route
                    Route information for ISIS(1)
                    -----------------------------
                    ISIS(1) Level-1 Forwarding Table
                    --------------------------------

IPV4 Destination   IntCost    ExtCost   ExitInterface   NextHop        Flags
-----------------------------------------------------------------------------
0.0.0.0/0          10         NULL      GE0/0/0         192.168.1.2    A/-/-/-
192.168.2.0/24     20         NULL      GE0/0/0         192.168.1.2    A/-/-/-
192.168.1.0/24     10         NULL      GE0/0/0         Direct         D/-/L/-
192.168.3.0/24     20         NULL      GE0/0/0         192.168.1.2    A/-/-/-
     Flags: D-Direct, A-Added to URT, L-Advertised in LSPs, S-IGP Shortcut,
            U-Up/Down Bit Set
<AR1>
```

6.3.3 IS-IS 协议与动态 BFD 联动功能配置实例

通常情况下，IS-IS 协议设定发送 Hello 报文的时间间隔为 10s，一般将邻居失效时间（即邻居的保持时间）配置为 Hello 报文时间间隔的 3 倍。若在邻居失效时间内没有收到邻居发来的 Hello 报文，则会删除邻居。由此可见，路由器能感知到邻居故障的时间最小为秒级，在高速的网络环境中可能会出现大量报文丢失的问题。

BFD 能够提供轻负荷、快速（毫秒级）的通道故障检测，解决了 IS-IS 协议现有检测机制不足的问题。使用 BFD 并不是以其代替 IS-IS 协议本身的 Hello 机制，而是配合 IS-IS 协议更快地发现邻接方面出现的故障，并及时通知 IS-IS 协议重新计算相关路由，以便正确指导报文的转发。

当主路径上的链路出现故障时，BFD 能够快速检测到故障并通告给 IS-IS 协议，IS-IS 协议会使故障链路的端口邻居下线并删除邻居对应的 IP 类型，从而触发拓扑计算，并更新 LSP，使得其他邻居及时收到更新的 LSP，实现网络拓扑的快速收敛。

如图 6.6 所示，路由器 AR1、路由器 AR2、路由器 AR3 和路由器 AR4 之间运行 IS-IS 协议。使能路由器 AR1、路由器 AR2、路由器 AR3 和路由器 AR4 的 IS-IS 协议进程的 BFD 特性；业务流量在主链路路由器 AR1→路由器 AR2→路由器 AR4 上传送，链路路由器 AR1→路由器 AR3→路由器 AR2→路由器 AR4 作为备份链路；在路由器 AR1 和路由器 AR2 之间的链路上创建端口的 BFD 特性，当路由器 AR1 和路由器 AR2 之间的链路出现故障时，BFD 能够快速检测到故障并通告给 IS-IS 协议，使业务流量使用备份链路传送，配置相关端口与 IP 地址，进行网络拓扑连接。

V6-2 配置 IS-IS 协议与动态 BFD 联动

图 6.6　IS-IS 协议与动态 BFD 联动功能配置实例

（1）对路由器 AR1 进行相关配置，相关实例代码如下。

```
<Huawei>system-view
[Huawei]sysname AR1
[AR1]isis 1
[AR1-isis-1]is-level level-1-2
[AR1-isis-1]network-entity 10.0000.0000.0001.00
[AR1-isis-1]quit
[AR1]interface GigabitEthernet 0/0/0
[AR1-GigabitEthernet0/0/0]ip address 192.168.1.1 24
[AR1-GigabitEthernet0/0/0]isis enable
[AR1-GigabitEthernet0/0/0]quit
[AR1]interface GigabitEthernet 0/0/1
[AR1-GigabitEthernet0/0/1]ip address 192.168.2.1 24
[AR1-GigabitEthernet0/0/1]isis enable
[AR1-GigabitEthernet0/0/1]quit
[AR1]
```

（2）对路由器 AR2 进行相关配置，相关实例代码如下。

```
<Huawei>system-view
[Huawei]sysname AR2
[AR2]isis 1
[AR2-isis-1]is-level level-1-2
[AR2-isis-1]network-entity 10.0000.0000.0002.00
[AR2-isis-1]quit
[AR2]interface GigabitEthernet 0/0/0
[AR2-GigabitEthernet0/0/0]ip address 192.168.2.2 24
[AR2-GigabitEthernet0/0/0]isis enable
[AR2-GigabitEthernet0/0/0]quit
```

```
[AR2]interface GigabitEthernet 0/0/1
[AR2-GigabitEthernet0/0/1]ip address 172.16.1.1 24
[AR2-GigabitEthernet0/0/1]isis enable
[AR2-GigabitEthernet0/0/1]quit
[AR2]interface GigabitEthernet 0/0/2
[AR2-GigabitEthernet0/0/1]ip address 192.168.3.2 24
[AR2-GigabitEthernet0/0/1]isis enable
[AR2-GigabitEthernet0/0/1]quit
[AR2]
```

（3）对路由器 AR3 进行相关配置，相关实例代码如下。

```
<Huawei>system-view
[Huawei]sysname AR3
[AR3]isis 1
[AR3-isis-1]is-level level-1-2
[AR3-isis-1]network-entity 10.0000.0000.0003.00
[AR3-isis-1]quit
[AR3]interface GigabitEthernet 0/0/0
[AR3-GigabitEthernet0/0/0]ip address 192.168.1.2 24
[AR3-GigabitEthernet0/0/0]isis enable
[AR3-GigabitEthernet0/0/0]quit
[AR3]interface GigabitEthernet 0/0/2
[AR3-GigabitEthernet0/0/1]ip address 192.168.3.1 24
[AR3-GigabitEthernet0/0/1]isis enable
[AR3-GigabitEthernet0/0/1]quit
[AR3]
```

（4）对路由器 AR4 进行相关配置，相关实例代码如下。

```
<Huawei>system-view
[Huawei]sysname AR4
[AR4]isis 1
[AR4-isis-1]is-level level-1-2
[AR4-isis-1]network-entity 10.0000.0000.0004.00
[AR4-isis-1]quit
[AR4]interface GigabitEthernet 0/0/1
[AR4-GigabitEthernet0/0/1]ip address 172.16.1.2 24
[AR4-GigabitEthernet0/0/1]isis enable
[AR4-GigabitEthernet0/0/1]quit
[AR4]
```

（5）配置完成后，使用 display isis peer 命令，可以看到路由器 AR1、路由器 AR2、路由器 AR3 和路由器 AR4 建立了邻居关系，这里以路由器 AR1 为例进行介绍。

```
<AR1>display isis peer
                      Peer information for ISIS(1)
  System Id       Interface    Circuit Id            State  HoldTime  Type  PRI
  --------------------------------------------------------------------------------
  0000.0000.0003  GE0/0/0      0000.0000.0001.01 Up   27s    L2    64
  0000.0000.0002  GE0/0/1      0000.0000.0001.02 Up   26s    L2    64
  Total Peer(s): 2
<AR1>
```

（6）跟踪数据流量走向，这里以路由器 AR1 为例进行介绍。

```
<AR1>tracert 172.16.1.2
 traceroute to    172.16.1.2(172.16.1.2), max hops: 30 ,packet length: 40,press CT
RL_C to break
 1 192.168.2.2 40 ms    10 ms    20 ms
 2 172.16.1.2 20 ms    30 ms    20 ms
<AR1>
```

从以上信息可以看出，业务流量在主链路路由器 AR1→路由器 AR2→路由器 AR4 上传送。

（7）路由器之间已经互相学习到路由，下面以查看路由器 AR1 的路由表为例进行介绍。

```
<AR1>display isis route
                    Route information for ISIS(1)
                    -----------------------------
                    ISIS(1) Level-2 Forwarding Table
                    --------------------------------
IPV4 Destination    IntCost    ExtCost    ExitInterface    NextHop        Flags

192.168.2.0/24      10         NULL       GE0/0/1          Direct         D/-/L/-
192.168.1.0/24      10         NULL       GE0/0/0          Direct         D/-/L/-
172.16.1.0/24       20         NULL       GE0/0/1          192.168.2.2    A/-/-/-
192.168.3.0/24      20         NULL       GE0/0/0          192.168.1.2    A/-/-/-
                                          GE0/0/1          192.168.2.2
   Flags: D-Direct, A-Added to URT, L-Advertised in LSPs, S-IGP Shortcut,
                              U-Up/Down Bit Set
<AR1>
```

从以上信息可以看出，去往 172.16.1.0/24 的路由下一跳地址为 192.168.2.2，流量在主链路路由器 AR1→路由器 AR2→路由器 AR4 上传送。

（8）配置 IS-IS 协议进程的 BFD 特性，在路由器 AR1 上使能 IS-IS 协议的 BFD 功能。

```
[AR1]bfd
[AR1-bfd]quit
[AR1]isis
[AR1-isis-1]bfd all-interfaces enable
[AR1-isis-1]quit
[AR1]
```

路由器 AR2、路由器 AR3 和路由器 AR4 与路由器 AR1 的配置相同，这里不再赘述。

（9）配置完成后，在路由器 AR1、路由器 AR2、路由器 AR3 或路由器 AR4 上使用 display isis bfd session all 命令，可以看到 BFD 参数已生效，这里以路由器 AR1 的显示为例进行介绍。

```
<AR1>display isis bfd session all
                    BFD session information for ISIS(1)
                    -----------------------------------
Peer System ID : 0000.0000.0003        Interface : GE0/0/0
TX : 1000            BFD State : up    Peer IP Address : 192.168.1.2
RX : 1000            LocDis : 8192     Local IP Address: 192.168.1.1
Multiplier : 3       RemDis : 8192     Type : L2
Diag : No diagnostic information
Peer System ID : 0000.0000.0002        Interface : GE0/0/1
TX : 1000            BFD State : up    Peer IP Address : 192.168.2.2
```

RX : 1000   LocDis : 8193   Local IP Address: 192.168.2.1
Multiplier : 3   RemDis : 8192   Type : L2
Diag : No diagnostic information
Total BFD session(s): 2
<AR1>

（10）配置端口的 BFD 特性，在路由器 AR1 的 GE 0/0/1 端口上配置 BFD 特性，并指定最小发送和接收时间间隔为 200ms，本地检测时间倍数为 4。

[AR1]interface GigabitEthernet 0/0/1
[AR1-GigabitEthernet0/0/1]isis bfd enable
[AR1-GigabitEthernet0/0/1]isis bfd min-rx-interval 200 min-tx-interval 200 detect-multiplier 4
[AR1-GigabitEthernet0/0/1]quit
[AR1]

（11）配置端口的 BFD 特性，在路由器 AR2 的 GE 0/0/0 端口上配置 BFD 特性，并指定最小发送和接收时间间隔为 200ms，本地检测时间倍数为 4。

[AR2]interface GigabitEthernet 0/0/0
[AR2-GigabitEthernet0/0/0]isis bfd enable
[AR2-GigabitEthernet0/0/0]isis bfd min-rx-interval 200 min-tx-interval 200 detect-multiplier 4
[AR2-GigabitEthernet0/0/0]quit
[AR2]

（12）配置完成后，在路由器 AR1 或路由器 AR2 上使用 display isis bfd session all 命令，可以看到 BFD 参数已生效，这里以路由器 AR2 的显示为例进行介绍。

<AR2>display isis bfd session all
                BFD session information for ISIS(1)

Peer System ID : 0000.0000.0001   Interface : GE0/0/0
TX : 200   BFD State : up   Peer IP Address : 192.168.2.1
RX : 200   LocDis : 8192   Local IP Address: 192.168.2.2
Multiplier : 4   RemDis : 8193   Type : L2
Diag : No diagnostic information
Peer System ID : 0000.0000.0004   Interface : GE0/0/1
TX : 1000   BFD State : up   Peer IP Address : 172.16.1.2
RX : 1000   LocDis : 8193   Local IP Address: 172.16.1.1
Multiplier : 3   RemDis : 8192   Type : L2
Diag : No diagnostic information
Peer System ID : 0000.0000.0003   Interface : GE0/0/2
TX : 1000   BFD State : up   Peer IP Address : 192.168.3.1
RX : 1000   LocDis : 8194   Local IP Address: 192.168.3.2
Multiplier : 3   RemDis : 8193   Type : L2
Diag : No diagnostic information
Total BFD session(s): 3
<AR2>

（13）验证配置结果，在路由器 AR2 的 GE 0/0/0 端口上使用 shutdown 命令，模拟主链路出现故障。

[AR2]interface GigabitEthernet 0/0/0
[AR2-GigabitEthernet0/0/0]shutdown
[AR2-GigabitEthernet0/0/0]quit

[AR2]

（14）在路由器 AR1 上查看路由表信息。

```
<AR1>display isis route
                  Route information for ISIS(1)
                  ------------------------------
                  ISIS(1) Level-2 Forwarding Table
                  ------------------------------
IPV4 Destination     IntCost    ExtCost   ExitInterface   NextHop        Flags
-------------------------------------------------------------------------------
192.168.1.0/24       10         NULL      GE0/0/0         Direct         D/-/L/-
172.16.1.0/24        30         NULL      GE0/0/0         192.168.1.2    A/-/-/-
192.168.3.0/24       20         NULL      GE0/0/0         192.168.1.2    A/-/-/-
           Flags: D-Direct, A-Added to URT, L-Advertised in LSPs, S-IGP Shortcut,
                            U-Up/Down Bit Set
<AR1>
```

从以上信息可以看出，在主链路失效后，备份链路路由器 AR1→路由器 AR3→路由器 AR2→路由器 AR4 生效，去往 172.16.1.0/24 的路由下一跳地址为 192.168.1.2。

（15）跟踪数据流量走向，这里以路由器 AR1 的显示结果为例进行介绍。

```
<AR1>tracert 172.16.1.2
  traceroute to  172.16.1.2(172.16.1.2), max hops: 30 ,packet length: 40,press CT
RL_C to break
 1 192.168.1.2 20 ms   20 ms   10 ms
 2 192.168.3.2 30 ms   30 ms   30 ms
 3 172.16.1.2 30 ms    40 ms   30 ms
<AR1>
```

从以上信息可以看出，业务流量在备份链路路由器 AR1→路由器 AR3→路由器 AR2→路由器 AR4 上传送。

（16）在路由器 AR1 上使用 display isis bfd session all 命令，只能看到路由器 AR1 和路由器 AR3 之间的 BFD 参数生效。

```
<AR1>display isis bfd session all
                  BFD session information for ISIS(1)
                  ------------------------------------
Peer System ID : 0000.0000.0003       Interface : GE0/0/0
TX : 1000            BFD State : up   Peer IP Address : 192.168.1.2
RX : 1000            LocDis : 8192    Local IP Address: 192.168.1.1
Multiplier : 3       RemDis : 8192    Type : L2
Diag : No diagnostic information
Total BFD session(s): 1
<AR1>
```

课后习题

1. 选择题

（1）以下不是 IS-IS 协议报文类型的是（　　）。
　　A. Hello　　　　B. LSP　　　　C. LSP ACK　　　　D. CSNP

（2）以下不是 Hello 报文的是（　　）。
　　　A．L1IIH　　　　B．L2IIH　　　　C．P-2-PIIH　　　D．L3IIH
（3）在广播网络中，IS-IS 协议通过（　　）机制保证邻居关系建立的可靠性。
　　　A．校验和　　　　B．三次握手　　　C．状态同步　　　D．老化计时器
（4）在广播网络中，DIS 默认发送 Hello 报文的时间间隔为（　　）。
　　　A．3.3s　　　　　B．5s　　　　　　C．10s　　　　　　D．20s
（5）在 IS-IS 协议的广播网络中，L2 路由器使用组播 MAC 地址（　　）作为发送 IIH 的目的地址。
　　　A．0180-C200-0014　　　　　　B．0180-C200-0015
　　　C．0180-C200-0016　　　　　　D．0180-C200-0017
（6）关于 IS-IS 协议所使用的 SNAP 地址，以下不是其组成部分的是（　　）。
　　　A．Area ID　　　B．DSCP　　　　C．System ID　　　D．SEL

2．简答题

（1）简述 IS-IS 协议的工作原理及其局限性。
（2）简述如何配置 IS-IS 协议与动态 BFD 联动。

项目7
BGP

【学习目标】

- 掌握BGP的基本概念。
- 理解BGP的工作原理。
- 掌握BGP的基本配置方法。
- 掌握BGP与BFD联动的配置方法。

7.1 项目陈述

BGP 是为取代最初的外部网关协议（Exterior Gateway Protocol，EGP）而设计的另一种外部网关协议。不同于最初的 EGP，BGP 能够进行路由优选、避免路由环路、更高效率地传递路由和维护大量的路由信息。虽然 BGP 用于在 AS 之间传递路由信息，但并不是所有 AS 之间传递路由信息都需要运行 BGP。例如，在数据中心上行的接入 Internet 的出口上，为了避免 Internet 海量路由对数据中心内部网络的影响，设备采用静态路由代替 BGP 与外部网络进行通信。

7.2 必备知识

7.2.1 BGP 概述

BGP 是一种实现 AS 之间的路由可达，并选择最佳路由的距离矢量路由协议。为方便管理规模不断扩大的网络，网络被分成了不同的 AS。1982 年，EGP 被用于在 AS 之间动态交换路由信息。但是 EGP 设计得比较简单，只发布网络可达的路由信息，而不对路由信息进行优选，同时没有考虑环路避免等问题，很快就无法满足网络管理的要求了。

1. AS

AS 是指在一个实体管辖下的拥有相同选路策略的 IP 网络。BGP 网络中的每个 AS 都被分配一个唯一的 AS 号，用于区分不同的 AS。AS 号分为 2 字节 AS 号和 4 字节 AS 号，其中，2 字节 AS 号为 1~65535，4 字节 AS 号为 1~4294967295。支持 4 字节 AS 号的设备能够与支持 2 字节 AS 号的设备兼容。

2. BGP 邻居类型

BGP 邻居类型按照运行方式分为外部边界网关协议（External Border Gateway Protocol，EBGP）和内部边界网关协议（Internal Border Gateway Protocol，IBGP），如图 7.1 所示。

（1）EBGP：运行于不同 AS 之间的 BGP 称为 EBGP。为了防止 AS 间产生环路，当 BGP 设备接收 EBGP 对等体发送的路由时，会将带有本地 AS 号的路由丢弃。

（2）IBGP：运行于同一 AS 内部的 BGP 称为 IBGP。为了防止 AS 内产生环路，BGP 设备不将从 IBGP 对等体学习到的路由通告给其他 IBGP 对等体，并与所有 IBGP 对等体建立全连接。

图 7.1 BGP 邻居类型

3. BGP 报文交互中的角色

BGP 报文交互中有 Speaker（发言者）和 Peer（对等体）两种角色。

（1）Speaker：发送 BGP 报文的设备称为 BGP 发言者，它接收或产生新的报文信息，并发布给其他 BGP Speaker。

（2）Peer：相互交换报文的 Speaker 之间互称对等体。若干相关的对等体可以构成对等体组（Peer Group）。

4. BGP 的 Router ID

BGP 的 Router ID 是一个用于标识 BGP 设备的 32 位值，通常是 IPv4 地址的形式，在 BGP 会话建立时发送的 Open 报文中携带。对等体之间建立 BGP 会话时，每个 BGP 设备都必须有唯一的 Router ID，否则对等体之间不能建立 BGP 连接。

BGP 的 Router ID 在 BGP 网络中必须是唯一的，可以手工配置，也可以让设备自动选取。默认情况下，BGP 选择设备上的 Loopback 端口的 IPv4 地址作为 BGP 的 Router ID。如果设备上没有配置 Loopback 端口，系统会选择端口中最大的 IPv4 地址作为 BGP 的 Router ID。一旦选出 Router ID，除非发生端口地址删除等事件，否则即使配置了更大的 IPv4 地址，也保持原来的 Router ID。

7.2.2 BGP 工作原理

1. BGP 的报文类型

BGP 对等体间通过以下 5 种报文进行交互，其中，Keepalive 报文为周期性发送，其余报文为触发式发送。

（1）Open 报文：用于建立 BGP 对等体连接。

（2）Update 报文：用于在对等体之间交换路由信息。

（3）Notification 报文：用于中断 BGP 连接。

（4）Keepalive 报文：用于保持 BGP 连接，默认时间间隔为 60s。

（5）Route-refresh 报文：用于在改变路由策略后请求对等体重新发送路由信息。只有支持路由刷新功能的 BGP 设备才会发送和响应此报文。

2. BGP 工作状态

BGP 对等体的交互过程中存在 6 种状态：空闲（Idle）、连接（Connect）、活跃（Active）、

Open 报文已发送(OpenSent)、Open 报文已确认(OpenConfirm)和连接已建立(Established)。在 BGP 对等体建立的过程中,通常可见的 3 个状态是 Idle、Active 和 Established,如图 7.2 所示。

图 7.2　BGP 对等体的交互过程

3. BGP 对等体之间的交互原则

BGP 设备将最优路由加入 BGP 路由表,形成 BGP 路由。BGP 设备与对等体建立邻居关系后,采取以下交互原则。

(1) 从 IBGP 对等体获得的 BGP 路由,BGP 设备只发布给其 EBGP 对等体。

(2) 从 EBGP 对等体获得的 BGP 路由,BGP 设备发布给其所有 EBGP 和 IBGP 对等体。

(3) 当存在多条到达同一目的地址的有效路由时,BGP 设备只将最优路由发布给对等体。

(4) 路由更新时,BGP 设备只发送更新的 BGP 路由。

(5) 所有对等体发送的路由,BGP 设备都会接收。

4. BGP 与 IGP 交互

BGP 与 IGP 在设备中使用不同的路由表,为了实现不同 AS 间的相互通信,BGP 需要与 IGP 进行交互,即 BGP 路由表和 IGP 路由表相互引入。

(1) BGP 引入 IGP 路由。

BGP 本身不发现路由,因此需要将其他路由引入 BGP 路由表中,实现 AS 间的路由互通。当一个 AS 需要将路由发布给其他 AS 时,AS 边缘路由器会在 BGP 路由表中引入 IGP 的路由。为了更好地规划网络,BGP 在引入 IGP 的路由时,可以使用路由策略进行路由过滤和路由属性设置,也可以设置 MED 值以指导 EBGP 对等体判断流量进入 AS 时选路。

BGP 引入路由时支持 Import 和 Network 两种方式。

① Import 方式是按协议类型将 RIP、OSPF 协议、IS-IS 协议等协议的路由引入 BGP 路由表中。为了保证引入的 IGP 路由的有效性,Import 方式还可以引入静态路由和直连路由。

② Network 方式是逐条将 IP 路由表中已经存在的路由引入 BGP 路由表中，比 Import 方式更精确。

（2）IGP 引入 BGP 路由。

当一个 AS 需要引入其他 AS 的路由时，AS 边缘路由器会在 IGP 路由表中引入 BGP 的路由。为了避免大量 BGP 路由对 AS 内的设备造成影响，当 IGP 引入 BGP 路由时，可以使用路由策略进行路由过滤和路由属性设置。

5. BGP 属性

在 BGP 路由表中，可能存在到达同一目的地的多条路由。此时，BGP 会选择其中一条路由作为最佳路由，并只把此路由发送给其对等体。BGP 为了选出最佳路由，会根据 BGP 的路由优选规则依次比较这些路由的 BGP 属性。

路由属性是对路由的特定描述，所有的 BGP 路由属性都可以分为以下 4 类。

（1）公认必须遵循：所有 BGP 设备都可以识别此类属性，且必须存在于 Update 报文中。如果缺少这类属性，路由信息就会出错。

（2）公认任意：所有 BGP 设备都可以识别此类属性，但不要求必须存在于 Update 报文中，即就算缺少这类属性，路由信息也不会出错。

（3）可选过渡：BGP 设备可以不识别此类属性，如果 BGP 设备不识别此类属性，则它仍然会接收这类属性并通告给其他对等体。

（4）可选非过渡：BGP 设备可以不识别此类属性，如果 BGP 设备不识别此类属性，则会忽略该属性且不会通告给其他对等体。

BGP 的常见属性类型如表 7.1 所示。

表 7.1 BGP 的常见属性类型

属性名	类型
Origin	公认必须遵循
AS_Path	公认必须遵循
Next_Hop	公认必须遵循
Local_Pref	公认任意
MED	可选非过渡
Originator_ID	可选非过渡
Cluster_List	可选非过渡
Community	可选过渡

下面介绍几种常用的 BGP 路由属性。

（1）Origin。

Origin 属性用来定义路径信息的来源，即标记一条路由是怎样成为 BGP 路由的。它有以下 3 种类型。

① IGP：具有最高的优先级，通过 network 命令注入 BGP 路由表的路由，其 Origin 属性为 IGP。

② EGP：优先级低于 IGP，通过 EGP 得到的路由信息，其 Origin 属性为 EGP。

③ Incomplete：优先级最低，通过其他方式学习到的路由信息，其 Origin 属性为 Incomplete，如 BGP 通过 import-route 命令引入的路由。

（2）AS_Path。

AS_Path 属性按矢量顺序记录了某条路由从本地到目的地址所要经过的所有 AS 号。在接收路由时，如果设备发现 AS_Path 列表中有本 AS 号，则不接收该路由，从而避免了 AS 间的路由环路。

当 BGP Speaker 传播自身引入的路由时，遵循下面的规则。

① 当 BGP Speaker 将这条路由通告给 EBGP 对等体时，会在 Update 报文中创建一个携带本地 AS 号的 AS_Path 列表。

② 当 BGP Speaker 将这条路由通告给 IBGP 对等体时，会在 Update 报文中创建一个空的 AS_Path 列表。

当 BGP Speaker 传播从其他 BGP Speaker 的 Update 报文中学习到的路由时，遵循下面的规则。

① 当 BGP Speaker 将这条路由通告给 EBGP 对等体时，会把本地 AS 号添加在 AS_Path 列表的最前面（最左侧）。收到此路由的 BGP 设备根据 AS_Path 属性就可以知道到达目的地址所要经过的 AS。距本地 AS 最近的相邻 AS 号排在前面，其他 AS 号按顺序依次排列。

② 当 BGP Speaker 将这条路由通告给 IBGP 对等体时，不会改变这条路由相关的 AS_Path 属性。

（3）Next_Hop。

Next_Hop 属性记录了路由的下一跳信息。BGP 的下一跳属性和 IGP 的有所不同，不一定就是邻居设备的 IP 地址。通常情况下，Next_Hop 属性遵循下面的规则。

① BGP Speaker 在向 EBGP 对等体发布某条路由时，会把该路由信息的 Next_Hop 属性设置为本地与对端建立 BGP 邻居关系的端口地址。

② BGP Speaker 将本地始发路由发布给 IBGP 对等体时，会把该路由信息的 Next_Hop 属性设置为本地与对端建立 BGP 邻居关系的端口地址。

③ BGP Speaker 在向 IBGP 对等体发布从 EBGP 对等体学习的路由时，并不改变该路由信息的 Next_Hop 属性。

（4）Local_Pref。

Local_Pref 属性表明了路由器的 BGP 优先级，用于判断流量离开 AS 时的最佳路由。当 BGP 的设备通过不同的 IBGP 对等体得到目的地址相同但下一跳不同的多条路由时，将优先选择 Local_Pref 属性值较高的路由。Local_Pref 属性仅在 IBGP 对等体之间有效，不通告给其他 AS。Local_Pref 属性可以手动配置，如果路由没有配置 Local_Pref 属性，则 BGP 选路时，该路由的 Local_Pref 值按默认值 100 处理。

（5）MED。

MED 属性用于判断流量进入 AS 时的最佳路由。当一个运行 BGP 的设备通过不同的 EBGP 对等体得到目的地址相同但下一跳不同的多条路由时，在其他条件相同的情况下，将优先选择 MED 值较小者作为最佳路由。MED 属性仅在相邻两个 AS 之间传递，收到此属性的 AS 一方不会再将其通告给第三方 AS。MED 属性可以手动配置，如果路由没有配置 MED 属性，则 BGP 选路时，该路由的 MED 值按默认值 0 处理。

（6）Community。

Community 属性用于标识具有相同特征的 BGP 路由，使路由策略的应用更加灵活，同时降低了维护管理的难度。Community 属性分为自定义 Community 属性和公认 Community 属性。

7.3 项目实施

7.3.1 BGP 基础配置

1. 启动 BGP 进程

（1）使用 system-view 命令，进入系统视图。

（2）使用 bgp {as-number-plain | as-number-dot }命令，启动 BGP，指定本地 AS 号，并进入 BGP 视图。

（3）使用 router-id ipv4-address 命令，配置 BGP 的 Router ID。

> **注意** 默认情况下，BGP 会自动选择系统视图下的 Router ID 作为 BGP 的 Router ID。如果选中的 Router ID 是物理端口的 IP 地址，则当 IP 地址发生变化时，会引起路由振荡。为了提高网络的稳定性，可以将 Router ID 手动配置为 Loopback 端口地址。

2. 配置 BGP 对等体

配置 BGP 对等体时，如果指定对等体所属的 AS 号与本地 AS 号相同，则表示配置 IBGP 对等体。如果指定对等体所属的 AS 号与本地 AS 号不同，则表示配置 EBGP 对等体。

为了增强 BGP 连接的稳定性，推荐使用路由可达的 Loopback 端口地址建立 BGP 连接。此时建议对等体两端同时使用 peerconnect-interface 命令，以保证两端 TCP 连接的端口和地址的正确性。如果仅有一端配置该命令，则可能导致 BGP 连接建立失败。

当使用 Loopback 端口的 IP 地址建立 EBGP 连接时，必须使用 peer ebgp-max-hop（其中，hop-count≥2）命令，否则 EBGP 连接将无法建立。

（1）使用 system-view 命令，进入系统视图。

（2）使用 bgp {as-number-plain | as-number-dot }命令，进入 BGP 视图。

（3）使用 peer { ipv4-address | ipv6-address } as-number { as-number-plain| as-number-dot}命令，创建 BGP 对等体。

默认情况下，不创建 BGP 对等体。

3. 配置 BGP 对等体组

在大型 BGP 网络中，对等体的数目众多，配置和维护极为不便。对于存在相同配置的 BGP 对等体，可以将它们加入一个 BGP 对等体组进行批量配置，以降低管理的难度，并提高路由发布效率。

（1）使用 system-view 命令，进入系统视图。

（2）使用 bgp { as-number-plain | as-number-dot }命令，进入 BGP 视图。

（3）使用 group group-name [external | internal]命令，创建对等体组。

（4）使用 peer group-name as-number { as-number-plain | as-number-dot }命令，配置 EBGP 对等体组的 AS 号。

（5）使用 peer { ipv4-address | ipv6-address } group group-name 命令，向对等体组中加入对等体。

4. 配置 BGP 引入路由

BGP 本身不发现路由，因此需要将其他路由（如 IGP 路由等）引入 BGP 路由表中，从而使这些路由在 AS 之内和 AS 之间传播。BGP 支持通过以下两种方式引入路由。

Import 方式：按照协议类型，将 RIP 路由、OSPF 协议路由、IS-IS 协议路由等引入 BGP 路由表中。为了保证引入的 IGP 路由的有效性，Import 方式还可以引入静态路由和直连路由。

Network 方式：逐条将 IP 路由表中已经存在的路由引入 BGP 路由表中。Network 方式比 Import 方式更精确。

（1）Import 方式引入路由的步骤。

① 使用 system-view 命令，进入系统视图。

② 使用 bgp {as-number-plain | as-number-dot }命令，进入 BGP 视图。

③ 根据网络类型，选择进入不同地址族视图，配置不同类型网络中的 BGP 设备。

使用 ipv4-family {unicast | multicast }命令，进入 IPv4 地址族视图。

使用 ipv6-family [unicast]命令，进入 IPv6 地址族视图。

④ 使用 import-route protocol [process-id] [med med | route-policy route-policy-name] * 命令，配置 BGP 引入其他协议的路由。

（2）Network 方式引入路由的步骤。

① 使用 system-view 命令，进入系统视图。

② 使用 bgp { as-number-plain | as-number-dot }命令，进入 BGP 视图。

③根据网络类型，选择进入不同地址族视图，配置不同类型网络中的 BGP 设备。

使用 ipv4-family { unicast | multicast }命令，进入 IPv4 地址族视图。

使用 ipv6-family [unicast]命令，进入 IPv6 地址族视图。

④ 使用 network ipv4-address [mask | mask-length] [route-policy route-policy-name]，或 network ipv6-address prefix-length [route-policy route-policy-name]命令，配置 BGP 逐条引入 IPv4 路由表或 IPv6 路由表中的路由。

5. 查看 BGP 的配置结果

（1）使用 display bgp peer [verbose]命令，查看所有 BGP 对等体的信息。

（2）使用 display bgp peer ipv4-address { log-info | verbose }命令，查看指定 BGP 对等体的信息。

（3）使用 display bgp routing-table [ipv4-address [{ mask | mask-length } [longer-prefixes]]命令，查看 BGP 路由的信息。

（4）使用 display bgp group [group-name]命令，查看对等体组信息。

（5）使用 display bgp multicast peer [[peer-address] verbose]命令，查看指定 MBGP 对等体的信息。

（6）使用 display bgp multicast group [group-name]命令，查看 MBGP 对等体组的信息。

（7）使用 display bgp multicast network 命令，查看 MBGP 发布的路由信息。

（8）使用 display bgp multicast routing-table [ip-address [mask-length [longer-prefixes] | mask [longer-prefixes]]]命令，查看 MBGP 路由表中指定网络的 MBGP 路由信息。

7.3.2 BGP 基本功能配置实例

如图 7.3 所示，需要在所有路由器之间运行 BGP，路由器 AR1 与路由器 AR3 之间建立 EBGP 连接；路由器 AR1、路由器 AR2 属于 AS 30001，之间建立 IBGP 全连接；路由器 AR3、路由器

AR4、路由器 AR5 属于 AS 30002，之间建立 IBGP 全连接，配置相关端口与 IP 地址，进行网络拓扑连接。

V7-1 配置 BGP 功能实例

图 7.3　BGP 基本功能配置实例

（1）配置路由器各端口的 IP 地址。

配置路由器 AR1 端口的 IP 地址。

```
<Huawei>system-view
[Huawei]sysname AR1
[AR1]interface GigabitEthernet 0/0/0
[AR1-GigabitEthernet0/0/0]ip address 100.1.1.1 24
[AR1-GigabitEthernet0/0/0]quit
[AR1]interface GigabitEthernet 0/0/1
[AR1-GigabitEthernet0/0/1]ip address 10.1.1.1 24
[AR1-GigabitEthernet0/0/1]quit
[AR1]
```

配置路由器 AR2 端口的 IP 地址。

```
<Huawei>system-view
[Huawei]sysname AR2
[AR2]interface GigabitEthernet 0/0/1
[AR2-GigabitEthernet0/0/1]ip address 10.1.1.2 24
[AR2-GigabitEthernet0/0/1]quit
[AR2]
```

配置路由器 AR3 端口的 IP 地址。

```
<Huawei>system-view
[Huawei]sysname AR3
[AR3]interface GigabitEthernet 0/0/0
[AR3-GigabitEthernet0/0/0]ip address 100.1.1.2 24
[AR3-GigabitEthernet0/0/0]quit
[AR3]interface GigabitEthernet 0/0/1
[AR3-GigabitEthernet0/0/1]ip address 20.1.1.2 24
[AR3-GigabitEthernet0/0/1]quit
[AR3]interface GigabitEthernet 0/0/2
[AR3-GigabitEthernet0/0/2]ip address 30.1.1.2 24
[AR3-GigabitEthernet0/0/2]quit
[AR3]
```

配置路由器 AR4 端口的 IP 地址。
```
<Huawei>system-view
[Huawei]sysname AR4
[AR4]interface GigabitEthernet 0/0/0
[AR4-GigabitEthernet0/0/0]ip address 40.1.1.1 24
[AR4-GigabitEthernet0/0/0]quit
[AR4]interface GigabitEthernet 0/0/1
[AR4-GigabitEthernet0/0/1]ip address 20.1.1.1 24
[AR4-GigabitEthernet0/0/1]quit
[AR4]
```
配置路由器 AR5 端口的 IP 地址。
```
<Huawei>system-view
[Huawei]sysname AR5
[AR5]interface GigabitEthernet 0/0/0
[AR5-GigabitEthernet0/0/0]ip address 40.1.1.2 24
[AR5-GigabitEthernet0/0/0]quit
[AR5]interface GigabitEthernet 0/0/2
[AR5-GigabitEthernet0/0/2]ip address 30.1.1.1 24
[AR5-GigabitEthernet0/0/2]quit
[AR5]
```
（2）配置 IBGP 连接。

配置路由器 AR1。
```
[AR1]bgp 30001
[AR1-bgp]peer 10.1.1.2 as-number 30001
[AR1-bgp]quit
[AR1]
```
配置路由器 AR2。
```
[AR2]bgp 30001
[AR2-bgp]peer 10.1.1.1 as-number 30001
[AR2-bgp]quit
[AR2]
```
配置路由器 AR3。
```
[AR3]bgp 30002
[AR3-bgp]peer 20.1.1.1 as-number 30002
[AR3-bgp]peer 30.1.1.1 as-number 30002
[AR3-bgp]quit
[AR3]
```
配置路由器 AR4。
```
[AR4]bgp 30002
[AR4-bgp]peer 20.1.1.2 as-number 30002
[AR4-bgp]peer 40.1.1.2 as-number 30002
[AR4-bgp]quit
[AR4]
```
配置路由器 AR5。
```
[AR5]bgp 30002
[AR5-bgp]peer 40.1.1.1 as-number 30002
[AR5-bgp]peer 30.1.1.2 as-number 30002
[AR5-bgp]quit
[AR5]
```

（3）配置 EBGP 连接。

配置路由器 AR1。

```
[AR1]bgp 30001
[AR1-bgp]peer 100.1.1.2 as-number 30002
[AR1-bgp]quit
[AR1]
```

配置路由器 AR3。

```
[AR3]bgp 30002
[AR3-bgp]peer 100.1.1.1 as-number 30001
[AR3-bgp]quit
[AR3]
```

（4）查看 BGP 对等体的连接状态，这里以路由器 AR3 为例进行介绍。

```
<AR3>display bgp peer
 BGP local router ID : 3.3.3.3
 Local AS number : 30002
 Total number of peers : 3         Peers in established state : 3
   Peer      V     AS    MsgRcvd   MsgSent   OutQ    Up/Down     State       Pre fRcv
   20.1.1.1  4     30002    16       17       0     00:14:47    Established    0
   30.1.1.1  4     30002    14       16       0     00:12:40    Established    0
   100.1.1.1 4     30001     7        7       0     00:05:19    Established    0
<AR3>
```

（5）配置路由器 AR1，发布路由 10.1.1.0/24。

```
[AR1]bgp 30001
[AR1-bgp]ipv4-family unicast              //配置单播发送方式
[AR1-bgp-af-ipv4]network 10.1.1.0 24      //发布路由信息
[AR1-bgp-af-ipv4]quit
[AR1-bgp]quit
[AR1]
```

（6）查看路由器 AR1 的路由表信息。

```
<AR1>display bgp routing-table
 BGP Local router ID is 100.1.1.1
 Status codes: * - valid, > - best, d - damped,
               h - history,  i - internal, s - suppressed, S - Stale
               Origin : i - IGP, e - EGP, ? - incomplete
 Total Number of Routes: 1
      Network         NextHop        MED        LocPrf      PrefVal Path/Ogn
 *>   10.1.1.0/24     0.0.0.0         0                       0      i
<AR1>
```

（7）查看路由器 AR4 的路由表信息。

```
<AR4>display bgp routing-table
 BGP Local router ID is 4.4.4.4
 Status codes: * - valid, > - best, d - damped,
               h - history,  i - internal, s - suppressed, S - Stale
               Origin : i - IGP, e - EGP, ? - incomplete
 Total Number of Routes: 1
      Network         NextHop        MED        LocPrf      PrefVal Path/Ogn
  i   10.1.1.0/24     100.1.1.1       0          100          0      30001i
```

<AR4>

从以上信息可以看出，路由器 AR4 虽然学习到了 AS 30001 中的 10.1.1.0 的路由，但因为下一跳 100.1.1.1 不可达，所以其不是有效路由。

（8）配置 BGP 引入直连路由。

```
[AR3]bgp 30002
[AR3-bgp]ipv4-family unicast            //配置单播发送方式
[AR3-bgp-af-ipv4]import-route direct    //引入直接路由
[AR3-bgp-af-ipv4]quit
[AR3-bgp]quit
[AR3]
```

（9）查看路由器 AR3 的路由表信息。

```
<AR3>display   bgp routing-table
 BGP Local router ID is 3.3.3.3
 Status codes: * - valid, > - best, d - damped,
               h - history,  i - internal, s - suppressed, S - Stale
               Origin : i - IGP, e - EGP, ? - incomplete

 Total Number of Routes: 9
   Network            NextHop         MED        LocPrf     PrefVal Path/Ogn
 *>  10.1.1.0/24      100.1.1.1       0                     0       30001i
 *>  20.1.1.0/24      0.0.0.0         0                     0       ?
 *>  20.1.1.2/32      0.0.0.0         0                     0       ?
 *>  30.1.1.0/24      0.0.0.0         0                     0       ?
 *>  30.1.1.2/32      0.0.0.0         0                     0       ?
 *>  100.1.1.0/24     0.0.0.0         0                     0       ?
 *>  100.1.1.2/32     0.0.0.0         0                     0       ?
 *>  127.0.0.0        0.0.0.0         0                     0       ?
 *>  127.0.0.1/32     0.0.0.0         0                     0       ?
<AR3>
```

（10）再次查看路由器 AR4 的路由表信息。

```
<AR4>display bgp routing-table
 BGP Local router ID is 4.4.4.4
 Status codes: * - valid, > - best, d - damped,
               h - history,  i - internal, s - suppressed, S - Stale
               Origin : i - IGP, e - EGP, ? - incomplete

 Total Number of Routes: 4
   Network            NextHop         MED        LocPrf     PrefVal Path/Ogn
 *>i 10.1.1.0/24      100.1.1.1       0          100        0       30001i
   i 20.1.1.0/24      20.1.1.2        0          100        0       ?
 *>i 30.1.1.0/24      20.1.1.2        0          100        0       ?
 *>i 100.1.1.0/24     20.1.1.2        0          100        0       ?
<AR4>
```

从以上信息可以看出，到 10.1.1.0 的路由变为有效路由，下一跳为路由器 AR1 的 GE 01010 端口的 IP 地址。

（11）验证配置结果，在路由器 AR4 上使用 ping 命令进行测试。

```
<AR4>ping 10.1.1.1
   PING 10.1.1.1: 56   data bytes, press CTRL_C to break
```

```
    Reply from 10.1.1.1: bytes=56 Sequence=1 ttl=254 time=20 ms
    Reply from 10.1.1.1: bytes=56 Sequence=2 ttl=254 time=30 ms
    Reply from 10.1.1.1: bytes=56 Sequence=3 ttl=254 time=20 ms
    Reply from 10.1.1.1: bytes=56 Sequence=4 ttl=254 time=30 ms
    Reply from 10.1.1.1: bytes=56 Sequence=5 ttl=254 time=30 ms
  --- 10.1.1.1 ping statistics ---
    5 packet(s) transmitted
    5 packet(s) received
    0.00% packet loss
    round-trip min/avg/max = 20/26/30 ms
<AR4>
```

7.3.3 BGP 与 BFD 联动功能配置实例

如图 7.4 所示，需要在所有路由器之间运行 BGP，路由器 AR1 属于 AS 30001，路由器 AR2、路由器 AR3、路由器 AR4 属于 AS 30002，它们之间建立 IBGP 全连接，路由器 AR1 与路由器 AR2 之间建立 EBGP 连接，路由器 AR1 与路由器 AR3 之间建立 EBGP 连接，业务流量在主链路路由器 AR1→路由器 AR3→路由器 AR4 上传送，链路路由器 AR1→路由器 AR2→路由器 AR3→路由器 AR4 为备份链路，配置相关端口与 IP 地址，进行网络拓扑连接。

图 7.4 BGP 与 BFD 联动功能配置实例

V7-2 配置 BGP 与 BFD 联动功能实例

V7-3 配置 BGP 与 BFD 联动功能实例——测试结果

（1）配置路由器各端口的 IP 地址，这里以路由器 AR1 为例进行介绍。

```
<Huawei>system-view
[Huawei]sysname AR1
[AR1]interface GigabitEthernet 0/0/0
[AR1-GigabitEthernet0/0/0]ip address 100.1.1.1 24
[AR1-GigabitEthernet0/0/0]quit
[AR1]interface GigabitEthernet 0/0/1
[AR1-GigabitEthernet0/0/1]ip address 100.1.2.1 24
[AR1-GigabitEthernet0/0/1]quit
[AR1]
```

其他路由器各端口的 IP 地址与此配置一致，这里不再赘述。

（2）配置 BGP 基本功能，在路由器 AR1 和路由器 AR2、路由器 AR1 和路由器 AR3 之间建立 EBGP 连接，在路由器 AR2 和路由器 AR3 之间建立 IBGP 连接。

配置路由器 AR1。

```
[AR1]bgp 30001
[AR1-bgp]router id 1.1.1.1
[AR1-bgp]peer 100.1.1.2 as-number 30002
[AR1-bgp]peer 100.1.1.2 ebgp-max-hop
[AR1-bgp]peer 100.1.2.2 as-number 30002
[AR1-bgp]peer 100.1.2.2 ebgp-max-hop
[AR1-bgp]quit
[AR1]
```

配置路由器 AR2。

```
[AR2]bgp 30002
[AR2-bgp]router id 2.2.2.2
[AR2-bgp]peer 100.1.1.1 as-number 30001
[AR2-bgp]peer 100.1.1.1 ebgp-max-hop
[AR2-bgp]peer 10.1.1.2 as-number 30002
[AR2-bgp]import-route direct
[AR2-bgp]quit
[AR2]
```

配置路由器 AR3。

```
[AR3]bgp 30002
[AR3-bgp]router id 3.3.3.3
[AR3-bgp]peer 100.1.2.1 as-number 30001
[AR3-bgp]peer 100.1.2.1 ebgp-max-hop
[AR3-bgp]peer 10.1.1.1 as-number 30002
[AR3-bgp]peer 20.1.1.2 as-number 30002
[AR3-bgp]import-route direct
[AR3-bgp]quit
[AR3]
```

配置路由器 AR4。

```
[AR4]bgp 30002
[AR4-bgp]router id 4.4.4.4
[AR4-bgp]peer 20.1.1.1 as-number 30002
[AR4-bgp]import-route direct
[AR4-bgp]ipv4-family unicast              //配置单播发送方式
[AR4-bgp-af-ipv4]network 20.1.1.0 24      //发布路由信息
[AR4-bgp-af-ipv4]quit
[AR4-bgp]quit
[AR4]
```

（3）在路由器 AR1 上查看 BGP 邻居关系。

```
<AR1>display bgp peer
 BGP local router ID : 1.1.1.1
 Local AS number : 30001
 Total number of peers : 2         Peers in established state : 2

  Peer        V    AS    MsgRcvd   MsgSent   OutQ   Up/Down       State Pre fRcv
  100.1.1.2   4    30002    21       24       0    00:13:22  Established   4
  100.1.2.2   4    30002    20       23       0    00:12:27  Established   4
```

<AR1>

（4）配置 MED 属性，通过策略配置路由器 AR2 和路由器 AR3 发送给路由器 AR1 的 MED 值。配置路由器 AR2。

```
[AR2]route-policy 10 permit node 10              //配置路由策略
[AR2-route-policy]apply cost 150                 //修改路径开销值为 150
[AR2-route-policy]quit
[AR2]bgp 30002
[AR2-bgp]peer 100.1.1.1 route-policy 10 export   //在出口上应用路由策略
[AR2-bgp]quit
[AR2]
```

配置路由器 AR3。

```
[AR3]route-policy 10 permit node 10              //配置路由策略
[AR3-route-policy]apply cost 120                 //修改路径开销值为 120
[AR3-route-policy]quit
[AR3]bgp 30002
[AR3-bgp]peer 100.1.2.1 route-policy 10 export   //在出口上应用路由策略
[AR3-bgp]quit
[AR3]
```

（5）查看路由器 AR1 上 BGP 的所有路由信息。

```
<AR1>display bgp routing-table
 BGP Local router ID is 1.1.1.1
 Status codes: * - valid, > - best, d - damped,
               h - history,  i - internal, s - suppressed, S - Stale
               Origin : i - IGP, e - EGP, ? - incomplete
Total Number of Routes: 8
    Network            NextHop         MED        LocPrf    PrefVal   Path/Ogn
 *>  10.1.1.0/24       100.1.2.2       120                  0         30002?
 *                     100.1.1.2       150                  0         30002?
 *>  20.1.1.0/24       100.1.2.2       120                  0         30002?
 *                     100.1.1.2       150                  0         30002?
 *>  100.1.1.0/24      100.1.2.2       120                  0         30002?
                      100.1.1.2       150                  0         30002?
 *>  100.1.2.0/24      100.1.1.2       150                  0         30002?
                      100.1.2.2       120                  0         30002?
<AR1>
```

从以上信息可以看出，去往 20.1.1.0/24 的路由下一跳地址为 100.1.2.2，流量在主链路路由器 AR1→路由器 AR3→路由器 AR4 上传送。

（6）跟踪数据流量走向，这里以路由器 AR1 为例进行介绍。

```
<AR1>tracert 20.1.1.1
 traceroute to   20.1.1.1(20.1.1.1), max hops: 30 ,packet length: 40,press CTRL_C
 to break
 1 100.1.2.2 30 ms   20 ms   20 ms
<AR1>
```

（7）配置 BFD 检测功能、发送和接收时间间隔、本地检测时间倍数。在路由器 AR1 上使能 BFD 功能，并指定最小发送和接收时间间隔为 200ms，本地检测时间倍数为 4。

```
[AR1]bfd
[AR1-bfd]quit
[AR1]bgp 30001
```

```
[AR1-bgp]peer 100.1.2.2 bfd enable
[AR1-bgp]peer 100.1.2.2 bfd min-rx-interval 200 min-tx-interval 200 detect-multiplier 4
[AR1-bgp]quit
[AR1]
```

（8）配置 BFD 检测功能、发送和接收时间间隔、本地检测时间倍数。在路由器 AR3 上使能 BFD 功能，并指定最小发送和接收时间间隔为 200ms，本地检测时间倍数为 4。

```
[AR3]bfd
[AR3-bfd]quit
[AR3]bgp 30002
[AR3-bgp]peer 100.1.2.1 bfd min-rx-interval 200 min-tx-interval 200 detect-multiplier 4
[AR3-bgp]quit
[AR3]
```

（9）在路由器 AR1 上显示 BGP 建立的所有 BFD 会话。

```
[AR1]display bgp bfd session all
  Local_Address    Peer_Address      LD/RD          Interface
  100.1.2.1        100.1.2.2         8192/8192      GigabitEthernet0/0/1
  Tx-interval(ms)  Rx-interval(ms)   Multiplier     Session-State
  200              200               4              Up
  Wtr-interval(m)
  0
[AR1]
```

（10）验证配置结果。在路由器 AR3 的 GE 0/0/1 端口上使用 shutdown 命令，模拟主链路出现故障。

```
[AR3]interface GigabitEthernet 0/0/1
[AR3-GigabitEthernet0/0/1]shutdown
[AR3-GigabitEthernet0/0/1]quit
[AR3]
```

（11）查看路由器 AR1 上 BGP 的所有路由信息。

```
[AR1]display bgp routing-table
 BGP Local router ID is 1.1.1.1
 Status codes: * - valid, > - best, d - damped,
               h - history,  i - internal, s - suppressed, S - Stale
               Origin : i - IGP, e - EGP, ? - incomplete
 Total Number of Routes: 3
   Network              NextHop          MED         LocPrf     PrefVal Path/Ogn
 *> 10.1.1.0/24         100.1.1.2        150                    0       30002?
 *> 20.1.1.0/24         100.1.1.2        150                    0       30002?
    100.1.1.0/24        100.1.1.2        150                    0       30002?
[AR1]
```

从以上信息可以看出，在主链路失效后，备份链路路由器 AR1→路由器 AR2→路由器 AR3→路由器 AR4 生效，去往 20.1.1.0/24 的路由下一跳地址为 100.1.1.2。

（12）再次跟踪数据流量走向，这里以 AR1 的显示结果为例进行介绍。

```
[AR1]tracert 20.1.1.1
 traceroute to  20.1.1.1(20.1.1.1), max hops: 30 ,packet length: 40,press CTRL_C
 to break
 1 100.1.1.2 30 ms   1 ms    10 ms
```

```
 2 10.1.1.2 < AS=30002 > 30 ms    30 ms    20 ms
[AR1]
```

（13）显示路由器 AR1 的配置信息，主要配置实例代码如下。

```
<AR1>display current-configuration
#
 sysname AR1
#
router id 1.1.1.1
#
bfd
#
interface GigabitEthernet0/0/0
 ip address 100.1.1.1 255.255.255.0
#
interface GigabitEthernet0/0/1
 ip address 100.1.2.1 255.255.255.0
#
bgp 30001
 peer 100.1.1.2 as-number 30002
 peer 100.1.1.2 ebgp-max-hop 255
 peer 100.1.2.2 as-number 30002
 peer 100.1.2.2 ebgp-max-hop 255
 peer 100.1.2.2 bfd min-tx-interval 200 min-rx-interval 200 detect-multiplier 4
 peer 100.1.2.2 bfd enable
 #
 ipv4-family unicast
  undo synchronization
  peer 100.1.1.2 enable
  peer 100.1.2.2 enable
#
return
<AR1>
```

（14）显示路由器 AR3 的配置信息，主要配置实例代码如下。

```
<AR3>display current-configuration
#
 sysname AR3
#
router id 3.3.3.3
#
bfd
#
interface GigabitEthernet0/0/0
 ip address 20.1.1.1 255.255.255.0
#
interface GigabitEthernet0/0/1
 shutdown
 ip address 100.1.2.2 255.255.255.0
#
interface GigabitEthernet0/0/2
 ip address 10.1.1.2 255.255.255.0
```

```
#
bgp 30002
 peer 10.1.1.1 as-number 30002
 peer 20.1.1.2 as-number 30002
 peer 100.1.2.1 as-number 30001
 peer 100.1.2.1 ebgp-max-hop 255
 peer 100.1.2.1 bfd min-tx-interval 200 min-rx-interval 200 detect-multiplier 4
 peer 100.1.2.1 bfd enable
#
ipv4-family unicast
  undo synchronization
  import-route direct
  peer 10.1.1.1 enable
  peer 20.1.1.2 enable
  peer 100.1.2.1 enable
  peer 100.1.2.1 route-policy 10 export
#
route-policy 10 permit node 10
 apply cost 120
#
return
<AR3>
```

课后习题

1. 选择题

（1）BGP 邻居间未建立连接且未尝试发起建立连接的状态是（　　）。
  A. Established　　　B. Idle　　　C. Active　　　D. Openconfirm

（2）下列关于 BGP 的说法中，不正确的是（　　）。
  A. BGP 是一种距离矢量协议　　　B. BGP 通过 UDP 发布路由信息
  C. BGP 支持路由聚合功能　　　　D. BGP 能够检测路由环路问题

（3）不属于 BGP 公认必遵属性的是（　　）。
  A. Origin 属性　　　B. AS-Path 属性　C. Next-Hop 属性　　D. MED 属性

（4）BGP 的作用是（　　）。
  A. 用于 AS 之间的路由器间交换路由信息
  B. 用于 AS 内部的路由器间交换路由信息
  C. 用于主干网中路由器之间交换路由信息
  D. 用于园区网中路由器之间交换路由信息

（5）BGP 路由的更新发送是基于 TCP 连接的，这种说法（　　）。
  A. 正确　　　　　　B. 错误

2. 简答题

（1）简述 BGP 的工作原理及其常见属性。
（2）简述 BGP 引入路由的方法。
（3）简述配置 BGP 与动态 BFD 联动的方法。

项目 8 路由策略

【学习目标】
- 掌握路由策略的基本原理。
- 掌握路由策略的配置方法。

8.1 项目陈述

路由策略主要实现了路由过滤和路由属性设置等功能，它通过改变路由属性（包括可达性）来改变网络流量所经过的路径。路由策略可以通过控制路由器的路由表规模，节约系统资源；通过控制路由的接收、发布和引入，提高网络安全性；通过修改路由属性，对网络数据流量进行合理规划，提高网络性能。

8.2 必备知识

8.2.1 Route-Policy 概述

路由协议在发布、接收和引入路由信息时需要根据实际组网需求实施一些路由策略（Route-Policy），以便对路由信息进行过滤和改变路由信息的属性。

（1）控制路由的接收和发布：只发布和接收必要、合法的路由信息，以控制路由表的容量，提高网络的安全性。

（2）控制路由的引入：在一种路由协议中引入其他路由协议发现的路由信息以丰富自己的路由信息时，只引入一部分满足条件的路由信息。

（3）设置特定路由的属性：通过修改 Route-Policy 过滤的路由的属性来满足自身需要。

8.2.2 Route-Policy 基本原理

Route-Policy 使用不同的匹配条件和匹配模式选择路由及改变路由属性，实现路由过滤，如图 8.1 所示。

图 8.1　Route-Policy 基本原理

一个 Route-Policy 中包含 N（N≥1）个节点（Node）。路由信息进入 Route-Policy 后，按节点序号从小到大依次检查各个节点是否匹配。匹配条件由 If-match 子句定义，涉及路由信息的属性和 Route-Policy 的 6 种过滤器。

Route-Policy 的 If-match 子句中，匹配的 6 种过滤器包括访问控制列表、地址前缀列表、AS 路径过滤器、团体属性过滤器、扩展团体属性过滤器和 RD 属性过滤器。这 6 种过滤器具有各自的匹配条件和匹配模式，因此，这 6 种过滤器在特定情况下可以单独使用，以实现路由过滤。

当路由与该节点的所有 If-match 子句都匹配成功后，会进入匹配模式选择，不再匹配其他节点。匹配模式有 permit 和 deny 两种。

（1）permit：路由将被允许通过，并执行该节点的 Apply 子句，对路由信息的一些属性进行设置。

（2）deny：路由将被拒绝通过。当路由与该节点的任意一个 If-match 子句匹配失败后，进入下一节点。如果其和所有节点都匹配失败，则路由信息将被拒绝通过。

8.2.3　配置 Route-Policy

Route-Policy 中可以包含多个匹配条件和动作。Route-Policy 中至少需要配置一个节点的匹配模式是 permit，否则所有路由都会被过滤。

1. 创建 Route-Policy

（1）使用 system-view 命令，进入系统视图。

（2）使用 route-policy route-policy-name { permit | deny } node node 命令，创建 Route-Policy，并进入 Route-Policy 视图。

当使用 Route-Policy 时，node 值小的节点先进行匹配。一个节点匹配成功后，路由将不再匹配其他节点；所有节点都匹配失败后，路由将被过滤。

（3）（可选）使用 description text 命令，配置 Route-Policy 的描述信息。

2.（可选）配置 if-match 子句

if-match 子句用来定义 Route-Policy 的匹配条件，匹配对象是 Route-Policy 过滤器和路由信息的一些属性。

在一个 Route-Policy 节点中，如果不配置 if-match 子句，则表示路由信息在该节点匹配成功；如果配置了一条或多条 if-match 子句，则各个 if-match 子句之间是"与"的关系，即路由信息必须同时满足所有 if-match 子句，才算该节点匹配成功。

> **注意** 当 if-match 子句匹配未配置的过滤器时，默认该 if-match 子句匹配成功。
> 对于同一个 Route-Policy 节点，if-match acl 命令和 if-match ip-prefix 命令不能同时配置，后配置的命令会覆盖先配置的命令。

配置 if-match 子句的步骤如下。

（1）使用 system-view 命令，进入系统视图。

（2）使用 route-policy route-policy-name { permit | deny } node node 命令，进入 Route-Policy 视图。

（3）根据实际情况配置 Route-Policy 中的 If-match 子句。

3.（可选）配置 Apply 子句

Apply 子句用来为 Route-Policy 指定动作，用来设置匹配成功的路由的属性。在一个节点中，如果没有配置 Apply 子句，则该节点仅起过滤路由的作用；如果配置了一个或多个 Apply 子句，则通过节点匹配的路由将执行所有 Apply 子句。

（1）使用 system-view 命令，进入系统视图。

（2）使用 route-policy route-policy-name { permit | deny } node node 命令，进入 Route-Policy 视图。

（3）根据实际情况配置 Route-Policy 中的 Apply 子句。

4. 检查 Route-Policy 的配置结果

使用 display route-policy [route-policy-name]命令，查看 Route-Policy 的详细配置信息。

8.3 项目实施

8.3.1 Route-Policy 过滤配置实例

如图 8.2 所示，主干网络中运行 OSPF 协议，路由器 AR1 从 Internet 接收路由信息，并为 OSPF 协议网络提供 Internet 路由。要求 OSPF 协议网络只能访问 172.16.1.0/24、172.16.2.0/24 和 172.16.3.0/24 三个网段的网络，其中，路由器 AR3 连接的网络只能访问 72.16.1.0/24 网段的网络。

V8-1 配置 Route-Policy 过滤实例

图 8.2 Route-Policy 过滤配置实例

在路由器 AR1 上配置 Route-Policy，在路由发布时应用 Route-Policy，使路由器 AR1 仅提供路由 172.16.1.0/24、172.16.2.0/24、172.16.3.0/24 给路由器 AR2，使 OSPF 协议网络只能访问 172.16.1.0/24、172.16.2.0/24 和 172.16.3.0/24 三个网段的网络。

在路由器 AR3 上配置 Route-Policy，在路由引入时应用 Route-Policy，使路由器 AR3 仅接收路由 172.16.1.0/24，即路由器 AR3 连接的网络只能访问 172.16.1.0/24 网段的网络，配置相关端口与 IP 地址，进行网络拓扑连接。

V8-2 配置 Route-Policy 过滤实例——测试结果

（1）配置路由器 AR1、路由器 AR2、路由器 AR3、路由器 AR4 各端口的 IP 地址，这里以路由器 AR2 为例，其他路由器各端口的 IP 地址与此配置一致，不再赘述。

```
<Huawei>system-view
[Huawei]sysname AR2
[AR2]interface GigabitEthernet 0/0/0
[AR2-GigabitEthernet0/0/0]ip address 192.168.1.2 24
[AR2-GigabitEthernet0/0/0]quit
[AR2]interface GigabitEthernet 0/0/1
[AR2-GigabitEthernet0/0/1]ip address 192.168.2.1 24
[AR2-GigabitEthernet0/0/1]quit
[AR2]interface GigabitEthernet 0/0/2
[AR2-GigabitEthernet0/0/2]ip address 192.168.3.1 24
[AR2-GigabitEthernet0/0/2]quit
[AR2]
```

（2）配置 OSPF 协议的基本功能。

配置路由器 AR1。

```
[AR1]ospf
[AR1-ospf-1]area 0
[AR1-ospf-1-area-0.0.0.0]network 192.168.1.0 0.0.0.255
[AR1-ospf-1-area-0.0.0.0]quit
[AR1-ospf-1]quit
[AR1]
```

配置路由器 AR2。

```
[AR2]ospf
[AR2-ospf-1-area-0.0.0.0]network 192.168.1.0 0.0.0.255
[AR2-ospf-1-area-0.0.0.0]network 192.168.2.0 0.0.0.255
[AR2-ospf-1-area-0.0.0.0]network 192.168.3.0 0.0.0.255
[AR2-ospf-1-area-0.0.0.0]quit
[AR2-ospf-1]quit
[AR2]
```

配置路由器 AR3。

```
[AR3]ospf
[AR3-ospf-1]area 0
[AR3-ospf-1-area-0.0.0.0]network 192.168.2.0 0.0.0.255
[AR3-ospf-1-area-0.0.0.0]quit
[AR3-ospf-1]quit
[AR3]
```

配置路由器 AR4。

```
[AR4]ospf
[AR4-ospf-1]area 0
[AR4-ospf-1-area-0.0.0.0]network 192.168.3.0 0.0.0.255
[AR4-ospf-1-area-0.0.0.0]quit
[AR4-ospf-1]quit
[AR4]
```

（3）在路由器 AR1 上配置 4 条静态路由，并将这些静态路由引入 OSPF 协议中。

```
[AR1]ip route-static 172.16.1.0 24 NULL 0
[AR1]ip route-static 172.16.2.0 24 NULL 0
[AR1]ip route-static 172.16.3.0 24 NULL 0
[AR1]ip route-static 172.16.4.0 24 NULL 0
[AR1]ospf
[AR1-ospf-1]import-route static
[AR1-ospf-1]quit
[AR1]
```

（4）在路由器 AR2 上查看 IP 路由表，可以看到 OSPF 协议引入的 4 条静态路由。

```
<AR2>display ip routing-table
Route Flags: R - relay, D - download to fib
------------------------------------------------------------------------
Routing Tables: Public
         Destinations : 17        Routes : 17
Destination/Mask    Proto    Pre    Cost    Flags    NextHop          Interface
      127.0.0.0/8   Direct   0      0       D        127.0.0.1        InLoopBack0
     127.0.0.1/32   Direct   0      0       D        127.0.0.1        InLoopBack0
127.255.255.255/32  Direct   0      0       D        127.0.0.1        InLoopBack0
     172.16.1.0/24  O_ASE    150    1       D        192.168.1.1      GigabitEthernet 0/0/0
     172.16.2.0/24  O_ASE    150    1       D        192.168.1.1      GigabitEthernet 0/0/0
     172.16.3.0/24  O_ASE    150    1       D        192.168.1.1      GigabitEthernet 0/0/0
     172.16.4.0/24  O_ASE    150    1       D        192.168.1.1      GigabitEthernet 0/0/0
    192.168.1.0/24  Direct   0      0       D        192.168.1.2      GigabitEthernet 0/0/0
   192.168.1.2/32   Direct   0      0       D        127.0.0.1        GigabitEthernet 0/0/0
 192.168.1.255/32   Direct   0      0       D        127.0.0.1        GigabitEthernet 0/0/0
    192.168.2.0/24  Direct   0      0       D        192.168.2.1      GigabitEthernet 0/0/1
   192.168.2.1/32   Direct   0      0       D        127.0.0.1        GigabitEthernet 0/0/1
 192.168.2.255/32   Direct   0      0       D        127.0.0.1        GigabitEthernet 0/0/1
    192.168.3.0/24  Direct   0      0       D        192.168.3.1      GigabitEthernet 0/0/2
   192.168.3.1/32   Direct   0      0       D        127.0.0.1        GigabitEthernet 0/0/2
 192.168.3.255/32   Direct   0      0       D        127.0.0.1        GigabitEthernet 0/0/2
255.255.255.255/32  Direct   0      0       D        127.0.0.1        InLoopBack0
<AR2>
```

（5）在路由器 AR1 上配置地址前缀列表 AR1-AR2。

```
[AR1]ip ip-prefix AR1-AR2 index 10 permit 172.16.1.0 24
[AR1]ip ip-prefix AR1-AR2 index 20 permit 172.16.2.0 24
[AR1]ip ip-prefix AR1-AR2 index 30 permit 172.16.3.0 24
[AR1]
```

（6）在路由器 AR1 上配置发布策略，引用地址前缀列表 AR1-AR2 进行过滤。

```
[AR1]ospf
[AR1-ospf-1]filter-policy ip-prefix AR1-AR2 export static
[AR1-ospf-1]quit
[AR1]
```

（7）在路由器 AR2 上查看 IP 路由表，可以看到路由器 AR2 仅接收了列表 AR1-AR2 中定义的 3 条路由。

```
<AR2>display ip routing-table
Route Flags: R - relay, D - download to fib
------------------------------------------------------------------------
Routing Tables: Public
         Destinations : 16      Routes : 16
Destination/Mask    Proto   Pre  Cost   Flags  NextHop        Interface
       127.0.0.0/8  Direct  0    0      D      127.0.0.1      InLoopBack0
      127.0.0.1/32  Direct  0    0      D      127.0.0.1      InLoopBack0
127.255.255.255/32  Direct  0    0      D      127.0.0.1      InLoopBack0
     172.16.1.0/24  O_ASE   150  1      D      192.168.1.1    GigabitEthernet 0/0/0
     172.16.2.0/24  O_ASE   150  1      D      192.168.1.1    GigabitEthernet 0/0/0
     172.16.3.0/24  O_ASE   150  1      D      192.168.1.1    GigabitEthernet 0/0/0
    192.168.1.0/24  Direct  0    0      D      192.168.1.2    GigabitEthernet 0/0/0
    192.168.1.2/32  Direct  0    0      D      127.0.0.1      GigabitEthernet 0/0/0
  192.168.1.255/32  Direct  0    0      D      127.0.0.1      GigabitEthernet 0/0/0
    192.168.2.0/24  Direct  0    0      D      192.168.2.1    GigabitEthernet 0/0/1
    192.168.2.1/32  Direct  0    0      D      127.0.0.1      GigabitEthernet 0/0/1
  192.168.2.255/32  Direct  0    0      D      127.0.0.1      GigabitEthernet 0/0/1
    192.168.3.0/24  Direct  0    0      D      192.168.3.1    GigabitEthernet 0/0/2
    192.168.3.1/32  Direct  0    0      D      127.0.0.1      GigabitEthernet 0/0/2
  192.168.3.255/32  Direct  0    0      D      127.0.0.1      GigabitEthernet 0/0/2
255.255.255.255/32  Direct  0    0      D      127.0.0.1      InLoopBack0
<AR2>
```

（8）配置路由接收策略，在路由器 AR3 上配置地址前缀列表 AR3-in。

```
[AR3]ip ip-prefix AR3-in index 10 permit 172.16.1.0 24
```

（9）在路由器 AR3 上配置接收策略，引用地址前缀列表 AR3-in 进行过滤。

```
[AR3]ospf
[AR3-ospf-1]filter-policy ip-prefix AR3-in import
[AR3-ospf-1]quit
[AR3]
```

（10）查看路由器 AR3 的 IP 路由表，可以看到路由器 AR3 的路由表中仅接收了列表 AR3-in 中定义的 1 条路由。

```
<AR3>display ip routing-table
Route Flags: R - relay, D - download to fib
------------------------------------------------------------------------
Routing Tables: Public
         Destinations : 8      Routes : 8
Destination/Mask    Proto   Pre  Cost   Flags  NextHop        Interface
       127.0.0.0/8  Direct  0    0      D      127.0.0.1      InLoopBack0
      127.0.0.1/32  Direct  0    0      D      127.0.0.1      InLoopBack0
127.255.255.255/32  Direct  0    0      D      127.0.0.1      InLoopBack0
```

172.16.1.0/24	O_ASE	150	1	D	192.168.2.1	GigabitEthernet 0/0/1
192.168.2.0/24	Direct	0	0	D	192.168.2.2	GigabitEthernet 0/0/1
192.168.2.2/32	Direct	0	0	D	127.0.0.1	GigabitEthernet 0/0/1
192.168.2.255/32	Direct	0	0	D	127.0.0.1	GigabitEthernet 0/0/1
255.255.255.255/32	Direct	0	0	D	127.0.0.1	InLoopBack0

\<AR3>

（11）查看路由器 AR4 的 IP 路由表，可以看到路由器 AR4 的本地路由表中接收了路由器 AR2 发送的所有路由信息。

```
<AR4>display ip routing-table
Route Flags: R - relay, D - download to fib
------------------------------------------------------------------
Routing Tables: Public
         Destinations : 12       Routes : 12
Destination/Mask    Proto   Pre  Cost   Flags    NextHop         Interface
     127.0.0.0/8    Direct  0    0      D        127.0.0.1       InLoopBack0
     127.0.0.1/32   Direct  0    0      D        127.0.0.1       InLoopBack0
127.255.255.255/32  Direct  0    0      D        127.0.0.1       InLoopBack0
    172.16.1.0/24   O_ASE   150  1      D        192.168.3.1     GigabitEthernet 0/0/2
    172.16.2.0/24   O_ASE   150  1      D        192.168.3.1     GigabitEthernet 0/0/2
    172.16.3.0/24   O_ASE   150  1      D        192.168.3.1     GigabitEthernet 0/0/2
   192.168.1.0/24   OSPF    10   2      D        192.168.3.1     GigabitEthernet 0/0/2
   192.168.2.0/24   OSPF    10   2      D        192.168.3.1     GigabitEthernet 0/0/2
   192.168.3.0/24   Direct  0    0      D        192.168.3.2     GigabitEthernet 0/0/2
   192.168.3.2/32   Direct  0    0      D        127.0.0.1       GigabitEthernet 0/0/2
 192.168.3.255/32   Direct  0    0      D        127.0.0.1       GigabitEthernet 0/0/2
255.255.255.255/32  Direct  0    0      D        127.0.0.1       InLoopBack0
<AR4>
```

（12）查看路由器 AR3 的 OSPF 协议路由表，可以看到 OSPF 协议路由表中接收了 3 条列表 AR1~AR2 中定义的路由。因为在链路状态协议中，filter-policy import 命令用于过滤从协议路由表加入本地路由表的路由。

```
<AR3>display ospf routing
        OSPF Process 1 with Router ID 192.168.2.2
                Routing Tables
Routing for Network
Destination        Cost    Type        NextHop         AdvRouter       Area
192.168.2.0/24     1       Transit     192.168.2.2     192.168.2.2     0.0.0.0
192.168.1.0/24     2       Transit     192.168.2.1     192.168.1.1     0.0.0.0
192.168.3.0/24     2       Transit     192.168.2.1     192.168.1.2     0.0.0.0

Routing for ASEs
Destination        Cost    Type        Tag             NextHop         AdvRouter
172.16.1.0/24      1       Type2       1               192.168.2.1     192.168.1.1
172.16.2.0/24      1       Type2       1               192.168.2.1     192.168.1.1
172.16.3.0/24      1       Type2       1               192.168.2.1     192.168.1.1
Total Nets: 6
Intra Area: 3  Inter Area: 0  ASE: 3  NSSA: 0
<AR3>
```

（13）查看路由器 AR1 的配置信息。

```
<AR1>display current-configuration
#
 sysname AR1
#
interface GigabitEthernet0/0/0
 ip address 192.168.1.1 255.255.255.0
#
ospf 1
 filter-policy ip-prefix AR1-AR2 export static
 import-route static
 area 0.0.0.0
  network 192.168.1.0 0.0.0.255
#
ip ip-prefix AR1-AR2 index 10 permit 172.16.1.0 24
ip ip-prefix AR1-AR2 index 20 permit 172.16.2.0 24
ip ip-prefix AR1-AR2 index 30 permit 172.16.3.0 24
#
ip route-static 172.16.1.0 255.255.255.0 NULL0
ip route-static 172.16.2.0 255.255.255.0 NULL0
ip route-static 172.16.3.0 255.255.255.0 NULL0
ip route-static 172.16.4.0 255.255.255.0 NULL0
#
return
<AR1>
```

8.3.2　路由引入时应用 Route-Policy 配置实例

如图 8.3 所示，路由器 AR1 与路由器 AR2 之间通过 OSPF 协议交换路由信息，路由器 AR3、路由器 AR4 与路由器 AR5 之间通过 IS-IS 协议交换路由信息。要求在路由器 AR1 上将 IS-IS 网络中的路由引入 OSPF 协议网络中后，在路由器 AR1 上配置 Route-Policy，将 172.16.1.0/24 的路由的开销设置为 100，并在 OSPF 网络中引入 IS-IS 协议路由时应用 Route-Policy，使 OSPF 协议网络中路由 172.16.1.0/24 的选路优先级较低；在路由器 AR1 上配置 Route-Policy，将 172.16.2.0/24 的路由的 Tag 属性设置为 30，并在 OSPF 网络中引入 IS-IS 协议路由时应用 Route-Policy，使路由 172.16.2.0/24 的优先级较高，配置相关端口与 IP 地址，进行网络拓扑连接。

图 8.3　路由引入时应用 Route-Policy 配置实例

V8-3　路由引入时应用 Route-Policy 配置实例

（1）配置路由器各端口的 IP 地址，这里以路由器 AR1 为例，其他路由器各端口的 IP 地址与此配置一致，不再赘述。

```
<Huawei>system-view
[Huawei]sysname AR1
[AR1]interface GigabitEthernet 0/0/0
[AR1-GigabitEthernet0/0/0]ip address 192.168.1.1 24
[AR1-GigabitEthernet0/0/0]quit
[AR1]interface GigabitEthernet 0/0/1
[AR1-GigabitEthernet0/0/1]ip address 192.168.2.1 24
[AR1-GigabitEthernet0/0/1]quit
[AR1]
```

V8-4 路由引入时应用 Route-Policy 配置实例——测试结果

（2）配置 OSPF 协议及路由引入。

配置路由器 AR1，启动 OSPF 协议，并引入 IS-IS 协议路由。

```
[AR1]router id 1.1.1.1
[AR1]ospf
[AR1-ospf-1]area 0
[AR1-ospf-1-area-0.0.0.0]network 192.168.2.0 0.0.0.255
[AR1-ospf-1-area-0.0.0.0]quit
[AR1-ospf-1]import-route isis 1
[AR1-ospf-1]quit
[AR1]
```

配置路由器 AR2，启动 OSPF 协议。

```
[AR2]router id 2.2.2.2
[AR2]ospf
[AR2-ospf-1]area 0
[AR2-ospf-1-area-0.0.0.0]network 192.168.2.0 0.0.0.255
[AR2-ospf-1-area-0.0.0.0]quit
[AR2-ospf-1]quit
[AR2]
```

（3）配置 IS-IS 协议。

配置路由器 AR1。

```
[AR1]isis
[AR1-isis-1]is-level level-2
[AR1-isis-1]network-entity 10.0000.0000.0001.00
[AR1-isis-1]quit
[AR1]interface GigabitEthernet 0/0/0
[AR1-GigabitEthernet0/0/0]isis enable
[AR1-GigabitEthernet0/0/0]ip address 192.168.1.1 24
[AR1-GigabitEthernet0/0/0]quit
[AR1]
```

配置路由器 AR3。

```
[AR3]router id 3.3.3.3
[AR3]isis
[AR3-isis-1]is-level level-2
[AR3-isis-1]network-entity 10.0000.0000.0003.00
[AR3-isis-1]quit
```

[AR3]interface GigabitEthernet 0/0/0
[AR3-GigabitEthernet0/0/0]isis enable
[AR3-GigabitEthernet0/0/0]ip address 192.168.1.2 24
[AR3-GigabitEthernet0/0/0]quit
[AR3]interface GigabitEthernet 0/0/1
[AR3-GigabitEthernet0/0/1]isis enable
[AR3-GigabitEthernet0/0/1]ip address 172.16.1.1 24
[AR3-GigabitEthernet0/0/1]quit
[AR3]interface GigabitEthernet 0/0/2
[AR3-GigabitEthernet0/0/2]isis enable
[AR3-GigabitEthernet0/0/2]ip address 172.16.2.1 24
[AR3-GigabitEthernet0/0/2]quit
[AR3]

配置路由器 AR4。

[AR4]router id 4.4.4.4
[AR4]isis
[AR4-isis-1]network-entity 10.0000.0000.0004.00
[AR4-isis-1]is-level level-2
[AR4-isis-1]quit
[AR4]interface GigabitEthernet 0/0/1
[AR4-GigabitEthernet0/0/1]isis enable
[AR4-GigabitEthernet0/0/1]ip address 172.16.1.2 24
[AR4-GigabitEthernet0/0/1]quit
[AR4]

配置路由器 AR5。

[AR5]router id 5.5.5.5
[AR5]isis
[AR5-isis-1]network-entity 10.0000.0000.0005.00
[AR5-isis-1]is-level level-2
[AR5-isis-1]quit
[AR5]interface GigabitEthernet 0/0/2
[AR5-GigabitEthernet0/0/2]isis enable
[AR5-GigabitEthernet0/0/2]ip address 172.16.2.2 24
[AR5-GigabitEthernet0/0/2]quit
[AR5]

（4）查看路由器 AR2 的 OSPF 协议路由表，可以看到引入的路由。

```
<AR2>display ospf routing
        OSPF Process 1 with Router ID 192.168.2.2
              Routing Tables
Routing for Network
Destination        Cost    Type      NextHop        AdvRouter       Area
192.168.2.0/24     1       Transit   192.168.2.2    192.168.2.2     0.0.0.0
Routing for ASEs
Destination        Cost    Type      Tag       NextHop        AdvRouter
172.16.1.0/24      1       Type2     1         192.168.2.1    192.168.1.1
172.16.2.0/24      1       Type2     1         192.168.2.1    192.168.1.1
192.168.1.0/24     1       Type2     1         192.168.2.1    192.168.1.1
```

Total Nets: 4
　Intra Area: 1　Inter Area: 0　ASE: 3　NSSA: 0
<AR2>

（5）配置访问控制列表，配置编号为 2000 的 ACL，允许 172.16.2.0/24 网段的数据通过。

[AR1]acl number 2000
[AR1-acl-basic-2000]rule permit source 172.16.2.0 0.0.0.255
[AR1-acl-basic-2000]quit
[AR1]

（6）配置名为 prefix-net 的地址前缀列表，允许 172.16.1.0/24 网段的数据通过。

[AR1]ip ip-prefix prefix-net index 10 permit 172.16.1.0 24
[AR1]

（7）配置 Route-Policy。

[AR1]route-policy isis-ospf-policy permit node 10
[AR1-route-policy]if-match ip-prefix prefix-net
[AR1-route-policy]apply cost 100
[AR1-route-policy]quit
[AR1]route-policy isis-ospf-policy permit node 20
[AR1-route-policy]if-match acl 2000
[AR1-route-policy]apply tag 30
[AR1-route-policy]quit
[AR1]route-policy isis-ospf-policy permit node 30
[AR1-route-policy]quit
[AR1]

（8）在路由引入时应用 Route-Policy，配置路由器 AR1。

[AR1]ospf
[AR1-ospf-1]import-route isis 1 route-policy isis-ospf-policy
[AR1-ospf-1]quit
[AR1]

（9）查看路由器 AR2 的 OSPF 协议路由表，可以看到目的地址为 172.16.1.0/24 的路由的开销为 100，目的地址为 172.16.2.0/24 的路由的标记（Tag）为 30，而其他路由的属性未发生变化。

<AR2>display ospf routing
　　　OSPF Process 1 with Router ID 192.168.2.2
　　　　Routing Tables
Routing for Network

Destination	Cost	Type	NextHop	AdvRouter	Area
192.168.2.0/24	1	Transit	192.168.2.2	192.168.2.2	0.0.0.0

Routing for ASEs

Destination	Cost	Type	Tag	NextHop	AdvRouter
172.16.1.0/24	100	Type2	1	192.168.2.1	192.168.1.1
172.16.2.0/24	1	Type2	30	192.168.2.1	192.168.1.1
192.168.1.0/24	1	Type2	1	192.168.2.1	192.168.1.1

Total Nets: 4
Intra Area: 1　Inter Area: 0　ASE: 3　NSSA: 0
<AR2>

（10）在路由器 AR1 上进行配置，使 IS-IS 协议引入 OSPF 协议路由，使全网连通。

[AR1]isis
[AR1-isis-1]import-route ospf

```
[AR1-isis-1]quit
[AR1]
```
（11）验证配置结果，使用 ping 命令进行测试，在路由器 AR2 上访问路由器 AR5 的 GE 0/0/2 端口的 IP 地址。

```
<AR2>ping 172.16.2.2
  PING 172.16.2.2: 56  data bytes, press CTRL_C to break
    Reply from 172.16.2.2: bytes=56 Sequence=1 ttl=253 time=40 ms
    Reply from 172.16.2.2: bytes=56 Sequence=2 ttl=253 time=50 ms
    Reply from 172.16.2.2: bytes=56 Sequence=3 ttl=253 time=30 ms
    Reply from 172.16.2.2: bytes=56 Sequence=4 ttl=253 time=30 ms
    Reply from 172.16.2.2: bytes=56 Sequence=5 ttl=253 time=30 ms
  --- 172.16.2.2 ping statistics ---
    5 packet(s) transmitted
    5 packet(s) received
    0.00% packet loss
    round-trip min/avg/max = 30/36/50 ms
<AR2>
```
（12）查看路由器 AR1 的配置信息。

```
<AR1>display current-configuration
#
 sysname AR1
#
router id 1.1.1.1
#
acl number 2000
 rule 10 permit source 172.16.2.0 0.0.0.255
#
isis 1
 is-level level-2
 network-entity 10.0000.0000.0001.00
 import-route ospf 1
#
interface GigabitEthernet0/0/0
 ip address 192.168.1.1 255.255.255.0
 isis enable 1
#
interface GigabitEthernet0/0/1
 ip address 192.168.2.1 255.255.255.0
#
ospf 1
 import-route isis 1 route-policy isis-ospf-policy
 area 0.0.0.0
  network 192.168.2.0 0.0.0.255
#
route-policy isis-ospf-policy permit node 10
 if-match ip-prefix prefix-net
 apply cost 100
```

```
#
route-policy isis-ospf-policy permit node 20
 if-match acl 2000
 apply tag 30
#
route-policy isis-ospf-policy permit node 30
#
ip ip-prefix prefix-net index 10 permit 172.16.1.0 24
#
return
<AR1>
```

课后习题

1. 选择题

（1）下列选项中，（　　）路由前缀无法满足以下 IP-Prefix 条件。

```
ip ip-prefix   prefix-test index 10 permit 10.1.1.0 16 greater-equal 24 less-equal 28
```

    A. 10.1.1.0/23   B. 10.1.1.0/24   C. 10.1.1.0/25   D. 10.1.1.0/26

（2）用于过滤路由信息以及为通过过滤的路由信息设置路由属性的是（　　）。

    A. AS-Path-Filter   B. IP-Prefix

    C. Route-Policy   D. Policy-Based-Route

（3）下列关于路由过滤的规则中，描述错误的是（　　）。

    A. 可以使用 filter-policy 命令进行过滤，也可以使用 ip-prefix 命令进行过滤

    B. 路由过滤可以过滤从其他路由协议引入的路由信息，但只能在出方向过滤

    C. 路由过滤可以在入方向过滤路由，对于链路状态路由协议，仅仅是不把路由信息加入路由表中

    D. 路由过滤可以在出方向过滤路由，且路由信息和链路状态信息均可被过滤

（4）定义 route-policy set-cost 如下，则下列描述正确的是（　　）。

```
ip ip-prefix prefix-net permit 10.1.1.0 16
route-policy set-cost permit node 10
if-match ip-prefix prefix-net
apply cost 100
route-policy set-cost permit node 20
apply cost 200
```

    A. 路由 10.1.1.0 /16 能够通过 node 10，其开销值被设置为 100

    B. 路由 10.1.1.0 /16 能够通过 node 10 后被继续匹配 node 20，最终开销值被设置为 200

    C. 所有路由的开销值都会被设置为 200

    D. 所有不通过 node 10 的路由都会被拒绝

2. 简答题

（1）简述 Route-Policy 的基本原理。

（2）简述配置 Route-Policy 的方法。

项目 9
策略路由

【学习目标】
- 掌握策略路由的基本原理。
- 掌握策略路由的配置方法。

9.1 项目陈述

策略路由（Policy-Based Routing，PBR）是一种依据用户制定的策略进行路由选择的机制，分为本地 PBR、端口 PBR 和智能策略路由（Smart Policy Routing，SPR）。

传统的路由转发原理是先根据报文的目的地址查找路由表，再进行报文转发，但是目前越来越多的用户希望能够在传统路由转发的基础上，根据自己定义的策略进行报文转发和选路。PBR 使网络管理者不仅能够根据报文的目的地址来制定 PBR，还能够根据报文的源地址、报文大小和链路质量等属性来制定 PBR，以改变数据包转发路径，满足用户需求。

9.2 必备知识

9.2.1 PBR 概述

PBR 与 Route-Policy 存在以下不同。

（1）PBR 的操作对象是数据包，在路由表已经产生的情况下，不按照路由表进行转发，而是根据需要，依照某种策略改变数据包转发路径。

（2）Route-Policy 的操作对象是路由信息。Route-Policy 主要实现了路由过滤和路由属性设置等功能，它通过改变路由属性（包括可达性）来改变网络流量所经过的路径。

（3）PBR 的查找优先级比 Route-Policy 高。当路由器接收到数据包并进行转发时，会优先根据 PBR 的规则进行匹配。如果能匹配上，则根据 PBR 进行转发，否则按照路由表中的路由条目来进行转发。其中，PBR 不改变路由表中的任何内容，它可以通过预先设置的规则来影响数据报文的转发。

1. 本地 PBR

本地 PBR 仅对本机下发的报文进行处理，对转发的报文不起作用。一条本地 PBR 可以配置多个策略点，且这些策略点具有不同的优先级，本机下发报文优先匹配优先级高的策略点。本机下发报文时，根据本地 PBR 节点的优先级，依次匹配各节点绑定的匹配规则。本地 PBR 支持基于 ACL 或报文长度的匹配规则。

如果没有找到匹配的本地 PBR 节点，则按照发送 IP 报文的一般流程，根据目的地址查找路由。

2. 端口 PBR

端口 PBR 只对转发的报文起作用，对本地下发的报文（如本地的 ping 报文）不起作用。端口 PBR 通过在数据流行为中配置重定向，实现只对端口入方向的报文生效。默认情况下，设备按照路由表的下一跳进行报文转发，如果配置了端口 PBR，则设备按照端口 PBR 指定的下一跳进行转发。

在按照端口 PBR 指定的下一跳进行报文转发时，如果设备上没有其下一跳 IP 地址对应的 ARP 表项，则设备会触发 ARP 学习，如果一直学习不到下一跳 IP 地址对应的 ARP 表项，则报文按照路由表指定的下一跳进行转发；如果设备上有或者学习到了此 ARP 表项，则按照端口 PBR 指定的下一跳 IP 地址进行报文转发。

3. SPR

SPR 是基于业务需求的 PBR，其通过匹配链路质量和网络业务对链路质量的需求，实现智能选路。

随着网络业务需求的多样化，以及业务数据的集中放置，链路质量对网络业务来说越来越重要。越来越多的用户把关注点从网络的连通性转移到业务的可用性上，如业务的可获得性、响应速度和业务质量等。这些复杂的业务需求对传统的基于逐跳的路由协议提出了挑战。它们无法感知链路的质量和业务的需求，所以无法保障用户的业务体验，即使路由可达，但链路质量可能已经很差甚至无法正常转发报文了。SPR 就是在这一背景下产生的一种 PBR，它可以主动探测链路质量并匹配业务的需求，从而选择一条最优链路转发业务数据，有效地避免网络黑洞、网络振荡等问题。

9.2.2 配置本地 PBR

1. 配置本地 PBR 的匹配规则

通过以下配置可以定义对何种报文进行本地 PBR。设置 IP 报文的 ACL 匹配条件前可以先配置 ACL。

（1）使用 system-view 命令，进入系统视图。

（2）使用 policy-based-route policy-name { deny | permit } node node-id 命令，创建 PBR 和策略点，若策略点已创建，则进入本地 PBR 视图。

默认情况下，本地 PBR 中未创建 PBR 或策略点。

> **注意** permit 表示对满足匹配条件的报文进行 PBR，deny 表示不对满足匹配条件的报文进行 PBR。
> 重复使用该命令可以在一条本地 PBR 下创建多个策略点，策略点由顺序号 node-id 来指定，顺序号的值越小表示优先级越高，相应策略优先执行。

（3）使用如下命令（可以只使用一条命令，也可以两条命令都使用）设置 IP 报文匹配条件。

① 使用 if-match acl acl-number 命令，设置 IP 报文的 ACL 匹配条件。默认情况下，本地 PBR 中未配置 IP 地址匹配条件。

② 使用 if-match packet-length min-length max-length 命令，设置 IP 报文的长度匹配条件。默认情况下，本地 PBR 中未配置 IP 报文长度匹配条件。

2. 配置本地 PBR 的动作

（1）使用 system-view 命令，进入系统视图。

（2）使用 policy-based-route policy-name { deny | permit } node node-id 命令，创建策略和策略点。若策略点已创建，则进入本地 PBR 视图。

默认情况下，本地 PBR 中未创建 PBR 或策略点。

（3）使用如下命令，配置 PBR 的动作。一个策略点中至少包含一条 Apply 子句，也可以多条 Apply 子句组合使用。

① 使用 apply output-interface interface-type interface-number 命令，指定本地 PBR 中报文的出端口。

② 使用 apply ip-address next-hop ip-address1 [ip-address2]命令，设置本地 PBR 中报文的下一跳地址。

③ 使用 apply ip-address next-hop { ip-address1 track ip-route ip-address2 { mask |mask-length } } &<1-2>命令，配置本地 PBR 的下一跳联动路由功能。

④ 使用 apply ip-address backup-nexthop ip-address 命令，配置本地 PBR 中报文转发的备份下一跳地址。

⑤ 使用 apply default output-interface interface-type interface-number 命令，配置本地 PBR 中报文的默认出端口。

在创建策略和策略点，配置本地 PBR 的动作时请注意以下几种情况。

① 如果策略中设置了两个下一跳地址，那么报文转发在两个下一跳地址之间负载均衡。

② 如果策略中设置了两个出端口，那么报文转发在两个出端口之间负载均衡。

③ 如果策略中同时设置了两个下一跳地址和两个出端口，那么报文转发仅在两个出端口之间负载均衡。

3. 应用本地 PBR

通过以下配置使能一条本地 PBR。本地 PBR 仅对本机下发的报文有效。

（1）使用 system-view 命令，进入系统视图。

（2）使用 ip local policy-based-route policy-name 命令，使能本地 PBR。默认情况下，本地 PBR 处于未使能状态。

4. 查看本地 PBR 的配置结果

（1）使用 display ip policy-based-route 命令，查看本地已使能的 PBR 的策略。

（2）使用 display ip policy-based-route setup local [verbose]命令，查看本地 PBR 的配置情况。

（3）使用 display ip policy-based-route statistics local 命令，查看本地 PBR 报文的统计信息。

（4）使用 display policy-based-route [policy-name [verbose]]命令，查看已创建的策略内容。

9.2.3 配置端口 PBR

配置端口 PBR 可以将到达端口的三层报文重定向到指定的下一跳地址。

通过配置重定向，设备可将符合流分类规则的报文重定向到指定的下一跳地址或指定端口上。当重定向不生效时，用户可以选择使报文按原路径转发或丢弃。包含重定向动作的流策略只能在端口的入方向上应用。

1. 配置流分类

（1）使用 system-view 命令，进入系统视图。

（2）使用 traffic classifier classifier-name [operator { and | or }]命令，创建一个流分类，进入流分类视图。

（3）根据实际情况配置流分类中的匹配规则。

（4）使用 quit 命令，退出流分类视图。

2. 配置流行为

（1）使用 traffic behavior behavior-name 命令，创建一个流行为并进入流行为视图，或进入已存在的流行为视图。

（2）根据实际需要进行如下配置。

① （可选）使用 statistic enable 命令，使能流量统计功能。

② 使用 quit 命令，退出流行为视图。

（3）使用 quit 命令，退出系统视图。

3. 配置流策略

（1）使用 system-view 命令，进入系统视图。

（2）使用 traffic policy policy-name 命令，创建一个流策略并进入流策略视图，或进入已存在的流策略视图。默认情况下，系统未创建任何流策略。

（3）使用 classifier classifier-name behavior behavior-name [precedence precedence-value]命令，在流策略中为指定的流分类配置所需流行为，即绑定流分类和流行为。默认情况下，流策略中没有绑定流分类和流行为。

（4）使用 quit 命令，退出流策略视图。

（5）使用 quit 命令，退出系统视图。

4. 应用流策略

（1）使用 system-view 命令，进入系统视图。

（2）使用 interface interface-type interface-number [subinterface-number]命令，进入端口视图或子端口视图。

（3）使用 traffic-policy policy-name inbound 命令，在端口或子端口的入方向应用流策略。目前，端口 PBR 仅支持在端口的入方向上应用。

5. 查看配置结果

（1）使用 display traffic classifier { system-defined | user-defined } [classifier-name]命令，查看已配置的流分类信息。

（2）使用 display traffic behavior { system-defined | user-defined } [behavior-name]命令，查看已配置的流行为信息。

（3）使用 display traffic policy user-defined [policy-name [classifier classifier-name]]命令，查看已配置的流策略信息。

（4）使用 display traffic-policy applied-record [policy-name]命令，查看流策略的应用记录。

9.3 项目实施

根据业务对链路质量的需求情况配置 SPR，以随链路质量变化情况动态切换业务数据的传输链路。

在配置 SPR 之前，需完成以下任务。

（1）配置端口的链路层协议参数及 IP 地址，使端口的链路层协议状态为 Up 且探测链路路由可达。

（2）配置区分业务流的 ACL。

（3）配置用于检测链路的 NQA 测试例。

1. 配置 SPR

（1）使用 system-view 命令，进入系统视图。

（2）使用 smart-policy-route 命令，创建 SPR 并进入 SPR 视图。

（3）配置 SPR 的切换周期、振荡抑制周期、从备份链路回切到主链路的时间、链路未被选中时可以自动关闭的端口、端口自动关闭的延迟时间、逃生链路、探测链路及链路组、使能重定向链路重定向增强功能等。

2. 配置 SPR 与业务关联

配置 SPR 的业务参数，包括指定需要 PBR 的业务流、配置业务的链路质量需求、绑定业务的探测链路等内容，绑定业务流 ACL 前需要先配置区分业务流的 ACL。

（1）使用 system-view 命令，进入系统视图。

（2）使用 smart-policy-route 命令，创建 SPR 并进入 SPR 视图。

（3）创建 SPR 的业务模板并进入业务模板视图。

使用 service-map name 命令，创建 SPR 的业务模板并进入业务模板视图。默认情况下，SPR 中未配置业务模板。

（4）配置 SPR 业务模板绑定 ACL。

① 绑定 IPv4 ACL。使用 match acl acl-number &<1-10>命令，配置 SPR 业务模板绑定 IPv4 ACL。

② 绑定 IPv6 ACL。使用 match acl ipv6 acl6-number &<1-10>命令，配置 SPR 业务模板绑定 IPv6 ACL。

3. 查看 SPR 的配置结果

（1）使用 display smart-policy-route 命令，查看 SPR 的路由配置信息。

（2）使用 display smart-policy-route service-map [name]命令，查看业务模板的配置信息。

（3）使用 display smart-policy-route link-state [interface-type interface-number]命令，查看探测链路的链路状态信息。

（4）使用 display smart-policy-route nqa-server link-state 命令，查看 NQA 服务器的链路状态信息。

4. 配置本地 PBR 实例

如图 9.1 所示，全网使用 OSPF 协议交换路由信息，要求在路由器 AR1 上使本机下发的不同长度的报文通过不同的下一跳地址进行转发。其中，报文长度为 64～1400 字节的报文设置 192.168.4.2 为下一跳地址；报文长度为 1401～1500 字节的报文设置 192.168.5.2 为下一跳地址；所有其他长度的报文都按基于目的地址的方法进行路由选路。同时，源地址为 192.168.1.0/24 网段的地址访问 Internet 时，192.168.4.2 作为下一跳地址；源地址为 192.168.2.0/24 网段的地址访问 Internet 时，192.168.5.2 作为下一跳地址；配置相关端口与 IP 地址，进行网络拓扑连接。

V9-1 配置本地 PBR 实例

图 9.1 配置本地 PBR 实例

（1）配置交换机 LSW1，相关实例代码如下。

V9-2 配置本地 PBR 实例——测试结果

```
<Huawei>system-view
[Huawei]sysname LSW1
[LSW1]vlan batch 10 20
[LSW1]interface GigabitEthernet 0/0/1
[LSW1-GigabitEthernet0/0/1]port link-type access
[LSW1-GigabitEthernet0/0/1]port default vlan 10
[LSW1-GigabitEthernet0/0/1]quit
[LSW1]interface GigabitEthernet 0/0/10
[LSW1-GigabitEthernet0/0/10]port link-type access
[LSW1-GigabitEthernet0/0/10]port default vlan 20
[LSW1-GigabitEthernet0/0/10]quit
[LSW1]interface GigabitEthernet 0/0/24
[LSW1-GigabitEthernet0/0/24]port link-type access
[LSW1-GigabitEthernet0/0/24]quit
[LSW1]interface Vlanif 1
[LSW1-Vlanif1]ip address 192.168.3.1 24
[LSW1-Vlanif1]quit
[LSW1]interface Vlanif 10
[LSW1-Vlanif10]ip address 192.168.1.254 24
[LSW1-Vlanif10]quit
[LSW1]interface Vlanif 20
[LSW1-Vlanif20]ip address 192.168.2.254 24
[LSW1-Vlanif20]quit
[LSW1]ospf
[LSW1-ospf-1]area 0
[LSW1-ospf-1-area-0.0.0.0]network 192.168.1.0 0.0.0.255
[LSW1-ospf-1-area-0.0.0.0]network 192.168.2.0 0.0.0.255
[LSW1-ospf-1-area-0.0.0.0]network 192.168.3.0 0.0.0.255
[LSW1-ospf-1-area-0.0.0.0]quit
[LSW1-ospf-1]quit
[LSW1]
```

（2）配置路由器 AR1，相关实例代码如下。

```
<Huawei>system-view
```

```
[Huawei]sysname AR1
[AR1]interface GigabitEthernet 0/0/0
[AR1-GigabitEthernet0/0/0]ip address 192.168.4.1 24
[AR1-GigabitEthernet0/0/0]quit
[AR1]interface GigabitEthernet 0/0/1
[AR1-GigabitEthernet0/0/1]ip address 192.168.5.1 24
[AR1-GigabitEthernet0/0/1]quit
[AR1]interface GigabitEthernet 0/0/2
[AR1-GigabitEthernet0/0/2]ip address 192.168.3.2 24
[AR1-GigabitEthernet0/0/2]quit
[AR1]ospf
[AR1-ospf-1]area 0
[AR1-ospf-1-area-0.0.0.0]network 192.168.3.0 0.0.0.255
[AR1-ospf-1-area-0.0.0.0]network 192.168.4.0 0.0.0.255
[AR1-ospf-1-area-0.0.0.0]network 192.168.5.0 0.0.0.255
[AR1-ospf-1-area-0.0.0.0]quit
[AR1-ospf-1]quit
[AR1]
```

（3）配置路由器 AR2，相关实例代码如下。

```
<Huawei>system-view
[Huawei]sysname AR2
[AR2]interface GigabitEthernet 0/0/0
[AR2-GigabitEthernet0/0/0]ip address 192.168.4.2 24
[AR2-GigabitEthernet0/0/0]quit
[AR2]interface GigabitEthernet 0/0/1
[AR2-GigabitEthernet0/0/1]ip address 172.16.1.1 24
[AR2-GigabitEthernet0/0/1]quit
[AR2]interface GigabitEthernet 0/0/2
[AR2-GigabitEthernet0/0/2]ip address 192.168.6.1 24
[AR2-GigabitEthernet0/0/2]quit
[AR2]ospf
[AR2-ospf-1]area 0
[AR2-ospf-1-area-0.0.0.0]network 192.168.4.0 0.0.0.255
[AR2-ospf-1-area-0.0.0.0]network 192.168.6.0 0.0.0.255
[AR2-ospf-1-area-0.0.0.0]network 172.16.1.0 0.0.0.255
[AR2-ospf-1-area-0.0.0.0]quit
[AR2-ospf-1]quit
[AR2]
```

（4）配置路由器 AR3，相关实例代码如下。

```
<Huawei>system-view
[Huawei]sysname AR3
[AR3]interface GigabitEthernet 0/0/0
[AR3-GigabitEthernet0/0/0]ip address 172.16.2.1 24
[AR3-GigabitEthernet0/0/0]quit
[AR3]interface GigabitEthernet 0/0/1
[AR3-GigabitEthernet0/0/1]ip address 192.168.5.2 24
[AR3-GigabitEthernet0/0/1]quit
```

```
[AR3]interface GigabitEthernet 0/0/2
[AR3-GigabitEthernet0/0/2]ip address 192.168.6.2 24
[AR3-GigabitEthernet0/0/2]quit
[AR3]ospf
[AR3-ospf-1]area 0
[AR3-ospf-1-area -0.0.0.0]network 192.168.5.0 0.0.0.255
[AR3-ospf-1-area -0.0.0.0]network 192.168.6.0 0.0.0.255
[AR3-ospf-1-area -0.0.0.0]network 172.16.2.0 0.0.0.255
[AR3-ospf-1-area -0.0.0.0]quit
[AR3-ospf-1]quit
[AR3]
```

(5) 配置路由器 AR1，配置名称为 packet-lab1 的 PBR。

```
[AR1]policy-based-route packet-lab1 permit node 10
[AR1-policy-based-route-packet-lab1-10]if-match packet-length 64 1400
[AR1-policy-based-route-packet-lab1-10]apply ip-address next-hop 192.168.4.2
[AR1-policy-based-route-packet-lab1-10]quit
[AR1]policy-based-route packet-lab1 permit node 20
[AR1-policy-based-route-packet-lab1-2if-match packet-length 1401 1500
[AR1-policy-based-route-packet-lab1-20]apply ip-address next-hop 192.168.5.2
[AR1-policy-based-route-packet-lab1-20]quit
[AR1]ip local policy-based-route packet-lab1     //使能本地 PBR
[AR1]
```

(6) 验证配置结果，清空路由器 AR2 的端口统计信息。

```
<AR2>reset counter interface GigabitEthernet 0/0/0
Info: Reset successfully.
<AR2>
```

(7) 在交换机 LSW1 上使用 ping 命令测试，访问路由器 AR2 的 GE 0/0/0 端口的 IP 地址，并将报文数据字段长度设为 100 字节。

```
<LSW1>ping -s 100 172.16.1.1
  PING 172.16.1.1: 100   data bytes, press CTRL_C to break
    Reply from 172.16.1.1: bytes=100 Sequence=1 ttl=254 time=10 ms
    Reply from 172.16.1.1: bytes=100 Sequence=2 ttl=254 time=30 ms
    Reply from 172.16.1.1: bytes=100 Sequence=3 ttl=254 time=50 ms
    Reply from 172.16.1.1: bytes=100 Sequence=4 ttl=254 time=40 ms
    Reply from 172.16.1.1: bytes=100 Sequence=5 ttl=254 time=10 ms
  --- 172.16.1.1 ping statistics ---
    5 packet(s) transmitted
    5 packet(s) received
    0.00% packet loss
    round-trip min/avg/max = 10/28/50 ms
<LSW1>
```

(8) 查看路由器 AR2 的端口统计信息。

```
<AR2>display interface GigabitEthernet 0/0/0
GigabitEthernet0/0/0 current state : UP
Line protocol current state : UP
Last line protocol up time : 2020-01-20 14:37:45 UTC-08:00
Description:HUAWEI, AR Series, GigabitEthernet0/0/0 Interface
```

```
     Route Port,The Maximum Transmit Unit is 1500
     Internet Address is 192.168.4.2/24
     IP Sending Frames' Format is PKTFMT_ETHNT_2, Hardware address is 00e0-fc06-6aca
     Last physical up time    : 2020-01-20 13:51:24 UTC-08:00
     Last physical down time : 2020-01-20 13:51:18 UTC-08:00
     Current system time: 2020-01-20 17:47:42-08:00
     Port Mode: FORCE COPPER
     Speed : 1000,  Loopback: NONE
     Duplex: FULL,   Negotiation: ENABLE
     Mdi    : AUTO
     Last 300 seconds input rate 24 bits/sec, 0 packets/sec
     Last 300 seconds output rate 24 bits/sec, 0 packets/sec
     Input peak rate 1880 bits/sec,Record time: 2020-01-20 14:37:55
     Output peak rate 1536 bits/sec,Record time: 2020-01-20 14:37:55
     Input:   9 packets, 1038 bytes
          Unicast:              5,       Multicast:              4
          Broadcast:            0,       Jumbo:                  0
          Discard:              0,       Total Error:            0
          CRC:                  0,       Giants:                 0
          Jabbers:              0,       Throttles:              0
          Runts:                0,       Symbols:                0
          Ignoreds:             0,       Frames:                 0
     Output:   9 packets, 1038 bytes
          Unicast:              5,       Multicast:              4
          Broadcast:            0,       Jumbo:                  0
          Discard:              0,       Total Error:            0
          Collisions:           0,       ExcessiveCollisions:    0
          Late Collisions:      0,       Deferreds:              0
            Input bandwidth utilization threshold : 100.00%
            Output bandwidth utilization threshold: 100.00%
            Input bandwidth utilization  :   0%
            Output bandwidth utilization :   0%
     <AR2>
```

（9）验证配置结果，清空路由器 AR2 的端口统计信息。

```
<AR3>reset counters interface GigabitEthernet 0/0/1
Info: Reset successfully.
<AR3>
```

（10）在交换机 LSW1 上使用 ping 命令测试，访问路由器 AR2 的 GE 0/0/0 端口的 IP 地址，并将报文数据字段长度设为 1410 字节。

```
<LSW1>ping -s 1410 172.16.2.1
  PING 172.16.2.1: 1410   data bytes, press CTRL_C to break
    Reply from 172.16.2.1: bytes=1410 Sequence=1 ttl=254 time=20 ms
    Reply from 172.16.2.1: bytes=1410 Sequence=2 ttl=254 time=50 ms
    Reply from 172.16.2.1: bytes=1410 Sequence=3 ttl=254 time=50 ms
    Reply from 172.16.2.1: bytes=1410 Sequence=4 ttl=254 time=50 ms
    Reply from 172.16.2.1: bytes=1410 Sequence=5 ttl=254 time=40 ms
  --- 172.16.2.1 ping statistics ---
```

```
    5 packet(s) transmitted
    5 packet(s) received
    0.00% packet loss
    round-trip min/avg/max = 20/42/50 ms
<LSW1>
```

（11）查看路由器 AR3 的端口统计信息。

```
<AR3>display interface GigabitEthernet 0/0/1
GigabitEthernet0/0/1 current state : UP
Line protocol current state : UP
Last line protocol up time : 2020-01-20 14:39:06 UTC-08:00
Description:HUAWEI, AR Series, GigabitEthernet0/0/1 Interface
Route Port,The Maximum Transmit Unit is 1500
Internet Address is 192.168.5.2/24
IP Sending Frames' Format is PKTFMT_ETHNT_2, Hardware address is 00e0-fc65-2a69
Last physical up time     : 2020-01-20 13:51:28 UTC-08:00
Last physical down time : 2020-01-20 13:51:21 UTC-08:00
Current system time: 2020-01-20 17:51:39-08:00
Port Mode: COMMON COPPER
Speed : 1000,   Loopback: NONE
Duplex: FULL,   Negotiation: ENABLE
Mdi    : AUTO
Last 300 seconds input rate 200 bits/sec, 0 packets/sec
Last 300 seconds output rate 200 bits/sec, 0 packets/sec
Input peak rate 11744 bits/sec,Record time: 2020-01-20 17:51:23
Output peak rate 11616 bits/sec,Record time: 2020-01-20 17:51:23
Input:   10 packets, 7670 bytes
    Unicast:                5,      Multicast:                5
    Broadcast:              0,      Jumbo:                    0
    Discard:                0,      Total Error:              0
    CRC:                    0,      Giants:                   0
    Jabbers:                0,      Throttles:                0
    Runts:                  0,      Symbols:                  0
    Ignoreds:               0,      Frames:                   0
Output:   9 packets, 7588 bytes
    Unicast:                5,      Multicast:                4
    Broadcast:              0,      Jumbo:                    0
    Discard:                0,      Total Error:              0
    Collisions:             0,      ExcessiveCollisions:      0
    Late Collisions:        0,      Deferreds:                0
       Input bandwidth utilization threshold : 100.00%
       Output bandwidth utilization threshold: 100.00%
       Input bandwidth utilization   :   0%
       Output bandwidth utilization :    0%
<AR3>
```

（12）定义 ACL。

```
[AR1]acl 2000
[AR1-acl-basic-2000]rule permit source 192.168.1.0 0.0.0.255
```

```
[AR1-acl-basic-2000]quit
[AR1]acl 2001
[AR1-acl-basic-2001]rule permit source 192.168.2.0 0.0.0.255
[AR1-acl-basic-2001]quit
```

（13）应用 PBR。

```
[AR1]policy-based-route rule-acl-1 permit node 10
[AR1-policy-based-route-rule-acl-1-10]if-match acl 2000
[AR1-policy-based-route-rule-acl-1-10]apply ip-address next-hop 192.168.4.2
[AR1-policy-based-route-rule-acl-1-10]quit
[AR1]policy-based-route rule-acl-1 permit node 20
[AR1-policy-based-route-rule-acl-1-20]if-match acl 2001
[AR1-policy-based-route-rule-acl-1-20]apply ip-address next-hop 192.168.5.2
[AR1-policy-based-route-rule-acl-1-20]quit
[AR1]ip local policy-based-route rule-acl
```

（14）配置主机 PC1 与主机 PC2 的 IP 地址，如图 9.2 所示。

图 9.2　配置主机 PC1 与主机 PC2 的 IP 地址

（15）验证配置结果，主机 PC1 访问路由器 AR2 GE 0/0/1 端口的 IP 地址，如图 9.3 所示。

（16）验证配置结果，主机 PC2 访问路由器 AR3 GE 0/0/0 端口的 IP 地址，如图 9.4 所示。

图 9.3　主机 PC1 访问路由器 AR2 GE 0/0/1 端口的 IP 地址

图 9.4　主机 PC2 访问路由器 AR3 GE 0/0/0 端口的 IP 地址

（17）显示交换机 LSW1 的配置信息，主要配置实例代码如下。

```
<LSW1>display current-configuration
```

```
#
sysname LSW1
#
vlan batch 10 20
#
interface Vlanif1
 ip address 192.168.3.1 255.255.255.0
#
interface Vlanif10
 ip address 192.168.1.254 255.255.255.0
#
interface Vlanif20
 ip address 192.168.2.254 255.255.255.0
#
interface GigabitEthernet0/0/1
 port link-type access
 port default vlan 10
#
interface GigabitEthernet0/0/10
 port link-type access
 port default vlan 20
#
ospf 1
 area 0.0.0.0
  network 192.168.1.0 0.0.0.255
  network 192.168.2.0 0.0.0.255
  network 192.168.3.0 0.0.0.255
#
return
<LSW1>
```

（18）显示路由器 AR1 的配置信息，主要配置实例代码如下。

```
<AR1>display current-configuration
#
 sysname AR1
#
ip local policy-based-route rule-acl
#
acl number 2000
 rule 5 permit source 192.168.1.0 0.0.0.255
acl number 2001
 rule 5 permit source 192.168.2.0 0.0.0.255
#
interface GigabitEthernet0/0/0
 ip address 192.168.4.1 255.255.255.0
#
interface GigabitEthernet0/0/1
 ip address 192.168.5.1 255.255.255.0
#
```

```
interface GigabitEthernet0/0/2
 ip address 192.168.3.2 255.255.255.0
#
ospf 1
 area 0.0.0.0
  network 192.168.3.0 0.0.0.255
  network 192.168.4.0 0.0.0.255
  network 192.168.5.0 0.0.0.255
#
policy-based-route packet-lab1 permit node 10
 if-match packet-length 64 1400
 apply ip-address next-hop 192.168.4.2
policy-based-route packet-lab1 permit node 20
 if-match packet-length 1401 1500
 apply ip-address next-hop 192.168.5.2
policy-based-route rule-acl-1 permit node 10
 if-match acl 2000
 apply ip-address next-hop 192.168.4.2
policy-based-route rule-acl-1 permit node 20
 if-match acl 2001
 apply ip-address next-hop 192.168.5.2
#
return
<AR1>
```

课后习题

1. 选择题

（1）【多选】在应用 PBR 时，以下描述正确的是（　　）。

　　A. 在端口视图下应用 PBR 时，PBR 只对本端口接收和发送的报文起作用

　　B. 在端口视图下应用 PBR 时，PBR 只对本端口接收的报文起作用

　　C. 在系统视图下应用 PBR 时，PBR 对通过本路由器收到的所有报文起作用

　　D. 在系统视图下应用 PBR 时，PBR 只对本地产生的报文起作用

（2）前缀列表可以用于路由信息过滤，这种说法是（　　）的。

　　A. 正确　　　　　　B. 错误

2. 简答题

（1）简述 PBR 的工作原理。

（2）简述 PBR 的配置方法。

项目10　GRE协议

【学习目标】
- 掌握GRE协议的基本原理。
- 掌握GRE协议的配置方法。

10.1　项目陈述

某公司由于规模扩大，需要建立分公司，现需要通过公共网络将总公司与分公司的网络连接起来。公司网络工程师小张负责组建网络，他通过建立 GRE 协议隧道（Tunnel）实现网络路由信息交换，本项目将对小张的组网方案进行介绍。

10.2　必备知识

10.2.1　GRE 协议概述

通用路由封装（Generic Routing Encapsulation，GRE）协议可以对某些网络层协议（如 IPX、ATM、IPv6、AppleTalk 等）的数据报文进行封装，使这些被封装的数据报文能够在另一个网络层协议（如 IPv4 协议）中传输。

GRE 协议提供了将一种协议的报文封装在另一种协议报文中的机制，是一种三层隧道封装技术，使报文可以通过 GRE 协议隧道进行透明的传输，解决了异种网络的传输问题。

1. GRE 协议特点

（1）GRE 协议实现机制简单，对隧道两端的设备负担小。

（2）GRE 协议隧道可以通过 IPv4 网络连通多种网络协议的本地网络，有效利用了原有的网络架构，降低了成本。

（3）GRE 协议隧道扩展了跳数受限网络协议的工作范围，支持企业灵活设计网络拓扑。

（4）GRE 协议隧道可以封装组播数据，和 IPSec 结合使用时可以保证语音、视频等组播业务的安全。

（5）GRE 协议隧道可使能 MPLS LDP，使用 GRE 协议隧道承载 MPLS LDP 报文，建立 LDP LSP，实现 MPLS 骨干网的互通。

（6）GRE 协议隧道可将不连续的子网连接起来，用于组建 VPN，实现企业总部和分部间的安全连接。

2．GRE 协议封装后的报文格式

GRE 协议封装后的报文格式如图 10.1 所示。

（1）乘客协议（Passenger Protocol）：封装前的报文称为净荷（Payload），净荷协议类型为乘客协议。

（2）封装协议（Encapsulation Protocol）：GRE Header 是由封装协议完成并填充的，封装协议也称为运载协议（Carrier Protocol）。

（3）传输协议（Delivery Protocol）：负责对封装后的报文进行转发的协议。

图 10.1　GRE 协议封装后的报文格式

10.2.2　GRE 协议基本原理

报文在 GRE 协议隧道中传输时包括封装和解封装两个过程。如图 10.2 所示，如果 X 协议报文从路由器 AR1 向路由器 AR2 传输，则封装在路由器 AR1 上完成，而解封装在路由器 AR2 上进行，封装后的数据报文在网络中传输的路径称为 GRE 协议隧道。

图 10.2　GRE 协议隧道实现 X 协议互通组网

（1）封装。

① 路由器 AR1 从连接 X 协议网络 1 的端口接收到 X 协议报文后，先交由 X 协议处理。

② X 协议根据报文头中的目的地址在路由表或转发表中查找出端口，确定转发此报文的方法。如果发现出端口是 GRE 协议 Tunnel 端口，则对报文进行 GRE 协议封装，即添加 GRE 协议头。

③ 因为骨干网传输协议为 IP，所以要为报文加上 IP 头。IP 头的源地址就是隧道源地址，目的地址就是隧道目的地址。

④ 根据该 IP 头的目的地址，在骨干网路由表中查找相应的出端口并发送报文，封装后的报文将在该骨干网中传输。

（2）解封装。

解封装过程与封装过程相反。

① 路由器 AR2 从 GRE 协议 Tunnel 端口收到该报文，分析 IP 头后，发现报文的目的地址为本设备，则路由器 AR2 剥掉 IP 头后交由 GRE 协议处理。

② GRE 协议剥掉 GRE 协议头，获取 X 协议报文，再交由 X 协议对此数据报文进行后续处理。

10.2.3　配置 GRE 协议隧道

（1）配置 Tunnel 端口。

GRE 协议隧道是通过隧道两端的 Tunnel 端口建立的，所以需要在隧道两端的设备上分别配置 Tunnel 端口。对于 GRE 协议的 Tunnel 端口，需要指定其协议类型为 GRE 协议，以及源地址/源端口、目的地址和 IP 地址。

Tunnel 的源地址：配置报文传输协议中的源地址。当配置地址类型时，直接将其作为源地址使用。当配置类型为源端口时，取该端口的 IP 地址作为源地址使用。

Tunnel 的目的地址：配置报文传输协议中的目的地址。

Tunnel 端口的 IP 地址：为了在 Tunnel 端口上启用动态路由协议，或使用静态路由协议发布 Tunnel 端口，需要为 Tunnel 端口分配 IP 地址。Tunnel 端口的 IP 地址可以不是公网地址，甚至可以借用其他端口的 IP 地址以节约 IP 地址。但是当 Tunnel 端口借用 IP 地址后，该地址不能直接与 Tunnel 端口互通，因此，在借用 IP 地址的情况下，必须配置静态路由或路由协议先实现借用地址的互通性，才能实现 Tunnel 端口的互通。

① 使用 system-view 命令，进入系统视图。

② 使用 interface tunnel interface-number 命令，创建 Tunnel 端口，并进入 Tunnel 端口视图。

③ 使用 tunnel-protocol gre 命令，配置 Tunnel 端口的协议类型为 GRE 协议。

④ 使用 source { source-ip-address | interface-type interface-number }命令，配置 Tunnel 的源地址或源端口。

⑤ 使用 destination [vpn-instance vpn-instance-name] dest-ip-address 命令，配置 Tunnel 的目的地址。

⑥ 使用 ip address ip-address { mask | mask-length } [sub]命令，配置 Tunnel 端口的 IP 地址。

（2）配置 Tunnel 端口的路由。

在保证本端设备和远端设备在骨干网中路由互通的基础上，本端设备和远端设备上必须存在经过 Tunnel 端口转发的路由，这样需要进行 GRE 协议封装的报文才能正确转发。经过 Tunnel 端口转发的路由可以是静态路由，也可以是动态路由。

① 使用 system-view 命令，进入系统视图。

② 配置经过 Tunnel 端口的路由，选择如下方法之一。

方法 1：使用 ip route-static ip-address { mask | mask-length } { nexthop-address | tunnel-interface-number [nexthop-address] } [description text]命令，配置静态路由。

方法 2：配置动态路由。可以使用 IGP 或 EGP，包括 OSPF 协议、RIP 等路由协议，此处不再详述其配置方法。

若采用 GRE 协议隧道实现 IPv6 协议的互通，则必须在 Tunnel 端口、与 IPv6 协议相连的物理端口上配置 IPv6 的路由协议。

（3）查看 GRE 协议配置结果。

① 使用 display interface tunnel [interface-number]命令，查看 Tunnel 端口的工作状态。

② 使用 display tunnel-info { tunnel-id tunnel-id | all | statistics [slots] }命令，查看隧道信息。

③ 使用 display ip routing-table 命令，查看 IPv4 路由表，到指定目的地址的路由出端口为 Tunnel 端口。

使用 display ipv6 routing-table 命令，查看 IPv6 路由表，到指定目的地址的路由出端口为 Tunnel 端口。

④ 使用 ping -a source-ip-address host 命令，ping 对端的 Tunnel 端口的 IP 地址，从本端 Tunnel 端口到对端 Tunnel 端口可以连通。

⑤ 使能 Keepalive 功能后，在 Tunnel 端口视图下使用 display keepalive packets count 命令，查看 GRE 协议 Tunnel 端口发送给对端及从对端接收的 Keepalive 报文的数量和 Keepalive 响应报文的数量。

10.3 项目实施

10.3.1 GRE 协议实现 IPv4 网络互通配置实例

如图 10.3 所示，路由器 AR1、路由器 AR2、路由器 AR3 使用 OSPF 协议实现公网互通，在主机 PC1 和主机 PC2 上运行 IPv4 私网协议，现需要主机 PC1 和主机 PC2 通过公网实现 IPv4 私网互通，同时需要保证私网数据传输的可靠性。其中，主机 PC1 和主机 PC2 分别指定了路由器 AR2 和路由器 AR3 为自己的默认网关，需要在路由器 AR2 和路由器 AR3 之间使用 GRE 协议隧道直连，而 Tunnel 端口和与私网相连端口上使用了 OSPF 协议，为了能够检测隧道链路状态，在 GRE 协议隧道两端的 Tunnel 端口上使能 Keepalive 功能，并配置与主机相连的网段运行 IGP，配置相关端口与 IP 地址，进行网络拓扑连接。

V10-1 配置 GRE 协议实现 IPv4 网络互通实例

图 10.3 GRE 协议实现 IPv4 网络互通配置实例

（1）配置路由器 AR1、路由器 AR2、路由器 AR3 各端口的 IP 地址，这里以路由器路由器 AR1 为例进行介绍，其他路由器各端口的 IP 地址与此配置一致，不再赘述。

```
<Huawei>system-view
[Huawei]sysname AR1
[AR1]interface GigabitEthernet 0/0/0
[AR1-GigabitEthernet0/0/0]ip address 172.16.1.2 24
[AR1-GigabitEthernet0/0/0]quit
[AR1]interface GigabitEthernet 0/0/1
[AR1-GigabitEthernet0/0/1]ip address 172.16.2.2 24
[AR1-GigabitEthernet0/0/1]quit
[AR1]
```

V10-2 配置 GRE 协议实现 IPv4 网络互通实例——测试结果

（2）配置各路由器的 OSPF 协议路由。

配置路由器 AR1 的 OSPF 协议路由。

```
[AR1]ospf 1
[AR1-ospf-1]area 0
[AR1-ospf-1-area-0.0.0.0]network 172.16.1.0 0.0.0.255
[AR1-ospf-1-area-0.0.0.0]network 172.16.2.0 0.0.0.255
[AR1-ospf-1-area-0.0.0.0]quit
[AR1-ospf-1]quit
[AR1]
```

配置路由器 AR2 的 OSPF 协议路由。

```
[AR2]ospf 1
[AR2-ospf-1]area 0
[AR2-ospf-1-area-0.0.0.0]network 172.16.1.0 0.0.0.255
[AR2-ospf-1-area-0.0.0.0]quit
[AR2-ospf-1]quit
[AR2]
```

配置路由器 AR3 的 OSPF 协议路由。

```
[AR3]ospf 1
[AR3-ospf-1]area 0
[AR3-ospf-1-area-0.0.0.0]network 172.16.2.0 0.0.0.255
[AR3-ospf-1-area-0.0.0.0]quit
[AR3-ospf-1]quit
[AR3]
```

（3）配置完成后，在路由器 AR2 和路由器 AR3 上使用 display ip routing-table protocol ospf 命令，可以看到它们能够学习到去往对端端口网段地址的 OSPF 协议路由，这里以路由器 AR2 为例进行介绍。

```
[AR2]display ip routing-table protocol ospf
Route Flags: R - relay, D - download to fib
------------------------------------------------------------------------

Public routing table : OSPF
         Destinations : 1        Routes : 1
OSPF routing table status : <Active>
         Destinations : 1        Routes : 1
Destination/Mask    Proto    Pre    Cost    Flags    NextHop         Interface
   172.16.2.0/24    OSPF     10     2       D        172.16.1.2      GigabitEthernet 0/0/0
```

```
OSPF routing table status : <Inactive>
         Destinations : 0        Routes : 0
[AR2]
```

（4）配置 Tunnel 端口。

配置路由器 AR2。

```
[AR2]interface Tunnel 0/0/1
[AR2-Tunnel0/0/1]tunnel-protocol gre
[AR2-Tunnel0/0/1]ip address 10.1.1.1 24
[AR2-Tunnel0/0/1]source 172.16.1.1
[AR2-Tunnel0/0/1]destination 172.16.2.1
[AR2-Tunnel0/0/1]keepalive
[AR2-Tunnel0/0/1]quit
[AR2]
```

配置路由器 AR3。

```
[AR3]interface Tunnel 0/0/1
[AR3-Tunnel0/0/1]tunnel-protocol gre
[AR3-Tunnel0/0/1]ip address 10.1.1.2 24
[AR3-Tunnel0/0/1]source 172.16.2.1
[AR3-Tunnel0/0/1]destination 172.16.1.1
[AR3-Tunnel0/0/1]keepalive
[AR3-Tunnel0/0/1]quit
[AR3]
```

（5）配置完成后，Tunnel 端口状态变为 Up，Tunnel 端口之间可以连通，这里以路由器 AR2 为例进行介绍。

```
<AR2>ping -a 10.1.1.1 10.1.1.2
  PING 10.1.1.2: 56  data bytes, press CTRL_C to break
    Reply from 10.1.1.2: bytes=56 Sequence=1 ttl=255 time=40 ms
    Reply from 10.1.1.2: bytes=56 Sequence=2 ttl=255 time=20 ms
    Reply from 10.1.1.2: bytes=56 Sequence=3 ttl=255 time=40 ms
    Reply from 10.1.1.2: bytes=56 Sequence=4 ttl=255 time=20 ms
    Reply from 10.1.1.2: bytes=56 Sequence=5 ttl=255 time=30 ms
  --- 10.1.1.2 ping statistics ---
    5 packet(s) transmitted
    5 packet(s) received
    0.00% packet loss
    round-trip min/avg/max = 20/30/40 ms
<AR2>
```

（6）查看 Keepalive 报文统计，这里以路由器 AR2 为例进行介绍。

```
[AR2-Tunnel0/0/1]display keepalive packets count
Send 59 keepalive packets to peers, Receive 59 keepalive response packets from peers
Receive 57 keepalive packets from peers, Send 57 keepalive response packets to peers.
[AR2-Tunnel0/0/1]
```

（7）配置 Tunnel 端口，使用 OSPF 协议路由。

配置路由器 AR2。

```
[AR2]ospf 2
```

```
[AR2-ospf-2]area 0
[AR2-ospf-2-area-0.0.0.0]network 10.1.1.0 0.0.0.255
[AR2-ospf-2-area-0.0.0.0]network 192.168.1.0 0.0.0.255
[AR2-ospf-2-area-0.0.0.0]quit
[AR2-ospf-2]quit
[AR2]
```

配置路由器 AR3。

```
[AR3]ospf 2
[AR3-ospf-2]area 0
[AR3-ospf-2-area-0.0.0.0]network 10.1.1.0 0.0.0.255
[AR3-ospf-2-area-0.0.0.0]network 192.168.2.0 0.0.0.255
[AR3-ospf-2-area-0.0.0.0]quit
[AR3-ospf-2]quit
[AR3]
```

（8）验证配置结果，配置完成后，在路由器 AR2 和路由器 AR3 上使用 display ip routing-table protocol ospf 命令，可以看到经过 Tunnel 端口去往对端用户侧网段的 OSPF 协议路由，且去往 Tunnel 目的端物理地址（10.1.1.0/24）的路由下一跳是 Tunnel 端口，这里以路由器 AR2 为例进行介绍。

```
[AR2]display ip routing-table protocol ospf
Route Flags: R - relay, D - download to fib
------------------------------------------------------------------------------
Public routing table : OSPF
         Destinations : 2       Routes : 2
OSPF routing table status : <Active>
         Destinations : 2       Routes : 2
Destination/Mask    Proto    Pre  Cost    Flags   NextHop         Interface
    172.16.2.0/24   OSPF     10   2       D       172.16.1.2      GigabitEthernet 0/0/0
    192.168.2.0/24  OSPF     10   1563    D       10.1.1.2        Tunnel0/0/1
OSPF routing table status : <Inactive>
         Destinations : 0       Routes : 0
[AR2]
```

（9）显示路由器 AR1、路由器 AR2、路由器 AR3 的配置信息，这里以路由器 AR3 为例进行介绍。

```
<AR3>display current-configuration
#
 sysname AR3
#
interface GigabitEthernet0/0/1
 ip address 172.16.2.1 255.255.255.0
#
interface GigabitEthernet0/0/2
 ip address 192.168.2.254 255.255.255.0
#
interface Tunnel0/0/1
 ip address 10.1.1.2 255.255.255.0
 tunnel-protocol gre
 keepalive
 source 172.16.2.1
```

```
    destination 172.16.1.1
#
ospf 1
 area 0.0.0.0
  network 172.16.2.0 0.0.0.255
#
ospf 2
 area 0.0.0.0
  network 10.1.1.0 0.0.0.255
  network 192.168.2.0 0.0.0.255
#
user-interface con 0
 authentication-mode password
user-interface vty 0 4
user-interface vty 16 20
#
return
<AR3>
```

（10）配置主机 PC1 与主机 PC2 的 IP 地址，如图 10.4 所示。

图 10.4　配置主机 PC1 与主机 PC2 的 IP 地址

（11）验证配置结果，使用 ping 命令，主机 PC1 访问主机 PC2，其测试结果如图 10.5 所示。

图 10.5　主机 PC1 访问主机 PC2 测试结果

10.3.2 GRE 协议隧道实现 IPv6 网络互通配置实例

如图 10.6 所示，两个 IPv6 网络分别通过路由器 AR2 和路由器 AR3 与 IPv4 网络中的路由器 AR1 连接，实现两个 IPv6 网络中的主机 PC1 和主机 PC2 的互通。其中，主机 PC1 和主机 PC2 分别指定了路由器 AR2 和路由器 AR3 为默认网关，在路由器 AR2 和路由器 AR3 之间建立直连链路，部署 GRE 协议隧道，通过静态路由指定到达对端的报文通过 Tunnel 端口转发，主干网络使用动态 OSPF 协议，配置相关端口与 IP 地址，进行网络拓扑连接。

V10-3 配置 GRE 协议隧道实现 IPv6 网络互通实例

图 10.6 GRE 协议隧道实现 IPv6 网络互通配置实例

（1）配置路由器 AR1、路由器 AR2、路由器 AR3、路由器 AR4 各端口的 IP 地址，这里以路由器 AR1 为例进行介绍，其他路由器各端口的 IP 地址与此配置一致，不再赘述。

```
<Huawei>system-view
[Huawei]sysname AR1
[AR1]interface GigabitEthernet 0/0/1
[AR1-GigabitEthernet0/0/1]ip address 10.1.1.1 24
[AR1-GigabitEthernet0/0/1]quit
[AR1]interface GigabitEthernet 0/0/2
[AR1-GigabitEthernet0/0/2]ip address 20.1.1.1 24
[AR1-GigabitEthernet0/0/2]quit
[AR1]
```

V10-4 配置 GRE 协议隧道实现 IPv6 网络互通实例 ——测试结果

（2）配置路由器的 OSPF 协议路由。
配置路由器 AR1 的 OSPF 协议路由。

```
[AR1]router id 1.1.1.1
[AR1]ospf 1
[AR1-ospf-1]area 0
[AR1-ospf-1-area-0.0.0.0]network 10.1.1.0 0.0.0.255
```

```
[AR1-ospf-1-area-0.0.0.0]network 20.1.1.0 0.0.0.255
[AR1-ospf-1-area-0.0.0.0]quit
[AR1-ospf-1]quit
[AR1]
```

配置路由器 AR2 的 OSPF 协议路由。

```
[AR2]router id 2.2.2.2
[AR2]ospf 1
[AR2-ospf-1]area 0
[AR2-ospf-1-area-0.0.0.0]network 10.1.1.0 0.0.0.255
[AR2-ospf-1-area-0.0.0.0]quit
[AR2-ospf-1]quit
[AR2]
```

配置路由器 AR3 的 OSPF 协议路由。

```
[AR3]router id 3.3.3.3
[AR3]ospf 1
[AR3-ospf-1]area 0
[AR3-ospf-1-area-0.0.0.0]network 20.1.1.0 0.0.0.255
[AR3-ospf-1-area-0.0.0.0]quit
[AR3-ospf-1]quit
[AR3]
```

（3）配置完成后，在路由器 AR2 和路由器 AR3 使用 display ip routing-table protocol ospf 命令，可以看到它们能够学习到去往对端端口网段地址的 OSPF 协议路由，这里以路由器 AR2 为例进行介绍。

```
[AR2]display ip routing-table protocol ospf
Route Flags: R - relay, D - download to fib
------------------------------------------------------------------------
Public routing table : OSPF
         Destinations : 1        Routes : 1
OSPF routing table status : <Active>
         Destinations : 1        Routes : 1
Destination/Mask    Proto    Pre  Cost     lags     NextHop          Interface
   20.1.1.0/24      OSPF     10   2        D        10.1.1.1         GigabitEthernet 0/0/1
OSPF routing table status : <Inactive>
         Destinations : 0        Routes : 0
[AR2]
```

（4）在路由器 AR2 上使用 ping 命令，访问路由器 AR3 的 GE 0/0/2 端口的 IP 地址。

```
[AR2]ping -a 10.1.1.2 20.1.1.2
  PING 20.1.1.2: 56   data bytes, press CTRL_C to break
    Reply from 20.1.1.2: bytes=56 Sequence=1 ttl=254 time=40 ms
    Reply from 20.1.1.2: bytes=56 Sequence=2 ttl=254 time=30 ms
    Reply from 20.1.1.2: bytes=56 Sequence=3 ttl=254 time=20 ms
    Reply from 20.1.1.2: bytes=56 Sequence=4 ttl=254 time=30 ms
    Reply from 20.1.1.2: bytes=56 Sequence=5 ttl=254 time=30 ms
  --- 20.1.1.2 ping statistics ---
    5 packet(s) transmitted
    5 packet(s) received
```

```
        0.00% packet loss
        round-trip min/avg/max = 20/30/40 ms
[AR2]
```
（5）配置 Tunnel 端口，建立 GRE 协议隧道，配置 IPv6 地址，启用 IPv6 协议。
配置路由器 AR2。
```
[AR2]ipv6
[AR2]interface GigabitEthernet 0/0/0
[AR2-GigabitEthernet0/0/0]ipv6 enable
[AR2-GigabitEthernet0/0/0]ipv6 address 2001::2/64
[AR2-GigabitEthernet0/0/0]quit
[AR2]interface Tunnel 0/0/1
[AR2-Tunnel0/0/1] ipv6 enable
[AR2-Tunnel0/0/1] ipv6 address 3001::1/64
[AR2-Tunnel0/0/1]tunnel-protocol gre
[AR2-Tunnel0/0/1]source 10.1.1.2
[AR2-Tunnel0/0/1]destination 20.1.1.2
[AR2-Tunnel0/0/1]quit
[AR2]
```
配置路由器 AR3。
```
[AR3]ipv6
[AR3]interface GigabitEthernet 0/0/0
[AR3-GigabitEthernet0/0/0]ipv6 enable
[AR3-GigabitEthernet0/0/0]ipv6 address 2002::2/64
[AR3-GigabitEthernet0/0/0]quit
[AR3]interface Tunnel 0/0/1
[AR3-Tunnel0/0/1] ipv6 enable
[AR3-Tunnel0/0/1] ipv6 address 3001::2/64
[AR3-Tunnel0/0/1]tunnel-protocol gre
[AR3-Tunnel0/0/1]source 20.1.1.2
[AR3-Tunnel0/0/1]destination 10.1.1.2
[AR3-Tunnel0/0/1]quit
[AR3]
```
（6）配置 Tunnel 端口的静态路由。
配置路由器 AR2。
```
[AR2]ipv6 route-static 2002:: 64 Tunnel0/0/1
```
配置路由器 AR3。
```
[AR3]ipv6 route-static 2001:: 64 Tunnel0/0/1
```
（7）在路由器 AR2 上使用 ping 命令，访问路由器 AR3 的 GE 0/0/0 端口的 IP 地址。
```
[AR2]ping ipv6 -a 2001::2 2002::2
    PING 2002::2 : 56   data bytes, press CTRL_C to break
        Reply from 2002::2      bytes=56 Sequence=1 hop limit=64    time = 30 ms
        Reply from 2002::2      bytes=56 Sequence=2 hop limit=64    time = 30 ms
        Reply from 2002::2      bytes=56 Sequence=3 hop limit=64    time = 20 ms
        Reply from 2002::2      bytes=56 Sequence=4 hop limit=64    time = 20 ms
```

```
            Reply from 2002::2    bytes=56 Sequence=5 hop limit=64    time = 20 ms
      --- 2002::2 ping statistics ---
      5 packet(s) transmitted
      5 packet(s) received
      0.00% packet loss
      round-trip min/avg/max = 20/24/30 ms
[AR2]
```

（8）显示路由器 AR1、路由器 AR2、路由器 AR3 的配置信息，这里以路由器 AR3 为例进行介绍，其主要配置实例代码如下。

```
<AR3>display current-configuration
#
 sysname AR3
#
ipv6
#
router id 3.3.3.3
#
interface GigabitEthernet0/0/0
 ipv6 enable
 ipv6 address 2002::2/64
#
interface GigabitEthernet0/0/2
 ip address 20.1.1.2 255.255.255.0
#
interface Tunnel0/0/1
 ipv6 enable
 ipv6 address 3001::2/64
 tunnel-protocol gre
 source 20.1.1.2
 destination 10.1.1.2
#
ospf 1
 area 0.0.0.0
  network 20.1.1.0 0.0.0.255
#
ipv6 route-static 2001:: 64 Tunnel0/0/1
#
user-interface con 0
 authentication-mode password
user-interface vty 0 4
user-interface vty 16 20
#
return
<AR3>
```

（9）配置主机 PC1 和主机 PC2 的 IPv6 地址，如图 10.7 所示。

图 10.7　配置主机 PC1 和主机 PC2 的 IPv6 地址

（10）验证配置结果，主机 PC1 访问主机 PC2，其测试结果如图 10.8 所示。

图 10.8　主机 PC1 访问主机 PC2 的测试结果

课后习题

简答题

（1）简述 GRE 协议的工作原理。
（2）简述 GRE 协议的配置方法。

项目 11
IPSec

【学习目标】
- 掌握IPSec的基本原理。
- 掌握IPSec的配置方法。

11.1 项目陈述

某公司由于规模扩大,需要建立分公司,现需要通过公共网络将总公司与分公司的网络连接起来,并需要加强公司网络的安全性。公司网络工程师小张负责组建网络,他通过建立 IPSec 隧道,实现网络路由信息交换,本项目将对小张的组网方案进行介绍。

11.2 必备知识

11.2.1 IPSec 概述

随着 Internet 的发展,越来越多的企业直接通过 Internet 进行互联,但由于 IP 未考虑安全性,而且 Internet 中有大量的不可靠用户和网络设备,用户业务数据要穿越这些未知网络,根本无法保证数据的安全性,数据易被伪造、篡改或窃取。因此,迫切需要一种兼容 IP 的通用的网络安全方案。

为了解决上述问题,互联网安全协议(Internet Protocol Security,IPSec)应运而生。IPSec 是对 IP 的安全性补充,其工作在 IP 层,为 IP 网络通信提供透明的安全服务。

IPSec 是 IETF 制定的一组开放的网络安全协议。它并不是一个单独的协议,而是一系列为 IP 网络提供安全性的协议和服务的集合,包括 AH 和 ESP 两个安全协议、密钥交换和用于认证及加密的一些算法,如图 11.1 所示。通过这些协议,在两个设备之间建立一条 IPSec 隧道,数据通过 IPSec 隧道进行数据转发,以保护数据的安全。

图 11.1 IPSec 体系

1. 安全协议

IPSec 使用 AH 和 ESP 两种 IP 传输层协议来提供认证和加密等安全服务。

（1）AH。

AH 仅支持认证功能，不支持加密功能。AH 在每一个数据包的标准 IP 头后面添加一个 AH 头，AH 对数据包和认证密钥进行 Hash 计算，接收方收到带有计算结果的数据包后，执行同样的 Hash 计算并与原计算结果进行比较，传输过程中对数据的任何更改都将使计算结果无效，这样就提供了数据来源认证和数据完整性校验，AH 协议的完整性验证范围为整个 IP 报文。

（2）ESP。

ESP 支持认证和加密功能。ESP 会在每一个数据包的标准 IP 头后面添加一个 ESP 头，并在数据包后面追加一个 ESP 尾（ESP Trailer 和 ESP Auth Data），与 AH 不同的是，ESP 对数据中的有效载荷进行加密后再封装到数据包中，以保证数据的机密性，但 ESP 没有对 IP 头的内容进行保护，除非 IP 头被封装在 ESP 内部（采用隧道模式）。

2. 封装模式

封装模式是指将 AH 或 ESP 相关的字段插入原始 IP 报文中，以实现对报文的认证和加密，封装模式有传输模式和隧道模式两种。

（1）传输模式。

在传输模式中，AH 头或 ESP 头被插入 IP 头与传输层协议头之间，以保护 TCP/UDP/ICMP 负载。由于传输模式未添加额外的 IP 头，所以原始报文中的 IP 地址在加密后报文的 IP 头中可见，以 TCP 报文为例，原始报文经过传输模式封装后，报文格式如图 11.2 所示。

图 11.2　TCP 报文在传输模式中封装后的报文格式

在传输模式中，与 AH 协议相比，ESP 的完整性验证范围不包括 IP 头，无法保证 IP 头的安全。

（2）隧道模式。

在隧道模式中，AH 头或 ESP 头被插入原始 IP 头之前，另外生成一个新的报文头放到 AH 头或 ESP 头之前，以保护 IP 头和负载。以 TCP 报文为例，原始报文经隧道模式封装后的报文格式如图 11.3 所示。

图 11.3　TCP 报文在隧道模式中封装后的报文格式

在隧道模式中，与 AH 协议相比，ESP 的完整性认证范围不包括新 IP 头，无法保证新 IP 头的安全。

传输模式和隧道模式的区别如下。

（1）从安全性来讲，隧道模式优于传输模式。它可以完全地对原始 IP 数据报进行认证和加密。隧道模式下可以隐藏内部 IP 地址、协议类型和端口。

（2）从性能来讲，隧道模式有一个额外的 IP 头，比传输模式占用更多带宽。

（3）从场景来讲，传输模式主要应用于两台主机或一台主机和一台 VPN 网关之间的通信；隧道模式主要应用于两台 VPN 网关之间或一台主机与一台 VPN 网关之间的通信。

当安全协议同时采用 AH 和 ESP 时，AH 和 ESP 必须采用相同的封装模式。

3. 加密和认证

IPSec 提供了两种安全机制：加密和认证。加密机制保证了数据的机密性，防止数据在传输过程中被窃听；认证机制能保证数据真实可靠，防止数据在传输过程中被仿冒和篡改。

（1）加密。

IPSec 采用对称加密算法对数据进行加密和解密。如图 11.4 所示，IPSec 发送方和 IPSec 接收方使用相同的密钥进行加密、解密。

用于加密和解密的对称密钥可以手工配置，也可以通过互联网密钥交换（Internet Key Exchange，IKE）协议自动协商生成。

图 11.4　IPSec 加密和解密的过程

常用的对称加密算法包括数据加密标准（Data Encryption Standard，DES）、三重数据加密标准（Triple Data Encryption Standard，3DES）、先进加密标准（Advanced Encryption Standard，AES）和算法（Senior Middle 1 和 Senior Middle 4，SM1 和 SM4）。其中，DES 和 3DES 算法安全性低，存在安全风险，不推荐使用。

（2）认证。

IPSec 的加密功能无法认证解密后的信息是否为原始发送的信息或是否完整。IPSec 采用了密钥散列消息认证码（Hash Message Authentication Code，HMAC）功能，比较数字签名以进行数据包完整性和真实性认证。

通常情况下，加密和认证需要配合使用。如图 11.5 所示，在 IPSec 发送方，加密后的报文通过认证算法和对称密钥生成数字签名，IP 报文和数字签名同时发给对端（数字签名填写在 AH 和 ESP 报文头的完整性校验值 ICV 字段中）；在 IPSec 接收方，使用相同的认证算法和对称密钥对加密报文进行处理，同样得到签名，再比较数字签名进行数据完整性和真实性认证，认证不通过的报文直接丢弃，认证通过的报文再进行解密。

同加密一样，用于认证的对称密钥也可以手工配置，或者通过 IKE 协议自动协商生成。

常用的认证算法包括消息摘要 5（Message Digest 5，MD5）、安全散列算法 1（Secure Hash Algorithm 1，SHA1）、安全散列算法 2（Secure Hash Algorithm 1，SHA2）、算法 SM3（Senior Middle 3）。其中，MD5、SHA1 安全性低，存在安全风险，不推荐使用。

4. 密钥交换

使用对称密钥进行加密、认证时，如何安全地共享密钥是一个很重要的问题。有以下两种方法可以解决这个问题。

图 11.5　IPSec 验证的过程

（1）带外共享密钥。

在发送、接收设备上手工配置静态的加密、验证密钥。双方通过带外共享的方式（如通过电话或邮件方式）保证密钥一致性。这种方式的缺点是安全性低，可扩展性差，在点对多点组网中配置密钥的工作量成倍增加。另外，为提升网络安全性需要周期性地修改密钥，而在这种方式下很难实施。

（2）使用安全的密钥分发协议。

通过 IKE 协议自动协商密钥。IKE 协议采用 DH（Diffie-Hellman）算法在不安全的网络中安全地分发密钥。这种方式配置简单，可扩展性好，在大型动态的网络环境中更能突出其优势。同时，通信双方通过交换密钥交换材料来计算共享的密钥，即使第三方截获了双方用于计算密钥的所有交换数据，也无法计算出真正的密钥，极大地提高了数据的安全性。

图 11.6　IKE 与 IPSec 的关系

IKE 协议建立在 Internet 安全联盟（Security Association，SA）和密钥管理协议定义的框架上，是基于用户数据报协议（User Datagram Protocol，UDP）的应用层协议。它为 IPSec 提供了自动协商密钥、建立 IPSec 安全联盟的服务，能够简化 IPSec 的配置和维护工作。

IKE 协议与 IPSec 的关系如图 11.6 所示，对等体之间建立一个 IKE SA 完成身份验证和密钥信息交换后，在 IKE SA 的保护下，根据配置的 AH/ESP 安全协议等参数协商出一对 IPSec SA。此后，对等体间的数据将在 IPSec 隧道中加密传输。IKE SA 是一个双向的逻辑连接，两个对等体间只建立一个 IKE SA。

IKE 协议具有一套自保护机制，可以在网络中安全地认证身份、分发密钥、建立 IPSec SA。

（1）身份认证。

身份认证指确认通信双方的身份（对等体的 IP 地址或名称），包括预共享密钥（Pre-Shared

Key，PSK）认证、RSA 数字证书（RSA-Signature）认证和数字信封认证。

① 在预共享密钥认证中，通信双方采用共享的密钥对报文进行 Hash 计算，判断双方的计算结果是否相同。如果相同，则认证通过，否则认证失败。当有一个对等体对应多个对等体时，需要为每个对等体配置预共享密钥，该方法在小型网络中容易建立，但安全性较低。

② 在 RSA 数字证书认证中，通信双方使用证书颁发机构（Certificate Authority，CA）颁发的证书进行数字证书合法性验证，双方各有自己的公钥（网络上传输）和私钥（自己持有）。发送方对原始报文进行 Hash 计算，并用自己的私钥对报文计算结果进行加密，生成数字签名。接收方使用发送方的公钥对数字签名进行解密，并对报文进行 Hash 计算，判断计算结果与解密后的结果是否相同。如果相同，则认证通过，否则认证失败。使用数字证书安全性高，但需要 CA 来颁发数字证书，适合在大型网络中使用。

③ 在数字信封认证中，发送方先随机产生一个对称密钥，使用接收方的公钥对此对称密钥进行加密（被公钥加密的对称密钥称为数字信封），发送方用对称密钥加密报文，同时用自己的私钥生成数字签名。接收方用自己的私钥解密数字信封得到对称密钥，再用对称密钥解密报文，同时根据发送方的公钥对数字签名进行解密，验证发送方的数字签名是否正确。如果正确，则认证通过，否则认证失败。数字信封认证在设备需要符合国家密码管理局要求时使用，此认证方法只能在 IKEv1 协议的主模式协商过程中支持。IKE 协议支持的认证算法包括 MD5、SHA1、SHA2-256、SHA2-384、SHA2-512、SM3。

（2）身份保护。

身份数据在密钥产生之后加密传送，实现了对身份数据的保护。

IKE 协议支持的加密算法包括 DES、3DES、AES-128、AES-192、AES-256、SM1 和 SM4。

注意 MD5 和 SHA1 认证算法不安全，建议使用 SHA2-256、SHA2-384、SHA2-512、SM3 算法。
DES 和 3DES 加密算法不安全，建议使用 AES 或 SM 算法。

11.2.2 配置采用虚拟隧道端口方式建立 IPSec 隧道

在采用虚拟隧道端口方式建立 IPSec 隧道之前，需完成以下任务。

（1）确定源端口和目的端口之间路由可达。

（2）确定需要 IPSec 保护的数据流，并将数据流引到虚拟隧道端口上。

（3）确定数据流被保护的强度，即确定使用的 IPSec 安全提议的参数。

1. 配置 IPSec 安全提议

IPSec 安全提议是安全策略或者安全框架的一个组成部分，它包括 IPSec 使用的安全协议、认证/加密算法及数据的封装模式，定义了 IPSec 的保护方法，为 IPSec 协商 SA 提供的各种安全参数。IPSec 隧道两端的设备需要配置相同的安全参数。

（1）使用 system-view 命令，进入系统视图。

（2）使用 ipsec proposal proposal-name 命令，创建 IPSec 安全提议并进入 IPSec 安全提议视图。

（3）使用 transform { ah | esp | ah-esp }命令，配置安全协议。默认情况下，IPSec 安全提议采用 ESP。

（4）配置安全协议的认证/加密算法。

① 安全协议采用 AH 协议时，AH 协议只能对报文进行认证，只能配置 AH 协议的认证算法。

使用 ah authentication-algorithm { md5 | sha1 | sha2-256 | sha2-384 | sha2-512 |sm3 } * 命令，配置 AH 协议使用的认证算法。默认情况下，AH 协议使用的认证算法为 SHA2-256。

② 安全协议采用 ESP 时，ESP 允许对报文同时进行加密和认证，也可以只加密或只认证，根据需要配置 ESP 的认证算法、加密算法即可。

使用 esp authentication-algorithm { md5 | sha1 | sha2-256 | sha2-384 |sha2-512 | sm3 } *命令，配置 ESP 使用的认证算法。默认情况下，ESP 使用的认证算法为 SHA2-256。

使用 esp encryption-algorithm { 3des | des | aes-128 | aes-192 | aes-256 |sm1 | sm4 } * 命令，配置 ESP 使用的加密算法。默认情况下，ESP 使用的加密算法为 AES-256。

③ 安全协议同时采用 AH 和 ESP 时，允许 AH 协议进行认证、ESP 对报文进行加密和认证，AH 协议的认证算法、ESP 的认证算法和加密算法均可选择配置。此时，设备先对报文进行 ESP 封装，再进行 AH 封装。

（5）使用 encapsulation-mode { transport | tunnel }命令，选择安全协议对数据的封装模式。默认情况下，安全协议对数据的封装模式为隧道模式。IKE 对等体采用 IKEv2 协议时，IKE 协议协商发起方配置的所有 IPSec 安全提议的报文封装模式必须一致，否则会导致 IKE 协议协商失败。

（6）使用 quit 命令，退出 IPSec 安全提议视图。

2. 配置 IPSec 安全策略

IPSec 安全策略是创建 SA 的前提，它规定了对哪些数据流采用哪种保护方法。配置 IPSec 安全策略时，通过引用 ACL 和 IPSec 安全提议，将 ACL 定义的数据流和 IPSec 安全提议定义的保护方法关联起来，并可以指定 SA 的协商方式、IPSec 隧道的起点和终点、所需要的密钥和 SA 的生存周期等。

一个 IPSec 安全策略由名称和序号共同唯一确定，相同名称的 IPSec 安全策略为一个 IPSec 安全策略组。IPSec 安全策略分为手工方式 IPSec 安全策略、ISAKMP 方式 IPSec 安全策略和策略模板方式 IPSec 安全策略 3 种。其中，ISAKMP 方式 IPSec 安全策略和策略模板方式 IPSec 安全策略均由 IKE 自动协商生成各参数。

（1）使用 system-view 命令，进入系统视图。

（2）使用 ipsec profile profile-name 命令，创建安全框架，并进入安全框架视图。默认情况下，系统中没有安全框架。

（3）使用 proposal proposal-name 命令，在安全框架中引用 IPSec 安全提议。默认情况下，安全框架没有引用 IPSec 安全提议。其中，proposal-name 是一个已创建的 IPSec 安全提议。

（4）使用 ike-peer peer-name 命令，在安全框架中引用 IKE 对等体。默认情况下，安全框架没有引用 IKE 对等体。其中，peer-name 是一个已创建的 IKE 对等体。

3. 配置虚拟隧道端口

虚拟隧道端口是一种三层逻辑端口，针对协议类型为 GRE 协议、mGRE 协议或 IPSec 的逻辑端口，设备可提供 IPSec 功能。它建立在 IKE 协商的基础上。通过配置虚拟隧道端口，并在虚拟隧道端口视图下应用 IPSec 安全框架建立 IPSec 隧道。

协议类型为 IPSec 的 Tunnel 端口的 IP 地址可以手工配置，也可以通过 IKEv2 协商动态申请。后者在大规模分支接入总部的场景中，可减少分支设备的配置和维护工作量。虚拟隧道模板（Tunnel-Template）端口与虚拟隧道端口类似，但配置了虚拟隧道模板的端口不能发起协商，只能作为协商响应方接收对端的协商请求，一般用于配置总部网关。每当增加一个分支网关接入时，总部网关就会动态生成一个虚拟隧道端口。

（1）配置虚拟隧道端口。

① 使用 system-view 命令，进入系统视图。

② 使用 interface tunnel interface-number 命令，进入 Tunnel 端口视图。

③ 使用 tunnel-protocol { gre [p2mp] | ipsec }命令，配置 Tunnel 端口的封装模式。

④ 根据实际需要选择如下配置之一。

使用 ip address ip-address { mask | mask-length } [sub]命令，手工配置 Tunnel 端口的 IPv4 私网地址。

（只针对 IPSec 类型的 Tunnel 端口）使用 ip address ike-negotiated 命令，配置通过 IKEv2 协商为 Tunnel 端口申请 IPv4 地址。

⑤ 使用 source { [vpn-instance vpn-instance-name] source-ip-address |interface-type interface-number [standby] }命令，配置 Tunnel 端口的源地址或源端口。其中，vpn-instance vpn-instance-name 参数只在 Tunnel 端口的封装模式为 IPSec 或 mGRE 方式时支持。

⑥ 使用 ipsec profile profile-name [shared]命令，在 Tunnel 端口上应用 IPSec 安全框架，使其具有 IPSec 的保护功能。默认情况下，Tunnel 端口上没有应用 IPSec 安全框架。

（2）配置虚拟隧道模板端口。

① 使用 system-view 命令，进入系统视图。

② 使用 interface tunnel-template interface-number 命令，进入 Tunnel-Template 端口视图。

③ 配置 Tunnel-Template 端口地址。

使用 ip address ip-address { mask | mask-length } [sub]命令，配置 Tunnel-Template 端口的 IPv4 私网地址。

使用 ip address unnumbered interface interface-type interface-number 命令，配置 Tunnel-Template 端口借用其他端口的 IP 地址。

④ 使用 tunnel-protocol ipsec 命令，配置 Tunnel-Template 端口的封装模式为 IPSec 方式。

⑤ 使用 source { [vpn-instance vpn-instance-name] source-ip-address |interface-type interface-number }命令，配置 Tunnel-Template 端口的源地址或源端口。

⑥ 使用 ipsec profile profile-name 命令，在 Tunnel-Template 端口上应用 IPSec 安全框架，使其具有 IPSec 的保护功能。默认情况下，Tunnel-Template 端口上没有应用 IPSec 安全框架。一个 Tunnel-Template 端口只能应用一个 IPSec 安全框架，一个 IPSec 安全框架也只能应用到一个 Tunnel-Template 端口上。

（3）查看采用虚拟隧道端口方式建立 IPSec 隧道的配置结果。

① 使用 display ipsec proposal [brief | name proposal-name]命令，查看 IPSec 安全提议的信息。

② 使用 display ipsec profile [brief | name profile-name]命令，查看 IPSec 安全框架的信息。

③ 使用 display ike identity [name identity-name]命令，查看身份过滤集的配置信息。

④ 使用 display ipsec sa [brief | duration | policy policy-name [seq-number] |remote ipv4-address]命令，查看 IPSec SA 相关信息。

（4）使用 display ipsec global config 命令，查看 IPSec 的全局配置信息。

11.2.3 配置 IKE 协议

在配置 IKE 协议之前，需完成以下任务。

（1）确定进行 IKE 协商时算法的强度，即确定使用的 IKE 安全提议的参数。

（2）当认证方法为数字证书公钥基础设施（Public Key Infrastructure，PKI）认证时，需要确定对等体所属的 PKI 域。

1. 配置 IKE 安全提议

IKE 安全提议是 IKE 对等体的一个组成部分，定义了对等体进行 IKE 协商时使用的参数，包括加密算法、认证方法、认证算法、DH 组和 IKE SA 的生存周期。

在进行 IKE 协商时，协商发起方会将自己的 IKE 安全提议发送给对端，由对端进行匹配，协商响应方从自己优先级最高的 IKE 安全提议开始，按照优先级顺序与对端进行匹配，直到找到一个匹配的 IKE 安全提议来使用。匹配的 IKE 安全提议将被用来建立 IKE 的安全隧道。

优先级由 IKE 安全提议的序号表示，数值越小，优先级越高。用户可以按照优先级创建多条 IKE 安全提议，但是协商双方必须至少有一条匹配的 IKE 安全提议才能协商成功。

IKE 安全提议的匹配原则如下：协商双方具有相同的加密算法、认证方法、认证算法和 DH 组。匹配的 IKE 安全提议的 IKE SA 的生存周期则取两端的最小值。

（1）使用 system-view 命令，进入系统视图。

（2）使用 ike proposal proposal-number 命令，创建一个 IKE 安全提议，并进入 IKE 安全提议视图。

（3）使用 authentication-method { pre-share | rsa-signature | digital-envelope }命令，配置认证方法。默认情况下，IKE 安全提议使用预共享密钥认证方法。

在进行 IKE 协商时，两端对等体使用的 IKE 安全提议中的认证方法必须保持一致，否则会导致 IKE 协商失败。

（4）使用 authentication-algorithm { md5 | sha1 | sha2-256 | sha2-384 | sha2-512 | sm3 }*命令，配置 IKEv1 协商时所使用的认证算法。默认情况下，IKEv1 协商时所使用的认证算法为 SHA2-256。

在进行 IKEv1 协商时，需要配置认证算法。认证算法安全级别由高到低为 SM3 > SHA2-512 > SHA2-384 > SHA2-256 > SHA1 > MD5。不建议使用 MD5 和 SHA1 算法，否则无法满足安全防御的要求。

（5）使用 encryption-algorithm { des | 3des | aes-128 | aes-192 | aes-256 | sm4 | sm1 } *命令，配置进行 IKE 协商时所使用的加密算法。默认情况下，IKE 协商时所使用的加密算法为 AES-256。

加密算法安全级别由高到低为 SM4 > SM1 > AES-256 > AES-192 > AES-128 >3DES > DES。不建议使用 DES 和 3DES 算法，否则无法满足安全防御的要求。

（6）使用 dh { group1 | group2 | group5 | group14 | group19 | group20 | group21 |group24 }*命令，配置 IKE 协商时采用的 DH 组。默认情况下，IKE 协商时采用的 DH 组为 group14。

DH 密钥交换组安全级别由高到低为 group24 > group21 > group20 > group19 > group14 > group5 > group2 > group1。不建议使用 group1、group2 和 group5，否则无法满足安全防御的要求。

（7）使用 prf { aes-xcbc-128 | hmac-md5 | hmac-sha1 | hmac-sha2-256 | hmac-sha2-384 |hmac-sha2-512 }*命令，配置 IKEv2 协商时所使用的伪随机数产生函数的算法。默认情况下，IKEv2 协商所使用的 PRF 算法为 HMAC-SHA2-256。不建议使用 HMAC-MD5 和 HMAC-SHA1 算法，否则无法满足安全防御的要求。

（8）使用 integrity-algorithm { aes-xcbc-96 | hmac-md5-96 | hmac-sha1-96 |

hmac-sha2-256 | hmac-sha2-384 | hmac-sha2-512 } *命令，配置 IKEv2 协商时所使用的完整性算法。默认情况下，IKEv2 协商时所使用的完整性算法为 HMAC-SHA2-256。

2. 配置 IKE 对等体

配置 IKE 协议以动态协商方式建立 IPSec 隧道时，需要引用 IKE 对等体，并配置 IKE 协商时对等体间的一系列属性。配置时需要注意以下几点。

（1）IKE 对等体两端使用相同的 IKE 版本。

（2）IKE 对等体两端使用 IKEv1 时必须采用相同的协商模式。

（3）IKE 对等体两端的身份认证参数必须匹配。

IKE 对等体使用的 IKE 协议有 IKEv1 协议和 IKEv2 协议两个版本。IKE 对等体使用 IKEv1 协议与使用 IKEv2 协议时相比，存在以下差异。

（1）IKEv1 协议需要配置第一阶段协商模式，而 IKEv2 协议不需要。

（2）采用数字证书认证时，IKEv1 协议不支持通过 IKE 协议进行数字证书的在线状态认证，但 IKEv2 协议支持。

（3）IKEv2 协议可以配置重认证时间间隔，以提高安全性，但 IKEv1 不支持此功能。

如果使用 RSA 签名认证，则要求被验证端已经导入本地证书和 CA 根证书，验证端已经导入 CA 根证书。

如果使用 RSA 认证，则要求被验证端已经生成 RSA 密钥对。当 IKEv1 使用证书方式协商时，如果配置的认证算法为 SHA2-512，则 RSA 密钥对长度必须为 1024 以上。

配置步骤如下。

① 使用 system-view 命令，进入系统视图。

② 使用 ike peer peer-name 命令，创建 IKE 对等体并进入 IKE 对等体视图。默认情况下，系统没有配置 IKE 对等体。

③ 使用 version { 1 | 2 }命令，配置 IKE 对等体使用的 IKE 协议版本号。默认情况下，一个 IKE 对等体同时支持 IKEv1 和 IKEv2 两个协议版本。

④ 使用 ike-proposal proposal-number 命令，引用 IKE 安全提议。proposal-number 是一个已创建的 IKE 安全提议。默认情况下，IKE 对等体没有引用 IKE 安全提议。

⑤ 配置身份认证参数。不同认证方法下身份认证参数的配置大不相同，要根据 IKE 安全提议选定的认证方法配置身份认证参数。

3. 查看 IKE 协议的配置结果

（1）使用 display ike identity [name identity-name]命令，查看 IKE 身份过滤集的配置信息。

（2）使用 display ike peer [brief | name peer-name]命令，查看 IKE 对等体配置的信息。

（3）使用 display ike proposal [number proposal-number]命令，查看 IKE 安全提议配置的参数。

（4）使用 display ike sa [remote ipv4-address]命令，查看当前 IKE SA 的摘要信息。

（5）使用 display ike sa [remote-id-type remote-id-type] remote-id remote-id 命令，根据对端 ID 查看 IKE SA 的摘要信息。

（6）使用 display ike sa verbose { remote ipv4-address | connection-id connection-id |[remote-id-type remote-id-type] remote-id remote-id }命令，查看当前 IKE SA 的详细信息。

（7）使用 display ike global config 命令，查看 IKE 协议的全局配置信息。

（8）使用 display ike user-table [number user-table-id [user-name user-name]]命令，查看 IKE 用户表的信息。

11.3 项目实施

11.3.1 虚拟隧道端口建立 IPSec 隧道配置实例

如图 11.7 所示，路由器 AR1 为企业分公司的网关，分公司的子网为 192.168.1.0/24，路由器 AR2 为企业总公司的网关，总公司的子网为 192.168.2.0/24，分公司与总公司通过公网建立通信。企业希望对分公司子网与总公司子网之间相互访问的流量进行安全保护，由于分公司与总公司通过公网建立通信，因此，可以在分公司的网关与总公司的网关之间建立一个 IPSec 隧道来实施安全保护，配置相关端口与 IP 地址，进行网络拓扑连接。

V11-1 虚拟隧道端口建立 IPSec 隧道配置实例——基础配置

V11-2 虚拟隧道端口建立 IPSec 隧道配置实例——IPSec 配置

图 11.7 虚拟隧道端口建立 IPSec 隧道配置实例

1. 配置思路

由于分公司的子网较为庞大，有大量需要 IPSec 保护的数据流，因此可基于虚拟隧道端口方式建立 IPSec 隧道，对 Tunnel 端口下的流量进行保护。

采用虚拟隧道端口方式建立 IPSec 隧道，相关思路如下。

（1）配置端口的 IP 地址和到对端的静态路由，保证两端路由可达。
（2）配置 IPSec 安全提议，定义 IPSec 的保护方法。
（3）配置 IKE 对等体，定义对等体间 IKE 协商时的属性。
（4）配置安全框架，并引用安全提议和 IKE 对等体，确定对何种数据流采取何种保护方法。
（5）在 Tunnel 端口上应用安全框架，使端口具有 IPSec 的保护功能。
（6）配置 Tunnel 端口的转发路由，将需要 IPSec 保护的数据流引到 Tunnel 端口上。

2. 配置步骤与方法

（1）配置路由器 AR1、路由器 AR2 各端口的 IP 地址，这里以路由器 AR1 为例进行介绍，路由器 AR2 各端口的 IP 地址与此配置一致，不再赘述。

```
<Huawei>system-view
[Huawei]sysname AR1
[AR1]interface GigabitEthernet 0/0/0
[AR1-GigabitEthernet0/0/0]ip address 172.16.1.1 24
[AR1-GigabitEthernet0/0/0]quit
[AR1]interface GigabitEthernet 0/0/1
```

V11-3 虚拟隧道端口建立 IPSec 隧道配置实例——测试结果

```
[AR1-GigabitEthernet0/0/1]ip address 192.168.1.254 24
[AR1-GigabitEthernet0/0/1]quit
[AR1]
```

（2）配置静态路由。

配置路由器 AR1。

```
[AR1]ip route-static 192.168.2.0 255.255.255.0 172.16.1.2
[AR1]
```

配置路由器 AR2。

```
[AR2]ip route-static 192.168.1.0 255.255.255.0 172.16.1.1
[AR2]
```

（3）分别在路由器 AR1 和路由器 AR2 上配置 IPSec 安全提议。

在路由器 AR1 上配置 IPSec 安全提议。

```
[AR1]ipsec proposal tran-sec-1              //创建 IPSec 安全提议，名称为 tran-sec-1
[AR1-ipsec-proposal-tran-sec-1]esp authentication-algorithm sha2-256
                                            //配置 ESP 使用的认证算法
[AR1-ipsec-proposal-tran-sec-1]esp encryption-algorithm aes-128
                                            //配置 ESP 使用的加密算法
[AR1-ipsec-proposal-tran-sec-1]quit
[AR1]
```

在路由器 AR2 上配置 IPSec 安全提议。

```
[AR2]ipsec proposal tran-sec-1              //创建 IPSec 安全提议，名称为 tran-sec-1
[AR2-ipsec-proposal-tran-sec-1]esp authentication-algorithm sha2-256
                                            //配置 ESP 使用的认证算法
[AR2-ipsec-proposal-tran-sec-1]esp encryption-algorithm aes-128
                                            //配置 ESP 使用的加密算法
[AR2-ipsec-proposal-tran-sec-1]quit
[AR2]
```

（4）此时，分别在路由器 AR1 和路由器 AR2 上使用 display ipsec proposal 命令，以显示所配置的信息，这里以路由器 AR1 为例进行介绍。

```
[AR1]display ipsec proposal
Number of proposals: 1
IPSec proposal name: tran-sec-1
  Encapsulation mode: Tunnel
  Transform         : esp-new
  ESP protocol      : Authentication SHA2-HMAC-256
                      Encryption     AES-128
[AR1]
```

（5）分别在路由器 AR1 和路由器 AR2 上配置 IKE 对等体。

在路由器 AR1 上配置 IKE 安全提议。

```
[AR1]ike proposal 1                                 //创建一个 IKE 安全提议
[AR1-ike-proposal-1]authentication-algorithm sm3    //配置 IKEv1 协商时所使用的认证算法
[AR1-ike-proposal-1]encryption-algorithm aes-cbc-128
                                                    //配置 IKE 协商时所使用的加密算法
[AR1-ike-proposal-1]dh group14                      //配置 IKE 协商时采用的 DH 组
[AR1-ike-proposal-1]quit
[AR1]
```

在路由器 AR1 上配置 IKE 对等体。

```
[AR1]ike peer spu- AR2 v1                              //创建 IKE 对等体
[AR1-ike-peer- spu- AR2]ike-proposal 1                 //引用 IKE 安全提议
[AR1-ike-peer- spu- AR2]pre-shared-key cipher Huawei123456
                                  //配置身份认证，预共享密钥认证，加密密码为 Huawei123456
[AR1-ike-peer- spu-AR1-AR2]quit
[AR1]
```

在路由器 AR2 上配置 IKE 安全提议。

```
[AR2]ike proposal 1                                    //创建一个 IKE 安全提议
[AR2-ike-proposal-1]authentication-algorithm sm3       //配置 IKEv1 协商时所使用的认证算法
[AR2-ike-proposal-1]encryption-algorithm aes-cbc-128
                                                       //配置 IKE 协商时所使用的加密算法
[AR2-ike-proposal-1]dh group14                         //配置 IKE 协商时采用的 DH 组
[AR2-ike-proposal-1]quit
[AR2]
```

在路由器 AR2 上配置 IKE 对等体。

```
[AR2]ike peer spu-AR1 v1                               //创建 IKE 对等体
[AR2-ike-peer- spu-AR1]ike-proposal 1                  //引用 IKE 安全提议
[AR2-ike-peer- spu-AR1]pre-shared-key cipher Huawei123456
                                  //配置身份认证，预共享密钥认证，加密密码为 Huawei123456
[AR2-ike-peer- spu-AR1-AR2]quit
[AR2]
```

（6）分别在路由器 AR1 和路由器 AR2 上创建安全框架。

在路由器 AR1 上创建安全框架。

```
[AR1]ipsec profile profile-1                           //创建安全框架，并进入安全框架视图
[AR1-ipsec-profile-profile-1]proposal tran-sec-1       //在安全框架中引用 IPSec 安全提议
[AR1-ipsec-profile-profile-1]ike-peer spu-AR2          //在安全框架中引用 IKE 对等体
[AR1-ipsec-profile-profile-1]quit
[AR1]
```

在路由器 AR2 上创建安全框架。

```
[AR2]ipsec profile profile-1                           //创建安全框架，并进入安全框架视图
[AR2-ipsec-profile-profile-1]proposal tran-sec-1       //在安全框架中引用 IPSec 安全提议
[AR2-ipsec-profile-profile-1]ike-peer spu-AR1          //在安全框架中引用 IKE 对等体
[AR2-ipsec-profile-profile-1]quit
[AR2]
```

（7）分别在路由器 AR1 和路由器 AR2 的端口上应用各自的安全框架。

在路由器 AR1 的端口上引用安全框架。

```
[AR1]interface Tunnel 0/0/0
[AR1-Tunnel0/0/0]ip address 10.1.1.1 255.255.255.0
[AR1-Tunnel0/0/0]tunnel-protocol ipsec                 //隧道启用 IPSec 协议
[AR1-Tunnel0/0/0]source 172.16.1.1
[AR1-Tunnel0/0/0]destination 172.16.1.2
[AR1-Tunnel0/0/0]ipsec profile profile-1               //引入 IPSec 安全框架
[AR1]
```

在路由器 AR2 的端口上引用安全框架。

```
[AR2]interface Tunnel 0/0/0
```

```
[AR2-Tunnel0/0/0]ip address 10.1.1.2 255.255.255.0
[AR2-Tunnel0/0/0]tunnel-protocol ipsec            //隧道启用 IPSec 协议
[AR2-Tunnel0/0/0]source 172.16.1.2
[AR2-Tunnel0/0/0]destination 172.16.1.1
[AR2-Tunnel0/0/0]ipsec profile profile-1          //引入 IPSec 安全框架
[AR2]
```

（8）在路由器 AR1 和路由器 AR2 上使用 display ipsec profile 命令，以显示所配置的信息，这里以路由器 AR1 为例进行介绍。

```
<AR1>display ipsec profile
==========================================
IPSec profile   : profile-1
Using interface: Tunnel0/0/0
==========================================
  IPSec Profile Name            :profile-1
  Peer Name                     : spu- AR2
  PFS    Group                  :0 (0:Disable 1:Group1 2:Group2 5:Group5 14:Group14)
  SecondsFlag                   :0 (0:Global 1:Local)
  SA Life Time Seconds          :3600
  KilobytesFlag                 :0 (0:Global 1:Local)
  SA Life Kilobytes             :1843200
  Anti-replay window size       :32
  Qos pre-classify              :0 (0:Disable 1:Enable)
  Number of IPSec Proposals :1
  IPSec Proposals Name          :tran-sec-1
<AR1>
```

（9）配置 Tunnel 端口的转发路由，将需要 IPSec 保护的数据流引到 Tunnel 端口上。

在路由器 AR1 上配置 Tunnel 端口的转发路由。

```
[AR1]ip route-static 192.168.2.0 255.255.255.0 Tunnel 0/0/0
[AR1]ip route-static 192.168.2.0 255.255.255.0 10.1.1.2
[AR1]
```

在路由器 AR2 上配置 Tunnel 端口的转发路由。

```
[AR2]ip route-static 192.168.1.0 255.255.255.0 Tunnel 0/0/0
[AR2]ip route-static 192.168.1.0 255.255.255.0 10.1.1.1
[AR2]
```

（10）查看配置结果，配置成功后，分别在路由器 AR1 和路由器 AR2 上使用 display ike sa 命令，以显示所配置的信息，这里以路由器 AR1 为例进行介绍。

```
<AR1>display ike sa
    Conn-ID  Peer        VPN     Flag(s)            Phase
  ---------------------------------------------------------------
       43    172.16.1.2   0      RD|ST                2
       42    172.16.1.2   0      RD|ST                1
  Flag Description:
  RD--READY   ST--STAYALIVE   RL--REPLACED   FD--FADING   TO--TIMEOUT
  HRT--HEARTBEAT   LKG--LAST KNOWN GOOD SEQ NO.   BCK--BACKED UP
<AR1>
```

（11）显示路由器 AR1、路由器 AR2 的配置信息，这里以路由器 AR1 为例进行介绍，其主要配置实例代码如下。

```
<AR1>display current-configuration
#
 sysname AR1
#
ipsec proposal tran-sec-1
 esp authentication-algorithm sha2-256
 esp encryption-algorithm aes-128
#
ike proposal 1
 encryption-algorithm aes-cbc-128
 dh group14
 authentication-algorithm sm3
#
ike peer spu-AR2 v1
 pre-shared-key cipher %$%$fU#m5_ep_4zcRC)R[Zk=,.2n%$%$
 ike-proposal 1
#
ipsec profile profile-1
 ike-peer spu-AR2
 proposal tran-sec-1
#
interface GigabitEthernet0/0/0
 ip address 172.16.1.1 255.255.255.0
#
interface GigabitEthernet0/0/1
 ip address 192.168.1.254 255.255.255.0
#
interface Tunnel0/0/0
 ip address 10.1.1.1 255.255.255.0
 tunnel-protocol ipsec
 source 172.16.1.1
 destination 172.16.1.2
 ipsec profile profile-1
#
ip route-static 192.168.2.0 255.255.255.0 172.16.1.2
ip route-static 192.168.2.0 255.255.255.0 Tunnel0/0/0
#
return
<AR1>
```

（12）配置主机 PC1 与主机 PC2 的 IP 地址，如图 11.8 所示。

（13）验证配置结果，使用 ping 命令，主机 PC1 访问主机 PC2，其测试结果如图 11.9 所示。

图 11.8　配置主机 PC1 与主机 PC2 的 IP 地址

图 11.9　主机 PC1 访问主机 PC2 的测试结果

11.3.2　虚拟隧道端口建立 GRE over IPSec 隧道配置实例

如图 11.10 所示，全网运行 OSPF 协议，路由器 AR2 为企业总公司的网关，路由器 AR3 为企业分公司的网关，总公司与分公司通过公网建立通信。

V11-4 虚拟隧道端口建立 GRE over IPSec 隧道配置实例——基础配置

V11-5 虚拟隧道端口建立 GRE over IPSec 隧道配置实例——GRE over IPSec 配置

图 11.10　虚拟隧道端口建立 GRE over IPSec 隧道配置实例

企业希望对总公司与分公司之间相互访问的流量进行安全保护，由于组播数据无法直接应用IPSec，所以基于虚拟隧道端口方式建立 GRE over IPSec，对 Tunnel 端口下的流量进行保护，配置相关端口与 IP 地址，进行网络拓扑连接。

1．配置思路

采用如下思路建立 GRE over IPSec 隧道。

（1）配置物理端口的 IP 地址和到对端的动态路由，保证两端路由可达。

（2）配置 GRE Tunnel 端口。

（3）配置 IPSec 安全提议，定义 IPSec 的保护方法。

（4）配置 IKE 对等体，定义对等体间进行 IKE 协商时的属性。

（5）配置安全框架，并引用安全提议和 IKE 对等体。

（6）在 Tunnel 端口上应用安全框架，使端口具有 IPSec 的保护功能。

（7）配置 Tunnel 端口的转发路由，将需要 IPSec 保护的数据流引到 Tunnel 端口上。

V11-6 虚拟隧道端口建立 GRE over IPSec 隧道配置实例——测试结果

2．配置步骤与方法

（1）配置路由器 AR1、路由器 AR2、路由器 AR3 各端口的 IP 地址，这里以路由器 AR1 为例进行介绍，其他路由器各端口的 IP 地址与此配置一致，不再赘述。

```
<Huawei>system-view
[Huawei]sysname AR1
[AR1]interface GigabitEthernet 0/0/1
[AR1-GigabitEthernet0/0/1]ip address 172.16.10.2 24
[AR1-GigabitEthernet0/0/1]quit
[AR1]interface GigabitEthernet 0/0/2
[AR1-GigabitEthernet0/0/2]ip address 172.16.20.2 24
[AR1-GigabitEthernet0/0/2]quit
[AR1]
```

（2）配置路由器的 OSPF 协议路由。

配置路由器 AR1 的 OSPF 协议路由。

```
[AR1]router id 1.1.1.1
[AR1]ospf 1
[AR1-ospf-1]area 0
[AR1-ospf-1-area-0.0.0.0]network 172.16.10.0 0.0.0.255
[AR1-ospf-1-area-0.0.0.0]network 172.16.20.0 0.0.0.255
[AR1-ospf-1-area-0.0.0.0]quit
[AR1-ospf-1]quit
[AR1]
```

配置路由器 AR2 的 OSPF 路由。

```
[AR2]router id 2.2.2.2
[AR2]ospf 1
[AR2-ospf-1]area 0
[AR2-ospf-1-area-0.0.0.0]network 172.16.10.0 0.0.0.255
[AR2-ospf-1-area-0.0.0.0]quit
[AR2-ospf-1]quit
[AR2]
```

配置路由器 AR3 的 OSPF 协议路由。

```
[AR3] router id 3.3.3.3
```

```
[AR3]ospf 1
[AR3-ospf-1]area 0
[AR3-ospf-1-area-0.0.0.0]network 172.16.20.0 0.0.0.255
[AR3-ospf-1-area-0.0.0.0]quit
[AR3-ospf-1]quit
[AR3]
```

（3）配置完成后，在路由器 AR2 和路由器 AR3 上使用 display ip routing-table protocol ospf 命令，可以看到它们能够学习到去往对端端口网段地址的 OSPF 协议路由，这里以路由器 AR2 为例进行介绍。

```
<AR2>display ip routing-table protocol ospf
Route Flags: R - relay, D - download to fib
------------------------------------------------------------------------------
Public routing table : OSPF
         Destinations : 1        Routes : 1
OSPF routing table status : <Active>
         Destinations : 1        Routes : 1
Destination/Mask    Proto   Pre  Cost    Flags   NextHop        Interface
172.16.20.0/24      OSPF    10   2       D       172.16.10.2    GigabitEthernet 0/0/1
OSPF routing table status : <Inactive>
         Destinations : 0        Routes : 0
<AR2>
```

（4）验证配置结果，使用 ping 命令，在路由器 AR2 上访问路由器 AR3 的 GE 0/0/2 端口的 IP 地址。

```
<AR2>ping -a 172.16.10.1 172.16.20.1
   PING 172.16.20.1: 56    data bytes, press CTRL_C to break
     Reply from 172.16.20.1: bytes=56 Sequence=1 ttl=254 time=30 ms
     Reply from 172.16.20.1: bytes=56 Sequence=2 ttl=254 time=30 ms
     Reply from 172.16.20.1: bytes=56 Sequence=3 ttl=254 time=30 ms
     Reply from 172.16.20.1: bytes=56 Sequence=4 ttl=254 time=20 ms
     Reply from 172.16.20.1: bytes=56 Sequence=5 ttl=254 time=30 ms
   --- 172.16.20.1 ping statistics ---
     5 packet(s) transmitted
     5 packet(s) received
     0.00% packet loss
     round-trip min/avg/max = 20/28/30 ms
<AR2>
```

（5）配置 GRE Tunnel 端口。

配置路由器 AR2。

```
[AR2]interface Tunnel 0/0/1
[AR2-Tunnel0/0/1]tunnel-protocol gre                    //配置 GRE 协议
[AR2-Tunnel0/0/1]ip address 10.1.1.1 24
[AR2-Tunnel0/0/1]source 172.16.10.1
[AR2-Tunnel0/0/1]destination 172.16.20.1
[AR2-Tunnel0/0/1]keepalive
[AR2-Tunnel0/0/1]quit
[AR2]
```

配置路由器 AR3。

[AR3]interface Tunnel 0/0/1
[AR3-Tunnel0/0/1]tunnel-protocol gre   //配置 GRE 协议
[AR3-Tunnel0/0/1]ip address 10.1.1.2 24
[AR3-Tunnel0/0/1]source 172.16.20.1
[AR3-Tunnel0/0/1]destination 172.16.10.1
[AR3-Tunnel0/0/1]keepalive
[AR3-Tunnel0/0/1]quit
[AR3]

（6）配置完成后，Tunnel 端口的状态变为 Up，Tunnel 端口之间可以连通，这里以路由器 AR2 为例进行介绍。

```
<AR2>ping –a 10.1.1.1 10.1.1.2
  PING 10.1.1.2: 56   data bytes, press CTRL_C to break
    Reply from 10.1.1.2: bytes=56 Sequence=1 ttl=255 time=50 ms
    Reply from 10.1.1.2: bytes=56 Sequence=2 ttl=255 time=20 ms
    Reply from 10.1.1.2: bytes=56 Sequence=3 ttl=255 time=20 ms
    Reply from 10.1.1.2: bytes=56 Sequence=4 ttl=255 time=30 ms
    Reply from 10.1.1.2: bytes=56 Sequence=5 ttl=255 time=30 ms
  --- 10.1.1.2 ping statistics ---
    5 packet(s) transmitted
    5 packet(s) received
    0.00% packet loss
    round-trip min/avg/max = 20/30/50 ms
<AR2>
```

（7）分别在路由器 AR2 和路由器 AR3 上创建 IPSec 安全提议。

在路由器 AR2 上创建 IPSec 安全提议。

[AR2]ipsec proposal tran-sec-1
[AR2-ipsec-proposal-tran-sec-1]esp authentication-algorithm sha2-256
[AR2-ipsec-proposal-tran-sec-1]esp encryption-algorithm aes-128
[AR2-ipsec-proposal-tran-sec-1]quit
[AR2]

在路由器 AR3 上创建 IPSec 安全提议。

[AR3]ipsec proposal tran-sec-1
[AR3-ipsec-proposal-tran-sec-1]esp authentication-algorithm sha2-256
[AR3-ipsec-proposal-tran-sec-1]esp encryption-algorithm aes-128
[AR3-ipsec-proposal-tran-sec-1]quit
[AR3]

（8）分别在路由器 AR2 和路由器 AR3 上配置 IKE 安全提议。

在路由器 AR2 上配置 IKE 安全提议。

[AR2]ike proposal 1
[AR2-ike-proposal-1]authentication-algorithm sha1
[AR2-ike-proposal-1]encryption-algorithm aes-cbc-128
[AR2-ike-proposal-1]dh group14
[AR2-ike-proposal-1]quit
[AR2]

在路由器 AR3 上配置 IKE 安全提议。

```
[AR3]ike proposal 1
[AR3-ike-proposal-1]authentication-algorithm sha1
[AR3-ike-proposal-1]encryption-algorithm aes-cbc-128
[AR3-ike-proposal-1]dh group14
[AR3-ike-proposal-1]quit
[AR3]
```

（9）分别在路由器 AR2 和路由器 AR3 上配置 IKE 对等体。

在路由器 AR2 上配置 IKE 对等体。

```
[AR2]ike peer spu-AR3 v1
[AR2-ike-peer-spu-AR3]ike-proposal 1
[AR2-ike-peer-spu-AR3]pre-shared-key cipher Huawei123456
[AR2-ike-peer-spu-AR3]quit
[AR2]
```

在路由器 AR3 上配置 IKE 对等体。

```
[AR3]ike peer spu-AR2 v1
[AR3-ike-peer-spu-AR2]ike-proposal 1
[AR3-ike-peer-spu-AR2]pre-shared-key cipher Huawei123456
[AR3-ike-peer-spu-AR2]quit
[AR3]
```

（10）分别在路由器 AR2 和路由器 AR3 上创建安全框架。

在路由器 AR2 上创建安全框架。

```
[AR2]ipsec profile profile-1
[AR2-ipsec-profile-profile-1]proposal tran-sec-1
[AR2-ipsec-profile-profile-1]ike-peer spu-AR3
[AR2-ipsec-profile-profile-1]quit
[AR2]
```

在路由器 AR3 上创建安全框架。

```
[AR3]ipsec profile profile-1
[AR3-ipsec-profile-profile-1]proposal tran-sec-1
[AR3-ipsec-profile-profile-1]ike-peer spu-AR2
[AR3-ipsec-profile-profile-1]quit
[AR3]
```

（11）分别在路由器 AR2 和路由器 AR3 的端口上应用各自的安全框架。

在路由器 AR2 的端口上应用安全框架。

```
[AR2]interface Tunnel 0/0/1
[AR2-Tunnel0/0/1]ipsec profile profile-1
[AR2-Tunnel0/0/1]quit
[AR2]
```

在路由器 AR3 的端口上应用安全框架。

```
[AR3]interface Tunnel 0/0/1
[AR3-Tunnel0/0/1]ipsec profile profile-1
[AR3-Tunnel0/0/1]quit
[AR3]
```

（12）在路由器 AR2 和路由器 AR3 上使用 display ipsec profile 命令，以显示所配置的信息。

在路由器 AR2 上使用 display ipsec profile 命令。

```
[AR2]display ipsec profile
```

```
========================================
IPSec profile   : profile-1
Using interface: Tunnel0/0/1
========================================
  IPSec Profile Name         :profile-1
  Peer Name                  :spu-AR3
  PFS    Group               :0 (0:Disable 1:Group1 2:Group2 5:Group5 14:Group14)
  SecondsFlag                :0 (0:Global 1:Local)
  SA Life Time Seconds       :3600
  KilobytesFlag              :0 (0:Global 1:Local)
  SA Life Kilobytes          :1843200
  Anti-replay window size    :32
  Qos pre-classify           :0 (0:Disable 1:Enable)
  Number of IPSec Proposals :1
  IPSec Proposals Name       :tran-sec-1
[AR2]
```

在路由器 AR3 上使用 display ipsec profile 命令。

```
[AR3]display ipsec profile
========================================
IPSec profile   : profile-1
Using interface: Tunnel0/0/1
========================================
  IPSec Profile Name         :profile-1
  Peer Name                  :spu-AR2
  PFS    Group               :0 (0:Disable 1:Group1 2:Group2 5:Group5 14:Group14)
  SecondsFlag                :0 (0:Global 1:Local)
  SA Life Time Seconds       :3600
  KilobytesFlag              :0 (0:Global 1:Local)
  SA Life Kilobytes          :1843200
  Anti-replay window size    :32
  Qos pre-classify           :0 (0:Disable 1:Enable)
  Number of IPSec Proposals :1
  IPSec Proposals Name       :tran-sec-1
[AR3]
```

（13）配置 Tunnel 端口的转发路由，将需要 IPSec 保护的数据流引到 Tunnel 端口上。

在路由器 AR2 上配置 Tunnel 端口的转发路由。

```
[AR2]ip route-static 192.168.20.0 255.255.255.0 tunnel 0/0/1
[AR2]
```

在路由器 AR3 上配置 Tunnel 端口的转发路由。

```
[AR3]ip route-static 192.168.10.0 255.255.255.0 tunnel 0/0/1
[AR3]
```

（14）查看配置结果，配置成功后，分别在路由器 AR2 和路由器 AR3 上使用 display ike sa 命令，以显示所配置的信息，这里以路由器 AR3 为例进行介绍。

```
[AR3]display ike sa
    Conn-ID  Peer          VPN    Flag(s)          Phase
    ------------------------------------------------------
    2        172.16.10.1   0      RD|ST            2
```

| | 1 | 172.16.10.1 | 0 | RD|ST | | 1 |

Flag Description:
RD—READY ST—STAYALIVE RL--REPLACED FD--FADING TO--TIMEOUT
HRT--HEARTBEAT LKG--LAST KNOWN GOOD SEQ NO. BCK--BACKED UP
[AR3]

（15）显示路由器 AR1、路由器 AR2 和路由器 AR3 的配置信息，这里以路由器 AR3 为例进行介绍，其主要配置实例代码如下。

```
<AR3>display current-configuration
#
 sysname AR3
#
router id 3.3.3.3
#
ipsec proposal tran-sec-1
 esp authentication-algorithm sha2-256
 esp encryption-algorithm aes-128
#
ike proposal 1
 encryption-algorithm aes-cbc-128
 dh group14
#
ike peer spu-AR2 v1
 pre-shared-key cipher %$%$fU#m5_ep_4zcRC)R[Zk=,.2n%$%$
 ike-proposal 1
#
ipsec profile profile-1
 ike-peer spu-AR2
 proposal tran-sec-1
#
interface GigabitEthernet0/0/0
 ip address 192.168.20.254 255.255.255.0
#
interface GigabitEthernet0/0/2
 ip address 172.16.20.1 255.255.255.0
#
interface Tunnel0/0/1
 ip address 10.1.1.2 255.255.255.0
 tunnel-protocol gre
 source 172.16.20.1
 destination 172.16.10.1
 ipsec profile profile-1
#
ospf 1
 area 0.0.0.0
  network 172.16.20.0 0.0.0.255
#
ip route-static 192.168.10.0 255.255.255.0 Tunnel0/0/1
```

```
#
return
<AR3>
```

（16）配置主机 PC1 和主机 PC2 的 IP 地址，如图 11.11 所示。

图 11.11　配置主机 PC1 和主机 PC2 的 IP 地址

（17）验证配置结果，使用 ping 命令，主机 PC1 访问主机 PC2，其测试结果如图 11.12 所示。

图 11.12　主机 PC1 访问主机 PC2 的测试结果

11.3.3　虚拟隧道端口建立 IPSec over GRE 隧道配置实例

如图 11.13 所示，全网运行 OSPF 协议，路由器 AR2 为企业总公司的网关，路由器 AR3 为企业分公司的网关，总公司与分公司通过公网建立通信。

企业希望对总公司与分公司之间相互访问的流量进行安全保护，由于组播数据无法直接应用 IPSec，所以需要基于虚拟隧道端口方式建立 IPSec over GRE 隧道，对 Tunnel 端口下的流量进行保护，配置相关端口与 IP 地址，进行网络拓扑连接。

1．配置思路

采用如下思路建立 IPSec over GRE 隧道。

（1）配置物理端口的 IP 地址和到对端的静态路由，保证两端路由可达。

（2）配置 GRE Tunnel 端口。

（3）配置 IPSec 安全提议，定义 IPSec 的保护方法。

（4）配置 IKE 对等体，定义对等体间进行 IKE 协商时的属性。

V11-7 虚拟隧道端口建立 IPSec over GRE 隧道配置实例——基础配置

图 11.13　虚拟隧道端口建立 IPSec over GRE 隧道配置实例

（5）配置安全框架，并引用安全提议和 IKE 对等体。

（6）配置 IPSec Tunnel 端口，将 IPSec Tunnel 的源端口配置为 GRE Tunnel 端口，且 IPSec Tunnel 的目的地址的路由必须从 GRE Tunnel 端口出去。

（7）在 IPSec Tunnel 端口上应用安全框架，使端口具有 IPSec 的保护功能。

V11-8 虚拟隧道端口建立 IPSec over GRE 隧道配置实例——IPSec over GRE 配置

（8）配置 IPSec Tunnel 端口的转发路由，将需要 IPSec 保护的数据流引到 IPSec Tunnel 端口上。

2. 配置步骤与方法

（1）配置路由器 AR1、路由器 AR2 和路由器 AR3 各端口的 IP 地址，这里以路由器 AR1 为例进行介绍，其他路由器各端口的 IP 地址与此配置一致，不再赘述。

```
<Huawei>system-view
[Huawei]sysname AR1
[AR1]interface GigabitEthernet 0/0/1
[AR1-GigabitEthernet0/0/1]ip address 172.16.10.2 24
[AR1-GigabitEthernet0/0/1]quit
[AR1]interface GigabitEthernet 0/0/2
[AR1-GigabitEthernet0/0/2]ip address 172.16.20.2 24
[AR1-GigabitEthernet0/0/2]quit
[AR1]
```

（2）配置路由器的 OSPF 协议路由。

配置路由器 AR1 的 OSPF 协议路由。

```
[AR1]router id 1.1.1.1
[AR1]ospf 1
[AR1-ospf-1]area 0
```

```
[AR1-ospf-1-area-0.0.0.0]network 172.16.10.0 0.0.0.255
[AR1-ospf-1-area-0.0.0.0]network 172.16.20.0 0.0.0.255
[AR1-ospf-1-area-0.0.0.0]quit
[AR1-ospf-1]quit
[AR1]
```

配置路由器 AR2 的 OSPF 协议路由。

```
[AR2]router id 2.2.2.2
[AR2]ospf 1
[AR2-ospf-1]area 0
[AR2-ospf-1-area-0.0.0.0]network 172.16.10.0 0.0.0.255
[AR2-ospf-1-area-0.0.0.0]quit
[AR2-ospf-1]quit
[AR2]
```

配置路由器 AR3 的 OSPF 协议路由。

```
[AR3] router id 3.3.3.3
[AR3]ospf 1
[AR3-ospf-1]area 0
[AR3-ospf-1-area-0.0.0.0]network 172.16.20.0 0.0.0.255
[AR3-ospf-1-area-0.0.0.0]quit
[AR3-ospf-1]quit
[AR3]
```

（3）配置完成后，在路由器 AR2 和路由器 AR3 上使用 display ip routing-table protocol ospf 命令，可以看到它们能够学习到去往对端端口网段地址的 OSPF 协议路由，这里以路由器 AR2 的显示为例进行介绍。

```
<AR2>display ip routing-table protocol ospf
Route Flags: R - relay, D - download to fib
------------------------------------------------------------------------
Public routing table : OSPF
         Destinations : 1       Routes : 1
OSPF routing table status : <Active>
         Destinations : 1       Routes : 1
Destination/Mask    Proto   Pre  Cost      Flags  NextHop         Interface
172.16.20.0/24      OSPF    10   2         D      172.16.10.2     GigabitEthernet 0/0/1
OSPF routing table status : <Inactive>
         Destinations : 0       Routes : 0
<AR2>
```

（4）验证配置结果，使用 ping 命令，在路由器 AR2 上访问路由器 AR3 的 GE 0/0/2 端口的 IP 地址。

```
<AR2>ping -a 172.16.10.1 172.16.20.1
    PING 172.16.20.1: 56   data bytes, press CTRL_C to break
    Reply from 172.16.20.1: bytes=56 Sequence=1 ttl=254 time=30 ms
    Reply from 172.16.20.1: bytes=56 Sequence=2 ttl=254 time=30 ms
    Reply from 172.16.20.1: bytes=56 Sequence=3 ttl=254 time=30 ms
    Reply from 172.16.20.1: bytes=56 Sequence=4 ttl=254 time=20 ms
    Reply from 172.16.20.1: bytes=56 Sequence=5 ttl=254 time=30 ms
    --- 172.16.20.1 ping statistics ---
    5 packet(s) transmitted
```

```
    5 packet(s) received
    0.00% packet loss
    round-trip min/avg/max = 20/28/30 ms
<AR2>
```

（5）配置 GRE Tunnel 端口。

配置路由器 AR2。

```
[AR2]interface Tunnel 0/0/1
[AR2-Tunnel0/0/1]tunnel-protocol gre          //配置 GRE 协议
[AR2-Tunnel0/0/1]ip address 10.1.1.1 24
[AR2-Tunnel0/0/1]source 172.16.10.1
[AR2-Tunnel0/0/1]destination 172.16.20.1
[AR2-Tunnel0/0/1]keepalive
[AR2-Tunnel0/0/1]quit
[AR2]
```

配置路由器 AR3。

```
[AR3]interface Tunnel 0/0/1
[AR3-Tunnel0/0/1]tunnel-protocol gre          //配置 GRE 协议
[AR3-Tunnel0/0/1]ip address 10.1.1.2 24
[AR3-Tunnel0/0/1]source 172.16.20.1
[AR3-Tunnel0/0/1]destination 172.16.10.1
[AR3-Tunnel0/0/1]keepalive
[AR3-Tunnel0/0/1]quit
[AR3]
```

（6）配置完成后，Tunnel 端口的状态变为 Up，Tunnel 端口之间可以连通，这里以路由器 AR2 为例进行介绍。

```
<AR2>ping -a 10.1.1.1 10.1.1.2
    PING 10.1.1.2: 56    data bytes, press CTRL_C to break
    Reply from 10.1.1.2: bytes=56 Sequence=1 ttl=255 time=50 ms
    Reply from 10.1.1.2: bytes=56 Sequence=2 ttl=255 time=20 ms
    Reply from 10.1.1.2: bytes=56 Sequence=3 ttl=255 time=20 ms
    Reply from 10.1.1.2: bytes=56 Sequence=4 ttl=255 time=30 ms
    Reply from 10.1.1.2: bytes=56 Sequence=5 ttl=255 time=30 ms
    --- 10.1.1.2 ping statistics ---
    5 packet(s) transmitted
    5 packet(s) received
    0.00% packet loss
    round-trip min/avg/max = 20/30/50 ms
<AR2>
```

（7）分别在路由器 AR2 和路由器 AR3 上创建 IPSec 安全提议。

在路由器 AR2 上创建 IPSec 安全提议。

```
[AR2]ipsec proposal tran-sec-1
[AR2-ipsec-proposal-tran-sec-1]esp authentication-algorithm sha2-256
[AR2-ipsec-proposal-tran-sec-1]esp encryption-algorithm aes-128
[AR2-ipsec-proposal-tran-sec-1]quit
[AR2]
```

在路由器 AR3 上创建 IPSec 安全提议。

```
[AR3]ipsec proposal tran-sec-1
```

```
[AR3-ipsec-proposal-tran-sec-1]esp authentication-algorithm sha2-256
[AR3-ipsec-proposal-tran-sec-1]esp encryption-algorithm aes-128
[AR3-ipsec-proposal-tran-sec-1]quit
[AR3]
```

（8）分别在路由器 AR2 和路由器 AR3 上配置 IKE 安全提议。

在路由器 AR2 上配置 IKE 安全提议。

```
[AR2]ike proposal 1
[AR2-ike-proposal-1]authentication-algorithm sha1
[AR2-ike-proposal-1]encryption-algorithm aes-cbc-128
[AR2-ike-proposal-1]dh group14
[AR2-ike-proposal-1]quit
[AR2]
```

在路由器 AR3 上配置 IKE 安全提议。

```
[AR3]ike proposal 1
[AR3-ike-proposal-1]authentication-algorithm sha1
[AR3-ike-proposal-1]encryption-algorithm aes-cbc-128
[AR3-ike-proposal-1]dh group14
[AR3-ike-proposal-1]quit
[AR3]
```

（9）分别在路由器 AR2 和路由器 AR3 上配置 IKE 对等体。

在路由器 AR2 上配置 IKE 对等体。

```
[AR2]ike peer spu-AR3 v1
[AR2-ike-peer-spu-AR3]ike-proposal 1
[AR2-ike-peer-spu-AR3]pre-shared-key cipher Huawei123456
[AR2-ike-peer-spu-AR3]quit
[AR2]
```

在路由器 AR3 上配置 IKE 对等体。

```
[AR3]ike peer spu-AR2 v1
[AR3-ike-peer-spu-AR2]ike-proposal 1
[AR3-ike-peer-spu-AR2]pre-shared-key cipher Huawei123456
[AR3-ike-peer-spu-AR2]quit
[AR3]
```

（10）分别在路由器 AR2 和路由器 AR3 上创建安全框架。

在路由器 AR2 上创建安全框架。

```
[AR2]ipsec profile profile-1
[AR2-ipsec-profile-profile-1]proposal tran-sec-1
[AR2-ipsec-profile-profile-1]ike-peer spu-AR3
[AR2-ipsec-profile-profile-1]quit
[AR2]
```

在路由器 AR3 上创建安全框架。

```
[AR3]ipsec profile profile-1
[AR3-ipsec-profile-profile-1]proposal tran-sec-1
[AR3-ipsec-profile-profile-1]ike-peer spu-AR2
[AR3-ipsec-profile-profile-1]quit
[AR3]
```

（11）分别在路由器 AR2 和路由器 AR3 上创建 IPSec Tunnel 端口，其中，IPSec Tunnel

的源端口为 GRE Tunnel 端口，且 IPSec Tunnel 的目的地址的路由必须从 GRE Tunnel 端口出去。

在路由器 AR2 上配置 IPSec Tunnel。

```
[AR2]interface Tunnel 0/0/2
[AR2-Tunnel0/0/2]ip address 20.1.1.1 24
[AR2-Tunnel0/0/2]tunnel-protocol ipsec        //配置 IPSec
[AR2-Tunnel0/0/2]source Tunnel 0/0/1
[AR2-Tunnel0/0/2]destination 10.1.1.2
[AR2-Tunnel0/0/2]quit
[AR2]
```

在路由器 AR3 上配置 IPSec Tunnel。

```
[AR3]interface Tunnel 0/0/2
[AR3-Tunnel0/0/2]ip address 20.1.1.2 24
[AR3-Tunnel0/0/2]tunnel-protocol ipsec        //配置 IPSec
[AR3-Tunnel0/0/2]source Tunnel 0/0/1
[AR3-Tunnel0/0/2]destination 10.1.1.1
[AR3-Tunnel0/0/2]quit
[AR3]
```

（12）分别在路由器 AR2 和路由器 AR3 的端口上应用各自的安全框架。

在路由器 AR2 的端口上应用安全框架。

```
[AR2]interface Tunnel 0/0/2
[AR2-Tunnel0/0/2]ipsec profile profile-1
[AR2-Tunnel0/0/2]quit
[AR2]
```

在路由器 AR3 的端口上应用安全框架。

```
[AR3]interface Tunnel 0/0/2
[AR3-Tunnel0/0/2]ipsec profile profile-1
[AR3-Tunnel0/0/2]quit
[AR3]
```

（13）在路由器 AR2 和路由器 AR3 上使用 display ipsec profile 命令，以显示所配置的信息。

在路由器 AR2 上使用 display ipsec profile 命令。

```
<AR2>display ipsec profile
===========================================
IPSec profile    : profile-1
Using interface: Tunnel0/0/2
===========================================
 IPSec Profile Name          :profile-1
 Peer Name                   :spu-AR3
 PFS    Group                :0 (0:Disable 1:Group1 2:Group2 5:Group5 14:Group14)
 SecondsFlag                 :0 (0:Global 1:Local)
 SA Life Time Seconds        :3600
 KilobytesFlag               :0 (0:Global 1:Local)
 SA Life Kilobytes           :1843200
 Anti-replay window size     :32
 Qos pre-classify            :0 (0:Disable 1:Enable)
 Number of IPSec Proposals :1
```

```
  IPSec Proposals Name        :tran-sec-1
<AR2>
```

在路由器 AR3 上使用 display ipsec profile 命令。

```
<AR3>display ipsec profile
==========================================
IPSec profile   : profile-1
Using interface: Tunnel0/0/2
==========================================
  IPSec Profile Name         :profile-1
  Peer Name                  :spu-AR2
  PFS    Group               :0 (0:Disable 1:Group1 2:Group2 5:Group5 14:Group14)
  SecondsFlag                :0 (0:Global 1:Local)
  SA Life Time Seconds       :3600
  KilobytesFlag              :0 (0:Global 1:Local)
  SA Life Kilobytes          :1843200
  Anti-replay window size    :32
  Qos pre-classify           :0 (0:Disable 1:Enable)
  Number of IPSec Proposals :1
  IPSec Proposals Name       :tran-sec-1
<AR3>
```

（14）配置 Tunnel 端口的转发路由，将需要 IPSec 保护的数据流引到 Tunnel 端口上。

在路由器 AR2 上配置 Tunnel 端口的转发路由。

```
[AR2]ip route-static 192.168.20.0 255.255.255.0 tunnel 0/0/2
[AR2]
```

在路由器 AR3 上配置 Tunnel 端口的转发路由。

```
[AR3]ip route-static 192.168.10.0 255.255.255.0 tunnel 0/0/2
[AR3]
```

（15）显示路由器 AR1、路由器 AR2、路由器 AR3 的配置信息，这里以路由器 AR3 为例进行介绍，其主要配置实例代码如下。

```
<AR3>display current-configuration
#
 sysname AR3
#
router id 3.3.3.3
#
ipsec proposal tran-sec-1
 esp authentication-algorithm sha2-256
 esp encryption-algorithm aes-128
#
ike proposal 1
 encryption-algorithm aes-cbc-128
 dh group14
#
ike peer spu-AR2 v1
 pre-shared-key cipher %$%$fU#m5_ep_4zcRC)R[Zk=,.2n%$%$
 ike-proposal 1
#
```

```
ipsec profile profile-1
 ike-peer spu-AR2
 proposal tran-sec-1
#
interface GigabitEthernet0/0/0
 ip address 192.168.20.254 255.255.255.0
#
interface GigabitEthernet0/0/2
 ip address 172.16.20.1 255.255.255.0
#
interface Tunnel0/0/1
 ip address 10.1.1.2 255.255.255.0
 tunnel-protocol gre
 source 172.16.20.1
 destination 172.16.10.1
#
interface Tunnel0/0/2
 ip address 20.1.1.2 255.255.255.0
 tunnel-protocol ipsec
 source Tunnel0/0/1
 destination 10.1.1.1
 ipsec profile profile-1
#
ospf 1
 area 0.0.0.0
  network 172.16.20.0 0.0.0.255
#
ip route-static 192.168.10.0 255.255.255.0 Tunnel0/0/2
#
user-interface con 0
 authentication-mode password
user-interface vty 0 4
user-interface vty 16 20
#
wlan ac
#
return
<AR3>
```

（16）配置主机 PC1 和主机 PC2 的 IP 地址，如图 11.11 所示。验证配置结果，主机 PC1 访问主机 PC2，其测试结果如图 11.12 所示。

课后习题

简答题

（1）简述 IPSec 的工作原理。

（2）简述 IPSec 的配置方法。

项目12
MPLS技术

【学习目标】
- 掌握MPLS技术的基本原理。
- 掌握MPLS技术的配置方法。

12.1 项目陈述

某公司由于规模较大,现行网络运行速度较慢,需要在公司骨干网络上实现快速转发。公司网络工程师小张负责组建网络,他通过 MPLS 协议实现公司网络的快速转发,本项目将对小张的组网方案进行介绍。

12.2 必备知识

12.2.1 MPLS 技术概述

多协议标签交换(Multi-Protocol Label Switching,MPLS)是由 IETF 提出的新一代的 IP 高速骨干网络交换标准,是一种在开放的通信网上利用标签引导数据高速、高效传输的新技术,不仅可以支持多种网络层协议,还可以兼容多种数据链路层技术。

MPLS 协议最初是为了提高转发速度而提出的,用于解决网络选址问题,如网络速度、可扩展性、服务质量管理及流量等,同时为下一代 IP 骨干网络解决了宽带管理及服务请求等问题。与传统 IP 路由方式相比,它利用标签(Label)进行数据转发,只在网络边缘分析 IP 头,而不用在每一跳中都分析 IP 头,从而节约了处理时间。

MPLS 独立于第二层和第三层协议,如 ATM 和 IP,它提供了一种方式,可将 IP 地址映射为简单的具有固定长度的标签,用于不同的包转发和包交换技术。它是现有路由和交换协议的端口,如 IP、ATM、帧中继、资源预留协议(Resource reSerVation Protocol,RSVP)、OSPF 协议等。在 MPLS 中,数据传输发生在 LSP 上,LSP 是每一个从源端到终端的路径上的节点的标签序列。

1. MPLS 常用术语

(1) Ingress LSR。

Ingress LSR(入站 LSR)处于 MPLS 域的边界,用于对 IP 报文进行处理,压入标签并产生标签报文。

（2）Transit LSR。

Transit LSR（中转LSR）用于对标签报文进行处理，如标签的置换，它们只会对标签进行操作，对IP头的信息不予处理。

（3）Egress LSR。

Egress LSR（出站LSR）是指将标签报文中的标签移除并将报文还原为IP报文的LSR。

（4）Label。

Label（标签）是指一个IP报文在进入MPLS域时被压入的一个标识符，其位于三层头和二层头的中间，长度为固定的32bit，分为4个字段，其格式如图12.1所示。

① 标签位：用于存储标签值，长度固定为20bit。

② Experimental：主要用于CoS，长度固定为3bit。

③ Bottom of Stack：栈底位，用于表示标签是否为该标签栈的栈底，1表示该标签后便是IP头，0表示非栈底，长度固定为1bit。

④ TTL：用于防止当网络中出现环路时标签被无限转发，长度固定为8bit。标签值的取值为0～15（特殊标签）、16～1023（静态LSP的标签值）、1024以上（标签分发协议所用）。

图12.1　MPLS标签的格式

（5）LSR。

开启了MPLS协议的路由器就是LSR，这些路由器维护着用于指导标签报文转发的信息，具有标签交换和标签分发的功能，并且能够依据这些信息对标签报文进行处理。

（6）LER。

标签边缘路由器（Label Edge Router，LER）负责对进入MPLS域的报文划分转发等价类（Forwarding Equivalence Class，FEC），并将这些FEC压入标签，进行MPLS转发。当报文离开MPLS域时会弹出标签，恢复为原来的报文，再进行相应转发。LER一定是LSR，但LSR不一定是LER。

（7）FEC。

FEC是指具有相同特征的报文，这些报文在LSR转发过程中采用相同的方式进行处理。

（8）LSP。

标签交换路径（Label Switched Path，LSP）是指报文在MPLS域中所经过的路径。

（9）LIB。

标签信息库（Label Information Base，LIB）用于存储本机对于某个网络的标签和邻居发给本机的对于某个网络的标签。标签信息库由标签分发协议生成，用于管理标签信息。

（10）FIB。

转发信息库（Forwarding Information Base，FIB）用于指明到达某网段或主机的报文应该通过路由器的哪个物理端口或逻辑端口发送，由 RIB 中提取的必要的路由信息生成，负责普通 IP 报文的转发。

（11）LFIB。

标签转发信息库（Label Forwarding Information Base，LFIB）用于指明入栈标签和出栈标签所对应的转发端口，由标签分发协议在 LSR 上建立，负责带 MPLS 标签报文的转发。

（12）LDP。

标签分发协议（Label Distribution Protocol，LDP）是一个公有协议，负责标签的分配、标签转发信息库的建立、标签交换路径的建立及拆除等，使用 UDP 发现邻居，再使用 TCP 建立邻居，目的端口号都为 646。LDP 为路由表中的每一条 IGP 前缀绑定了一个标签，LSR 和邻居交换这些标签，并将其存储在 LIB 中。

（13）RIB。

路由信息库（Routing Information Base，RIB）由 IP 路由协议（IP Routing Protocol）生成，用于选择路由。

2. MPLS 网络的运行过程

（1）路由器运行 IGP，生成 FIB 和路由信息库（Routing Information Base，RIB）。

（2）路由器运行 LDP，为每一个路由前缀捆绑一个标签，LDP 邻居相互交换标签信息，生成 LIB。

（3）路由器通过 LIB 和 FIB 中的数据生成 LFIB。

（4）路由器通过 LFIB 转发数据。

3. 传统的 IP 与 MPLS 网络转发的区别

（1）IP 转发是基于目的 IP 地址和 RIB 的，而 MPLS 转发是基于 MPLS 标签和 LFIB 的。

（2）MPLS 转发和 IP 都是逐跳转发的，但 IP 转发在每一跳都对分组进行了识别分类，而 MPLS 转发只在入口路由器上进行分类，分好类后即可通过 FEC 进行转发。

12.2.2　MPLS 体系结构

IP 报文进入 MPLS 网络时，MPLS 入口的 LER 分析 IP 报文的内容并为这些 IP 报文添加合适的标签，所有 MPLS 网络中的 LSR 根据标签转发数据。当 IP 报文离开 MPLS 网络时，标签由出口 LER 弹出。

IP 报文在 MPLS 网络中经过的路径称为 LSP。LSP 是一个单向路径，与数据流的方向一致。

LSP 的入口 LER 称为入节点（Ingress），位于 LSP 中间的 LSR 称为中间节点（Transit），LSP 的出口 LER 称为出节点（Egress）。一条 LSP 可以有 0 个、1 个或多个中间节点，但有且只有一个入节点和出节点。

根据 LSP 的方向，MPLS 报文由 Ingress 发往 Egress，则 Ingress 是 Transit 的上游节点，Transit 是 Ingress 的下游节点。同理，Transit 是 Egress 的上游节点，Egress 是 Transit 的下游节点。

MPLS 体系结构被分为两个独立的单元，即控制平面（Control Plane）和转发平面（Forwarding Plane），如图 12.2 所示。

图 12.2　MPLS 体系结构

（1）控制平面：负责产生和维护路由信息及标签信息。

（2）转发平面：即数据平面（Data Plane），负责普通 IP 报文的转发，以及带 MPLS 标签报文的转发。

12.2.3　MPLS 转发过程

标签操作类型包括标签压入（Push）、标签交换（Swap）和标签弹出（Pop），它们是标签转发的基本动作。

（1）Push：当 IP 报文进入 MPLS 域时，MPLS 边界设备在报文二层头和 IP 头之间插入一个新标签；或者，MPLS 中间设备根据需要在标签栈顶增加一个新的标签（即标签嵌套封装）。

（2）Swap：当报文在 MPLS 域内转发时，根据 LFIB，以下一跳分配的标签替换 MPLS 报文的栈顶标签。

（3）Pop：当报文离开 MPLS 域时，将 MPLS 报文的标签剥掉。

在最后一跳的节点上，标签已经没有使用价值了。这时可以利用倒数第二跳弹出（Penultimate Hop Popping，PHP）特性，在倒数第二跳节点处将标签弹出，减少最后一跳的负担。在最后一跳的节点上直接进行 IP 转发或者下一层标签转发。

默认情况下，设备支持 PHP 特性，支持 PHP 特性的 Egress 节点分配给倒数第二跳节点的标签值为 3，下面以支持 PHP 特性的 LSP 为例，说明 MPLS 的基本转发过程，如图 12.3 所示。

MPLS 标签已分发完成，建立了一条 LSP，其目的地址为 20.1.1.1/32，则 MPLS 的基本转发过程如下。

（1）Ingress 节点收到目的地址为 20.1.1.1/32 的 IP 报文，压入标签 X 并转发。

（2）Transit 节点收到该标签报文，进行标签交换，将标签 X 换成标签 Y。

（3）倒数第二跳的 Transit 节点收到带标签 Y 的报文，因为 Egress 分给它的标签值为 3，所以进行 PHP 操作，弹出标签 Y 并转发报文。从倒数第二跳转发给 Egress 的报文以 IP 报文形式传输。

（4）Egress 节点收到该 IP 报文，将其转发给目的地 20.1.1.1/32。

图 12.3　MPLS 的基本转发过程

12.3　项目实施

如图 12.4 所示配置 MPLS 实例，全网运行 OSPF 协议，为解决 IP 网络转发性能低下问题，决定使用 MPLS 技术来提高路由器的转发速度，配置相关端口与 IP 地址，进行网络拓扑连接。

图 12.4　MPLS 配置实例

（1）配置交换机 LSW1、交换机 LSW2，以及路由器 AR1、路由器 AR2 和路由器 AR3 各端口的 IP 地址，这里以交换机 LSW1、路由器 AR1 为例进行介绍，其他交换机和路由器各端口的 IP 地址配置与此一致，不再赘述。

配置交换机 LSW1 各端口的 IP 地址。

V12-1 MPLS 配置实例——基础配置

```
<Huawei>system-view
[Huawei]sysname LSW1
[LSW1]interface GigabitEthernet 0/0/1
[LSW1-GigabitEthernet0/0/1]port link-type access
[LSW1-GigabitEthernet0/0/1]quit
```

[LSW1]interface Vlanif 1
[LSW1-Vlanif1]ip address 10.1.1.1 24
[LSW1-Vlanif1]quit
[LSW1]

配置路由器 AR1 各端口的 IP 地址。

<Huawei>system-view
[Huawei]sysname AR1
[AR1]interface GigabitEthernet 0/0/0
[AR1-GigabitEthernet0/0/0]ip address 20.1.1.1 24
[AR1-GigabitEthernet0/0/0]quit
[AR1]interface GigabitEthernet 0/0/1
[AR1-GigabitEthernet0/0/1]ip address 10.1.1.2 24
[AR1-GigabitEthernet0/0/1]quit
[AR1]

V12-2 MPLS 配置
实例——MPLS
配置

V12-3 MPLS 配置
实例——测试结果

（2）配置 OSPF 协议。

配置交换机 LSW1 的 OSPF 协议。

[LSW1]router id 1.1.1.1
[LSW1]ospf 1
[LSW1-ospf-1]area 0
[LSW1-ospf-1-area-0.0.0.0]network 10.1.1.0 0.0.0.255
[LSW1-ospf-1-area-0.0.0.0]quit
[LSW1-ospf-1]quit
[LSW1]

配置交换机 LSW2 的 OSPF 协议。

[LSW2]router id 2.2.2.2
[LSW2]ospf 1
[LSW2-ospf-1]area 0
[LSW2-ospf-1-area-0.0.0.0]network 40.1.1.0 0.0.0.255
[LSW2-ospf-1-area-0.0.0.0]quit
[LSW2-ospf-1]quit
[LSW2]

配置路由器 AR1 的 OSPF 协议。

[AR1]router id 3.3.3.3
[AR1]ospf 1
[AR1-ospf-1]area 0
[AR1-ospf-1-area-0.0.0.0]network 10.1.1.0 0.0.0.255
[AR1-ospf-1-area-0.0.0.0]network 20.1.1.0 0.0.0.255
[AR1-ospf-1-area-0.0.0.0]quit
[AR1-ospf-1]quit
[AR1]

配置路由器 AR2 的 OSPF 协议。

[AR2]router id 4.4.4.4
[AR2]ospf 1
[AR2-ospf-1]area 0
[AR2-ospf-1-area-0.0.0.0]network 20.1.1.0 0.0.0.255
[AR2-ospf-1-area-0.0.0.0]network 30.1.1.0 0.0.0.255
[AR2-ospf-1-area-0.0.0.0]quit

```
[AR2-ospf-1]quit
[AR2]
```

配置路由器 AR3 的 OSPF 协议。

```
[AR3]router id 5.5.5.5
[AR3]ospf 1
[AR3-ospf-1]area 0
[AR3-ospf-1-area-0.0.0.0]network 30.1.1.0 0.0.0.255
[AR3-ospf-1-area-0.0.0.0]network 40.1.1.0 0.0.0.255
[AR3-ospf-1-area-0.0.0.0]quit
[AR3-ospf-1]quit
[AR3]
```

(3) 配置完成后,使用 ping 命令测试全网的连通性。

```
<LSW1>ping -a 10.1.1.1 40.1.1.1
  PING 40.1.1.1: 56   data bytes, press CTRL_C to break
    Reply from 40.1.1.1: bytes=56 Sequence=1 ttl=252 time=60 ms
    Reply from 40.1.1.1: bytes=56 Sequence=2 ttl=252 time=50 ms
    Reply from 40.1.1.1: bytes=56 Sequence=3 ttl=252 time=50 ms
    Reply from 40.1.1.1: bytes=56 Sequence=4 ttl=252 time=50 ms
    Reply from 40.1.1.1: bytes=56 Sequence=5 ttl=252 time=70 ms
  --- 40.1.1.1 ping statistics ---
    5 packet(s) transmitted
    5 packet(s) received
    0.00% packet loss
    round-trip min/avg/max = 50/56/70 ms
<LSW1>
```

(4) 查看设备路由表信息,这里以交换机 LSW1 为例进行介绍。

```
<LSW1>display ip routing-table
Route Flags: R - relay, D - download to fib
------------------------------------------------------------------------
Routing Tables: Public
         Destinations : 7        Routes : 7
Destination/Mask    Proto    Pre    Cost    Flags    NextHop      Interface
      10.1.1.0/24  Direct   0      0       D        10.1.1.1     Vlanif1
      10.1.1.1/32  Direct   0      0       D        127.0.0.1    Vlanif1
      20.1.1.0/24  OSPF     10     2       D        10.1.1.2     Vlanif1
      30.1.1.0/24  OSPF     10     3       D        10.1.1.2     Vlanif1
      40.1.1.0/24  OSPF     10     4       D        10.1.1.2     Vlanif1
      127.0.0.0/8  Direct   0      0       D        127.0.0.1    InLoopBack0
     127.0.0.1/32  Direct   0      0       D        127.0.0.1    InLoopBack0
<LSW1>
```

(5) 在各路由器的端口上配置 MPLS 和 LDP。

在路由器 AR1 的各端口上配置 MPLS 和 LDP。

```
[AR1]mpls lsr-id 3.3.3.3
[AR1]mpls
[AR1]interface LoopBack 0
[AR1-LoopBack0]ip address 3.3.3.3 24
[AR1-LoopBack0]quit
```

```
[AR1]ospf 1
[AR1-ospf-1-area-0.0.0.0]network 3.3.3.0 0.0.0.255
[AR1-ospf-1-area-0.0.0.0]quit
[AR1-ospf-1]quit
[AR1]
[AR1]interface GigabitEthernet 0/0/0
[AR1-GigabitEthernet0/0/0]mpls
[AR1-GigabitEthernet0/0/0]mpls ldp
[AR1-GigabitEthernet0/0/0]quit
[AR1]interface GigabitEthernet 0/0/1
[AR1-GigabitEthernet0/0/1]mpls
[AR1-GigabitEthernet0/0/1]mpls ldp
[AR1-GigabitEthernet0/0/1]quit
[AR1]
```

在路由器 AR2 的各端口上配置 MPLS 和 LDP。

```
[AR2]mpls lsr-id 4.4.4.4
[AR2]mpls
[AR2]interface LoopBack 0
[AR2-LoopBack0]ip address 4.4.4.4 24
[AR2-LoopBack0]quit
[AR2]ospf 1
[AR2-ospf-1-area-0.0.0.0]network 4.4.4.0 0.0.0.255
[AR2-ospf-1-area-0.0.0.0]quit
[AR2-ospf-1]quit
[AR2]interface GigabitEthernet 0/0/0
[AR2-GigabitEthernet0/0/0]mpls
[AR2-GigabitEthernet0/0/0]mpls ldp
[AR2-GigabitEthernet0/0/0]quit
[AR2]interface GigabitEthernet 0/0/2
[AR2-GigabitEthernet0/0/2]mpls
[AR2-GigabitEthernet0/0/2]mpls ldp
[AR2-GigabitEthernet0/0/2]quit
[AR2]
```

在路由器 AR3 的各端口上配置 MPLS 和 LDP。

```
[AR3]mpls lsr-id 5.5.5.5
[AR3]mpls
[AR3]interface LoopBack 0
[AR3-LoopBack0]ip address 5.5.5.5 24
[AR3-LoopBack0]quit
[AR3]ospf 1
[AR3-ospf-1-area-0.0.0.0]network 5.5.5.0 0.0.0.255
[AR3-ospf-1-area-0.0.0.0]quit
[AR3-ospf-1]quit
[AR3]interface GigabitEthernet 0/0/1
[AR3-GigabitEthernet0/0/1]mpls
[AR3-GigabitEthernet0/0/1]mpls ldp
[AR3-GigabitEthernet0/0/1]quit
```

```
[AR3]interface GigabitEthernet 0/0/2
[AR3-GigabitEthernet0/0/2]mpls
[AR3-GigabitEthernet0/0/2]mpls ldp
[AR3-GigabitEthernet0/0/2]quit
[AR3]
```

（6）配置完成后，在各节点上使用 display mpls ldp session 命令，可以查看路由器 AR1、路由器 AR2 和路由器 AR3 之间的本地 LDP 会话状态。

查看路由器 AR1 的本地 LDP 会话状态。

```
<AR1>display mpls ldp session
 LDP Session(s) in Public Network
 Codes: LAM(Label Advertisement Mode), SsnAge Unit(DDDD:HH:MM)
 A '*' before a session means the session is being deleted.
 ------------------------------------------------------------------
 PeerID           Status      LAM   SsnRole   SsnAge       KASent/Rcv
 ------------------------------------------------------------------
 4.4.4.4:0        Operational DU    Passive   0000:00:29   118/118
 ------------------------------------------------------------------
 TOTAL: 1 session(s) Found.
<AR1>
```

查看路由器 AR2 的本地 LDP 会话状态。

```
<AR2>display mpls ldp session
 LDP Session(s) in Public Network
 Codes: LAM(Label Advertisement Mode), SsnAge Unit(DDDD:HH:MM)
 A '*' before a session means the session is being deleted.
 ------------------------------------------------------------------
 PeerID           Status      LAM   SsnRole   SsnAge       KASent/Rcv
 ------------------------------------------------------------------
 3.3.3.3:0        Operational DU    Active    0000:00:31   126/126
 5.5.5.5:0        Operational DU    Passive   0000:00:30   124/124
 ------------------------------------------------------------------
 TOTAL: 2 session(s) Found.
<AR2>
```

查看路由器 AR3 的本地 LDP 会话状态。

```
<AR3>display mpls ldp session
 LDP Session(s) in Public Network
 Codes: LAM(Label Advertisement Mode), SsnAge Unit(DDDD:HH:MM)
 A '*' before a session means the session is being deleted.
 ------------------------------------------------------------------
 PeerID           Status      LAM   SsnRole   SsnAge       KASent/Rcv
 ------------------------------------------------------------------
 4.4.4.4:0        Operational DU    Active    0000:00:31   128/128
 ------------------------------------------------------------------
 TOTAL: 1 session(s) Found.
<AR3>
```

（7）在各 MPLS 路由器上将 LDP LSP 的触发策略修改为 all，使路由表中的所有静态路由和 IGP 表项都可以触发并建立 LDP LSP。

配置路由器 AR1。

[AR1]mpls
[AR1-mpls]lsp-trigger all
[AR1-mpls]quit
[AR1]

配置路由器 AR2。

[AR2]mpls
[AR2-mpls]lsp-trigger all
[AR2-mpls]quit
[AR2]

配置路由器 AR3。

[AR3]mpls
[AR3-mpls]lsp-trigger all
[AR3-mpls]quit
[AR3]

（8）配置完成后，在各节点上使用 display mpls ldp lsp 命令，可以看到 LDP LSP 的建立情况。查看路由器 AR1 的 LDP LSP 的建立情况。

```
<AR1>display mpls ldp lsp
 LDP LSP Information
-------------------------------------------------------------------------------
 DestAddress/Mask    In/OutLabel      UpstreamPeer      NextHop         OutInterface
-------------------------------------------------------------------------------
 3.3.3.0/24          3/NULL           4.4.4.4           3.3.3.3         Loop0
 3.3.3.3/32          3/NULL           4.4.4.4           127.0.0.1       InLoop0
 *3.3.3.3/32         Liberal/1024                       DS/4.4.4.4
 *4.4.4.0/24         Liberal/3                          DS/4.4.4.4
 4.4.4.4/32          NULL/3           -                 20.1.1.2        GE0/0/0
 4.4.4.4/32          1024/3           4.4.4.4           20.1.1.2        GE0/0/0
 5.5.5.5/32          NULL/1027        -                 20.1.1.2        GE0/0/0
 5.5.5.5/32          1027/1027        4.4.4.4           20.1.1.2        GE0/0/0
 10.1.1.0/24         3/NULL           4.4.4.4           10.1.1.2        GE0/0/1
 *10.1.1.0/24        Liberal/1025                       DS/4.4.4.4
 20.1.1.0/24         3/NULL           4.4.4.4           20.1.1.1        GE0/0/0
 *20.1.1.0/24        Liberal/3                          DS/4.4.4.4
 30.1.1.0/24         NULL/3           -                 20.1.1.2        GE0/0/0
 30.1.1.0/24         1025/3           4.4.4.4           20.1.1.2        GE0/0/0
 40.1.1.0/24         NULL/1026        -                 20.1.1.2        GE0/0/0
 40.1.1.0/24         1026/1026        4.4.4.4           20.1.1.2        GE0/0/0
-------------------------------------------------------------------------------
 TOTAL: 12 Normal LSP(s) Found.
 TOTAL: 4 Liberal LSP(s) Found.
 TOTAL: 0 Frr LSP(s) Found.
 A '*' before an LSP means the LSP is not established
 A '*' before a Label means the USCB or DSCB is stale
 A '*' before a UpstreamPeer means the session is stale
 A '*' before a DS means the session is stale
 A '*' before a NextHop means the LSP is FRR LSP
```

\<AR1\>

查看路由器 AR2 的 LDP LSP 的建立情况。

```
<AR2>display mpls ldp lsp
   LDP LSP Information
-------------------------------------------------------------------------
DestAddress/Mask    In/OutLabel    UpstreamPeer    NextHop      OutInterface
-------------------------------------------------------------------------
*3.3.3.0/24         Liberal/3                      DS/3.3.3.3
 3.3.3.3/32         NULL/3         -               20.1.1.1     GE0/0/0
 3.3.3.3/32         1024/3         3.3.3.3         20.1.1.1     GE0/0/0
 3.3.3.3/32         1024/3         5.5.5.5         20.1.1.1     GE0/0/0
*3.3.3.3/32         Liberal/1024                   DS/5.5.5.5
 4.4.4.0/24         3/NULL         3.3.3.3         4.4.4.4      Loop0
 4.4.4.0/24         3/NULL         5.5.5.5         4.4.4.4      Loop0
 4.4.4.4/32         3/NULL         3.3.3.3         127.0.0.1    InLoop0
 4.4.4.4/32         3/NULL         5.5.5.5         127.0.0.1    InLoop0
*4.4.4.4/32         Liberal/1024                   DS/3.3.3.3
*4.4.4.4/32         Liberal/1025                   DS/5.5.5.5
*5.5.5.0/24         Liberal/3                      DS/5.5.5.5
 5.5.5.5/32         NULL/3         -               30.1.1.2     GE0/0/2
 5.5.5.5/32         1027/3         3.3.3.3         30.1.1.2     GE0/0/2
 5.5.5.5/32         1027/3         5.5.5.5         30.1.1.2     GE0/0/2
*5.5.5.5/32         Liberal/1027                   DS/3.3.3.3
 10.1.1.0/24        NULL/3         -               20.1.1.1     GE0/0/0
 10.1.1.0/24        1025/3         3.3.3.3         20.1.1.1     GE0/0/0
 10.1.1.0/24        1025/3         5.5.5.5         20.1.1.1     GE0/0/0
*10.1.1.0/24        Liberal/1026                   DS/5.5.5.5
 20.1.1.0/24        3/NULL         3.3.3.3         20.1.1.2     GE0/0/0
 20.1.1.0/24        3/NULL         5.5.5.5         20.1.1.2     GE0/0/0
*20.1.1.0/24        Liberal/3                      DS/3.3.3.3
*20.1.1.0/24        Liberal/1027                   DS/5.5.5.5
 30.1.1.0/24        3/NULL         3.3.3.3         30.1.1.1     GE0/0/2
 30.1.1.0/24        3/NULL         5.5.5.5         30.1.1.1     GE0/0/2
*30.1.1.0/24        Liberal/1025                   DS/3.3.3.3
*30.1.1.0/24        Liberal/3                      DS/5.5.5.5
 40.1.1.0/24        NULL/3         -               30.1.1.2     GE0/0/2
 40.1.1.0/24        1026/3         3.3.3.3         30.1.1.2     GE0/0/2
 40.1.1.0/24        1026/3         5.5.5.5         30.1.1.2     GE0/0/2
*40.1.1.0/24        Liberal/1026                   DS/3.3.3.3
-------------------------------------------------------------------------
TOTAL: 20 Normal LSP(s) Found.
TOTAL: 12 Liberal LSP(s) Found.
TOTAL: 0 Frr LSP(s) Found.
A '*' before an LSP means the LSP is not established
A '*' before a Label means the USCB or DSCB is stale
A '*' before a UpstreamPeer means the session is stale
A '*' before a DS means the session is stale
A '*' before a NextHop means the LSP is FRR LSP
```

\<AR2\>
查看路由器 AR3 的 LDP LSP 的建立情况。
```
<AR3>display mpls ldp lsp
  LDP LSP Information
-------------------------------------------------------------------------------
DestAddress/Mask    In/OutLabel    UpstreamPeer    NextHop      OutInterface
-------------------------------------------------------------------------------
3.3.3.3/32          NULL/1024      -               30.1.1.1     GE0/0/2
3.3.3.3/32          1024/1024      4.4.4.4         30.1.1.1     GE0/0/2
*4.4.4.0/24         Liberal/3                      DS/4.4.4.4
4.4.4.4/32          NULL/3         -               30.1.1.1     GE0/0/2
4.4.4.4/32          1025/3         4.4.4.4         30.1.1.1     GE0/0/2
5.5.5.0/24          3/NULL         4.4.4.4         5.5.5.5      Loop0
5.5.5.5/32          3/NULL         4.4.4.4         127.0.0.1    InLoop0
*5.5.5.5/32         Liberal/1027                   DS/4.4.4.4
10.1.1.0/24         NULL/1025      -               30.1.1.1     GE0/0/2
10.1.1.0/24         1026/1025      4.4.4.4         30.1.1.1     GE0/0/2
20.1.1.0/24         NULL/3         -               30.1.1.1     GE0/0/2
20.1.1.0/24         1027/3         4.4.4.4         30.1.1.1     GE0/0/2
30.1.1.0/24         3/NULL         4.4.4.4         30.1.1.2     GE0/0/2
*30.1.1.0/24        Liberal/3                      DS/4.4.4.4
40.1.1.0/24         3/NULL         4.4.4.4         40.1.1.2     GE0/0/1
*40.1.1.0/24        Liberal/1026                   DS/4.4.4.4
-------------------------------------------------------------------------------
TOTAL: 12 Normal LSP(s) Found.
TOTAL: 4 Liberal LSP(s) Found.
TOTAL: 0 Frr LSP(s) Found.
A '*' before an LSP means the LSP is not established
A '*' before a Label means the USCB or DSCB is stale
A '*' before a UpstreamPeer means the session is stale
A '*' before a DS means the session is stale
A '*' before a NextHop means the LSP is FRR LSP
<AR3>
```
（9）查看路由器 AR1、路由器 AR2、路由器 AR3 的配置信息，这里以路由器 AR1 为例进行介绍，其主要配置实例代码如下。
```
<AR1>display current-configuration
#
 sysname AR1
#
router id 3.3.3.3
#
mpls lsr-id 3.3.3.3
mpls
 lsp-trigger all
#
mpls ldp
#
```

```
interface GigabitEthernet0/0/0
 ip address 20.1.1.1 255.255.255.0
 mpls
 mpls ldp
#
interface GigabitEthernet0/0/1
 ip address 10.1.1.2 255.255.255.0
 mpls
 mpls ldp
#
interface LoopBack0
 ip address 3.3.3.3 255.255.255.0
#
ospf 1
 area 0.0.0.0
  network 3.3.3.0 0.0.0.255
  network 10.1.1.0 0.0.0.255
  network 20.1.1.0 0.0.0.255
#
return
<AR1>
```

课后习题

1. 选择题

（1）推入一层 MPLS 标签的报文比原来的 IP 报文多（　　）字节。
　　A. 4　　　　　　　　B. 8　　　　　　　　C. 16　　　　　　　　D. 32

（2）MPLS 支持多层标签和转发平面面向连接的特性已经在很多方面得到了应用，部署 MPLS 的原因不包括（　　）。
　　A. MPLS 支持流量工程功能
　　B. MPLS 有能力使用 VPN 服务
　　C. MPLS 在基于软件的路由器上简化了路由查找
　　D. 各厂商标准普遍认可 MPLS

（3）在一个 MPLS 网络中，设备会根据 MPLS 数据帧的标签进行转发，其中 MPLS 的标签位于（　　）。
　　A. 二层头后　　　　　　　　　　　　B. 二层头前及三层头后
　　C. 三层头前　　　　　　　　　　　　D. 一层头前

（4）MPLS 又称为多协议标签交换技术，以下特性或者协议中，MPLS 不支持（　　）。
　　A. MULTICAST　　　B. OSPF 协议　　　C. BGP　　　　　D. IS-IS 协议

（5）MPLS 技术以标签交换代替 IP 转发，当 MPLS 运行在以太网中时，它使用的封装模式是（　　）。
　　A. 包模式　　　　　B. 传输模式　　　　C. 帧模式　　　　D. 通道模式

（6）在二层和三层网络中有很多防止环路产生的方法，MPLS 也可以通过多种方式来防止环路，下面不是 MPLS 防止环路方法的是（   ）。

  A．LDP 环路检测机制

  B．硬件环路检测机制

  C．在 MPLS 网络中可以使用 TTL 防止报文的无限循环转发

  D．MPLS 要依靠 IGP 建立 LSP，IGP 本身的一些机制可以避免自由环路发生

（7）IETF 定义的多协议标签交换技术是一种第三层交换技术，它由不同的设备组成，其中，负责为网络添加或删除标记的设备是（   ）。

  A．标签分发路由器   B．标签边缘路由器

  C．标签交换路由器   D．标签传送路由器

（8）【多选】多协议标签交换技术用于向 IP 层提供连接服务，MPLS 网络由（   ）组成。

  A．标签分发路由器   B．标签边缘路由器

  C．标签交换路由器   D．标签传送路由器

（9）MPLS 依据标签对数据进行转发，如果没有标签，则通过 MPLS 域的 IP 报文通过（   ）转发。

  A．ATM B．多标签 MPLS C．普通 IP D．单标签 MPLS

（10）【多选】在 MPLS 域中建立 LSP 时要防止环路产生，LDP 的环路检测包括（   ）。

  A．路径向量法 B．STP C．硬件检测环路 D．最大跳数法

（11）MPLS 技术的核心是标签交换，这种说法是（   ）的。

  A．正确   B．错误

（12）下列关于传统 IP 网络在转发数据时的描述中，错误的是（   ）。

  A．传统 IP 转发是基于连接的，效率比较低

  B．传统 IP 网络基于 IGP Metric 计算最优路径

  C．传统 IP 转发采用的是逐跳转发

  D．传统 IP 网络无法提供好的 QoS

2．简答题

（1）简述 MPLS 的工作原理。

（2）简述 MPLS 的配置方法。

项目13
VXLAN技术

【学习目标】
- 掌握VXLAN技术的基本原理。
- 掌握VXLAN技术的配置方法。

13.1 项目陈述

在传统二层网络环境下,网络隔离技术使用 VLAN 技术,于 IEEE 802.1q 中定义的 VLAN Tag 域只有 12 比特,仅能表示 4096 个 VLAN,无法满足大二层网络中标识大量租户或租户群的需求。传统二层网络中的 VLAN 无法满足网络动态调整需求,虚拟机迁移范围也受到网络架构的限制,虚拟可扩展局域网(Virtual eXtensible LAN,VXLAN)技术的提出很好地解决了上述问题。

13.2 必备知识

13.2.1 VXLAN 技术概述

1. VXLAN 的含义

VXLAN 是一种 Overlay 网络技术,其对原始二层以太网帧进行 UDP 封装(MAC-in-UDP),增加了 8 字节 VXLAN 头部、8 字节 UDP 头部、20 字节 IP 头部和 14 字节以太网头部,共 50 字节,如图 13.1 所示。

VXLAN 采用了 MAC-in-UDP 封装方式,是第三层网络虚拟化(Network Virtualization Over layer 3,NVO3)中的一种网络虚拟化技术。作为云计算的核心技术之一,服务器虚拟化可大幅降低 IT 成本、提高业务部署灵活性、降低运维成本等优势已经得到越来越多的用户的认可和部署。

如图 13.2 所示,一台服务器可虚拟出多台虚拟机,而一台虚拟机相当于一台主机,主机的数量发生了数量级的变化,这也为虚拟网络带来了如下问题。

(1)虚拟机规模受网络规格限制。

在传统二层网络环境下,数据报文是通过查询 MAC 地址表进行二层转发的,而 MAC 地址表的容量限制了虚拟机的数量。

图 13.1　VXLAN 报文格式

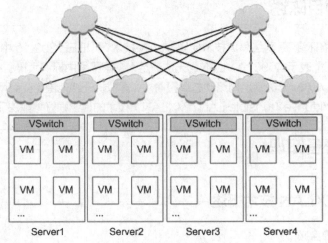

图 13.2　服务器虚拟化示意图

（2）网络隔离能力限制。

当前主流的网络隔离技术是 VLAN，其仅支持 4096 个 VLAN，因此无法满足大二层网络中标识大量租户或租户群的需求，同时也无法满足网络动态调整的需求。

（3）虚拟机迁移范围受网络架构限制。

虚拟机启动后，由于服务器资源等问题（如 CPU 过高、内存不够等），可能需要将虚拟机迁移到新的服务器上。为了保证虚拟机迁移过程中业务不中断，需要保证虚拟机的 IP 地址保持不变，这就要求业务网络是一个二层网络，且要求网络本身具备多路径的冗余备份和可靠性。

针对大二层网络，VXLAN 技术的提出很好地解决了上述问题。

（1）针对虚拟机规模受网络规格限制。

VXLAN 将虚拟机发出的数据包封装在 UDP 中，并使用物理网络的 IP、MAC 地址作为外层头进行封装，对网络只表现为封装后的参数，因此，极大地降低了大二层网络对 MAC 地址规格的需求。

（2）针对网络隔离能力限制。

VXLAN 引入了类似 VLAN ID 的用户标识，称为 VXLAN 网络标识（VXLAN Network Identifier,VNI），由 24bit 组成，支持多达 16M 的 VXLAN 段，从而满足了可能使用大量用户标识的需求。

（3）针对虚拟机迁移范围受网络架构限制。

VXLAN 通过采用 MAC-in-UDP 封装来延伸二层网络，将以太报文封装在 IP 报文之上，通过路由在网络中传输，无须关注虚拟机的 MAC 地址，且路由网络无网络结构限制，具备大规模扩展能力、故障自愈能力、负载均衡能力。通过路由网络，虚拟机迁移不受网络架构的限制。

2. VXLAN 的优点

与 VLAN 相比，VXLAN 具有更高的扩展性和灵活性，其主要有以下优点。

（1）应用部署更灵活。

通过 VXLAN 封装后的二层以太网帧可以跨三层网络边界，使组网及应用部署变得更加灵活，同时解决了多租户网络环境中 IP 地址冲突的问题。

（2）更好的扩展性。

传统网络通过 VLAN 对客户网络进行逻辑隔离，VNI 字段为 12bit，VLAN 数量最大为 4096。VXLAN 使用了 24bit 的 VNI，最大支持 16000000 个逻辑网络，扩展性得到了极大的增强。

（3）提高了网络利用率。

传统以太网使用 STP 预防环路，STP 导致网络冗余路径处于阻塞状态，VXLAN 报文基于 3 层 IP 报头传输，能有效利用网络路径，支持等价多路径（Equal-Cost Multi-Path，ECMP）和链路聚合协议，传统以太网帧无法穿越三层网络，部署 VXLAN 后，VXLAN 隧道端点（VXLAN Tunnel End Point，VTEP）之间的数据基于三层寻址，网络互联端口不再是二层端口，可以将交换机之间的互联端口部署为三层模式。

3. VXLAN 常用术语

（1）VXLAN 隧道端点。

VXLAN 使用 VTEP 设备对 VXLAN 报文进行封装与解封装，包括 ARP 请求报文和正常的 VXLAN 数据报文，VTEP 将原始以太网帧通过 VXLAN 封装后发送至对端 VTEP 设备，对端 VTEP 设备接收到 VXLAN 报文后进行解封装，并根据原始 MAC 地址进行转发，VTEP 可以是物理交换机、物理服务器或者其他支持 VXLAN 的硬件设备或软件。

（2）VNI。

VNI 类似于 VLAN ID，用于区分 VXLAN 段，不同 VXLAN 段的虚拟机不能直接通过二层相互通信。一个 VNI 表示一个租户，即使多个终端用户属于同一个 VNI，其也表示一个租户。

在分布式网关部署场景下，VNI 分为二层 VNI 和三层 VNI。二层 VNI 是普通的 VNI，以 1∶1 的方式映射到广播域，实现 VXLAN 报文与子网的转发。三层 VNI 和 VPN 实例进行关联，用于 VXLAN 报文的跨子网转发。

（3）VXLAN 网关。

和 VLAN 类似，不同 VNI 之间的 VXLAN，以及 VXLAN 和非 VXLAN 之间不能直接相互通信。为了使 VXLAN 之间，以及 VXLAN 和非 VXLAN 之间能够进行通信，VXLAN 引入了 VXLAN 网关。

VXLAN 网关分为二层网关和三层网关。二层网关可用于解决租户接入 VXLAN 虚拟网络的问题，也可用于同一 VXLAN 虚拟网络的子网通信。三层网关用于 VXLAN 虚拟网络的跨子网通信及外部网络的访问。

（4）桥接域。

桥接域（Bridge Domain，BD）是 VXLAN 网络中转发数据报文的二层广播域。在 VXLAN 网络中，将 VNI 以 1：1 的方式映射到 BD 中，BD 成为 VXLAN 网络转发数据报文的实体。

（5）虚拟接入点。

虚拟接入点（Virtual Access Point，VAP）即 VXLAN 业务接入点，可以是二层子端口或 VLAN。

当接入节点是二层子端口时，通过在二层子端口上配置流封装类型使不同的端口接入不同的数据报文，将二层子端口关联 BD 后，可使数据报文通过 BD 转发。

当业务接入点是 VLAN 时，将 VLAN 绑定到 BD，也可以使数据报文通过 BD 转发。

（6）Underlay 网络和 Overlay 网络。

VXLAN 技术将已有的物理网络作为 Underlay 网络，在其上构建出虚拟的二层或三层网络，即 Overlay 网络。Overlay 网络通过封装技术，利用 Underlay 网络提供的三层转发路径，使租户报文在不同站点间传递。对于租户来说，Underlay 网络是透明的，他们只能感知到 Overlay 网络。

（7）VBDIF 端口。

基于 BD 创建的三层逻辑端口就是 VBDIF 端口。通过 VBDIF 端口配置 IP 地址可实现不同网段的 VXLAN 间及 VXLAN 和非 VXLAN 间的通信，也可实现二层网络接入三层网络。

（8）网络虚拟边缘节点。

网络虚拟边缘节点（Network Virtualization Edge，NVE）是实现网络虚拟化功能的网络实体。报文经过 NVE 封装转换后，NVE 间即可基于三层基础网络建立二层虚拟化网络。

13.2.2　VXLAN 网关划分

根据三层网关部署方式的不同，VXLAN 三层网关又可以分为集中式网关和分布式网关。

（1）集中式网关部署。

集中式网关是指将三层网关集中部署在一台设备上，如图 13.3 所示，所有跨子网的流量都经过三层网关进行转发，实现子流量的集中管理。

图 13.3　VXLAN 集中式网关部署

部署集中式网关的优点和缺点如下。

优点：对跨子网流量进行集中管理，网关的部署和管理比较简单。

缺点：转发路径不是最优的，同一二层网关下跨子网的数据中心的三层流量都需要经过集中三层网关转发；ARP 表项规格瓶颈，由于采用了集中式三层网关，因此通过三层网关转发的终端租户的 ARP 表项都需要在三层网关上生成，而三层网关上的 ARP 表项规格有限，不利于数据中心网络的扩展。

（2）分布式网关部署。

分布式网关可以解决集中式网关的缺点。VXLAN 分布式网关是指在典型的 Spine-Leaf 组网结构下，将 Leaf 节点作为 VTEP，每个 Leaf 节点都可作为 VXLAN 的三层网关，Spine 节点不感知 VXLAN 隧道，只作为 VXLAN 报文的转发节点。如图 13.4 所示，服务器 Server1 和服务器 Server2 不在同一个网段中，但是都连接在一个 Leaf 节点上。服务器 Server1 和服务器 Server2 通信时，流量只需要在 Leaf1 节点上进行转发即可，不再需要经过 Spine 节点。

图 13.4　VXLAN 分布式网关部署

Spine 节点关注于高速 IP 转发，强调的是设备的高速转发能力。

Leaf 节点作为 VXLAN 网络中的二层网关设备，与物理服务器或虚拟机（Virtual Machine，VM）对接，用于解决终端租户接入 VXLAN 虚拟网络的问题；作为 VXLAN 网络中的三层网关设备，进行 VXLAN 报文封装/解封装，实现了跨子网的终端租户通信及外部网络的访问。

VXLAN 的分布式网关具有如下特点。

同一个 Leaf 节点既可以做 VXLAN 二层网关，也可以做 VXLAN 三层网关，部署灵活；Leaf 节点只需要学习自身连接服务器的 ARP 表项即可，而不必像集中式三层网关一样学习所有服务器的 ARP 表项，解决了集中式三层网关带来的 ARP 表项瓶颈问题，网络规模扩展能力强。

13.2.3　VXLAN 运行机制

在设备上部署 VXLAN 网络时，需要分别对下行接入业务选择部署及上行 VXLAN 隧道建立部署，两者部署完成后，报文即可在 VXLAN 中进行转发。因此，VXLAN 运行机制可以概括为报文识别、隧道建立及 VXLAN 报文转发。

1. 报文识别

在 VXLAN 中，VNI 以 1：1 的方式映射到 BD 中。当报文到达 VTEP 后，VTEP 只要能够识别出报文所属的 BD，就能够选择正确的 VXLAN 隧道进行转发。VTEP 有以下两种方式来识别报文所属的 VXLAN。

（1）基于 VLAN 识别报文所属的 VXLAN。

基于网络规划，在 VTEP 上建立 VLAN 与 BD 的一对一或多对一的映射。当 VTEP 收到业务侧的报文后，根据 VLAN 与 BD 及 BD 与 VNI 的对应关系即可选择相应的 VXLAN 隧道进行转发。

如图 13.5 所示，VLAN 10 与 VLAN 20 同属于 BD 10，VTEP 上存在 BD 10 与 VLAN 10 及 VLAN 20 的对应关系，同时 BD 10 对应的 VXLAN 的 VNI 为 1000。当 VTEP 接收到主机 PC_1 或主机 PC_2 的报文时，即可选择正确的 VXLAN 隧道进行转发。

图 13.5　基于 VLAN 识别报文所属的 VXLAN

（2）基于报文流封装类型识别报文所属的 VXLAN。

报文的流封装类型可概括地分为携带指定 VLAN Tag 与不携带 VLAN Tag 两种。基于此，可在 VTEP 连接下行业务的物理端口上创建二层子端口，并配置二层子端口对报文的不同处理方式，同时将二层子端口与 BD 进行一一映射。业务侧报文到达 VTEP 后，即可进入指定的二层子端口。VTEP 即可根据二层子端口与 BD 的映射关系，以及 BD 与 VNI 的映射关系，来选择正确的 VXLAN 隧道进行报文转发，如图 13.6 所示。

图 13.6　基于报文流封装类型识别报文所属的 VXLAN

2. 隧道建立

VXLAN 隧道由一对 VTEP IP 地址确定，创建 VXLAN 隧道实际上是两端 VTEP 获取对端 VTEP IP 地址的过程，只要对端 VTEP IP 地址是三层路由可达的，VXLAN 隧道就可以成功建立。

静态 VXLAN 隧道的创建需要通过手工配置源端和目的端的 VNI 与 VTEP 的 IP 地址，配置工作量大，灵活性较差，不适用于大规模的组网场景。

通过 BGP EVPN（Ethernet VPN）方式动态建立 VXLAN 隧道，就是在两端 VTEP 之间建

立 BGP EVPN 对等体，对等体之间利用 BGP EVPN 路由来互相传递 VNI 和 VTEP IP 地址信息，从而动态建立 VXLAN 隧道。该方式通过部署 EVPN 协议实现 VTEP 自动发现和 VXLAN 隧道的动态创建，灵活性高，适用于大规模的 VXLAN 组网场景。

VXLAN 隧道建立情况分为以下两种。

（1）租户分支 1 与租户分支 2 位于同一子网，且两者属于同一 VNI，属于同一 VNI 的租户处于同一个逻辑二层网络，彼此可以直接通过 VXLAN 隧道实现二层互联。

（2）租户分支与租户总部属于不同子网，且两者属于不同 VNI，属于不同 VNI 的租户之间不能直接进行 VXLAN 二层互联，需要经过 VXLAN 网关。

3. VXLAN 报文转发

VXLAN 将二层网络报文封装为 VXLAN 格式的报文，可以跨传统的三层网络进行传输，使得用户能够在三层网络之上构建逻辑的大二层网络。

（1）MAC 地址学习。

在 VXLAN 网络中，为了实现用户的互通，支持 MAC 地址动态学习。

（2）同子网报文转发。

根据报文中包含的目的 MAC 地址类型，报文转发流程分为已知单播报文转发和广播&未知单播&组播（Broadcast&Unknown-Unicast&Multicast，BUM）报文转发两部分。

同子网已知单播报文转发与 BUM 报文转发只在 VXLAN 二层网关之间进行，三层网关无须感知。

当 BUM 报文进入 VXLAN 隧道时，源端 VTEP 依据获取的隧道列表对报文进行复制，并分别进行封装。

（3）跨子网报文转发。

VXLAN 为站点内的租户提供三层业务时，需要在网络中部署 VXLAN 三层网关，以便站点内的租户通过三层网关与外界网络或其他 VXLAN 网络内的租户进行三层通信。

13.2.4 配置 VXLAN——静态方式

1. 配置 VXLAN 接入业务部署方式

在设备上部署 VXLAN 时，需要对下行接入业务进行选择部署，针对接入业务侧的部署，有以下两种方式。

基于 VLAN 方式：将一个或多个 VLAN 与 BD 进行关联，将指定 VLAN 内的用户加入 BD。该方式控制粒度较粗，但配置简单，适合在现网上直接进行部署。

基于报文流封装类型：根据报文携带 VLAN Tag 的情况，将不同流封装类型的报文送到不同的二层子端口，并将二层子端口与 BD 绑定，以将指定用户加入 BD。该方式控制粒度较细且较为灵活，但配置复杂，适合在新建网上进行部署。

（1）使用 system-view 命令，进入系统视图。

（2）使用 bridge-domain bd-id 命令，创建 BD，并进入 BD 视图。默认情况下，没有创建 BD。

（3）（可选）使用 description description 命令，配置 BD 的描述信息。默认情况下，未配置 BD 的描述信息。

（4）使用 quit 命令，退出 BD 视图，返回系统视图。

（5）配置业务接入点接入业务。

基于 VLAN 方式进行部署的方法如下。

① 使用 vlan vlan-id 命令，创建 VLAN 并进入 VLAN 视图。

② 使用 quit 命令，退出 VLAN 视图，返回系统视图。

③ 使用 bridge-domain bd-id 命令，进入已经创建的 BD 视图。

④ 使用 l2 binding vlan vlan-id 命令，将指定 VLAN 与 BD 相关联，实现数据报文在 BD 内的转发。默认情况下，VLAN 与 BD 无关联。

基于报文流封装类型进行部署的方法如下。

① 使用 interface interface-type interface-number.subnum mode l2 命令，创建二层子端口，并进入二层子端口视图。

② 使用 encapsulation { dot1q vid pe-vid | default | untag | qinq vid vlan-vid [ce-vid cevid] }命令，配置二层子端口允许通过的流封装类型，实现不同的端口接入不同的数据报文。默认情况下，二层子端口没有配置允许通过的流封装类型。

③ 使用 bridge-domain bd-id 命令，将指定二层子端口与 BD 相关联，实现数据报文在 BD 内的转发。默认情况下，二层子端口与 BD 无关联。

2. 配置 VXLAN 隧道

在设备上部署 VXLAN 时，需要对上行 VXLAN 隧道建立部署。VXLAN 隧道是通过隧道两端的 VTEP 的 IP 地址建立的，所以需要在隧道两端的设备上分别配置源端 VTEP 与目的端 VTEP 的 IP 地址。

（1）使用 system-view 命令，进入系统视图。

（2）使用 bridge-domain bd-id 命令，进入 BD 视图。

（3）使用 vxlan vni vni-id 命令，配置 BD 所对应的 VXLAN 的 VNI。默认情况下，没有配置 BD 所对应的 VXLAN 的 VNI。

（4）使用 quit 命令，退出 BD 视图，返回系统视图。

（5）使用 interface nve nve-number 命令，创建 NVE 端口，并进入 NVE 端口视图。

（6）使用 source ip-address 命令，配置 VXLAN 隧道源端 VTEP 的 IP 地址。默认情况下，VXLAN 隧道源端 VTEP 没有配置 IP 地址。

（7）使用 vni vni-id head-end peer-list ip-address &<1-10>命令，配置头端复制列表。默认情况下，没有配置 VNI 头端复制列表。

（8）使用 quit 命令，退出 NVE 端口视图，返回系统视图。

3. 配置 VXLAN 三层网关

VBDIF 端口在 VXLAN 三层网关上配置，用于跨网段报文的转发，因此同网段通信时不需要配置此端口。

VXLAN 为站点内的租户提供三层业务时，需要在网络中部署 VXLAN 三层网关，以便站点内的租户通过 VXLAN 三层网关与外界网络或其他 VXLAN 网络内的租户进行三层通信。

基于 BD 可创建三层逻辑端口——VBDIF 端口，每个 BD 对应一个 VBDIF 端口，在为 VBDIF 端口配置 IP 地址后，该端口即可作为本 BD 内租户的网关，对需要进行通信的报文进行基于 IP 地址的三层转发。为了实现不同网段的终端用户的互通，终端用户的默认网关地址必须对应 VXLAN 三层网关上的 VBDIF 端口的 IP 地址。

（1）配置步骤。

① 使用 system-view 命令，进入系统视图。

② 使用 interface vbdif bd-id 命令，创建 VBDIF 端口，并进入 VBDIF 端口视图。VBDIF 端口的编号必须对应一个已经存在的 ID 的 BD。

③ 使用 ip address ip-address { mask | mask-length } [sub]命令，配置 VBDIF 端口的 IP

地址，实现三层互通。默认情况下，VBDIF 端口上没有配置 IP 地址。

（2）查看 VXLAN 的配置结果。

① 使用 display bridge-domain [bd-id [brief | verbose]]命令，查看 BD 的配置信息。

② 使用 display vxlan tunnel [tunnel-id] [verbose]命令，查看 VXLAN 隧道的信息。

③ 使用 display vxlan vni [vni-id [verbose]]命令，查看 VNI 的 VXLAN 的配置信息。

④ 使用 display vxlan peer [vni vni-id]命令，查看 VNI 的目的端 VTEP 的 IP 地址。

⑤ 使用 display vxlan encapsulation interface interface-type interface-number [bridge-domain bd-id | default | dot1q [vid pe-vid] | qinq [vid vlan-vid [ce-vid cevid]] | untag]命令，查看主端口下二层子端口 VXLAN 封装信息。

13.2.5　配置 VXLAN——BGP EVPN 方式

BGP/MPLS IP VPN 的基本模型由三部分组成：用户边缘（Customer Edge，CE）设备、运营商边缘（Provider Edge，PE）设备和运营商（Provider，P）骨干设备。

CE 设备：有端口直接与服务提供商网络相连。CE 设备可以是路由器或交换机，也可以是一台主机。通常情况下，CE 设备"感知"不到 VPN 的存在，也不需要支持 MPLS。

PE 设备：与 CE 设备直接相连。在 MPLS 网络中，对 VPN 的所有处理都发生在 PE 设备上，对 PE 设备的性能要求较高。

P 设备：不与 CE 设备直接相连。P 骨干设备只具备基本 MPLS 的转发能力，不维护 VPN 信息。

PE 设备和 P 骨干设备仅由服务提供商管理；CE 设备仅由用户管理，除非用户把管理权委托给服务提供商。一台 PE 设备可以接入多台 CE 设备，一台 CE 设备也可以连接属于相同或不同服务提供商的多台 PE 设备。

1. 配置 VXLAN 接入业务选择部署方式

在设备上部署 VXLAN 时，需要对下行接入业务进行选择部署。针对接入业务侧的部署，有以下两种方式。

基于 VLAN 方式：将一个或多个 VLAN 与 BD 进行关联，将指定 VLAN 内的用户加入 BD。该方式控制粒度较粗，但配置简单，适合在现网上直接进行部署。

基于报文流封装类型：根据报文携带 VLAN Tag 的情况，将不同流封装类型的报文送到不同的二层子端口，并将二层子端口与 BD 绑定，以将指定用户加入 BD。该方式控制粒度较细且较为灵活，但配置复杂，适合在新建网上进行部署。

（1）使用 system-view 命令，进入系统视图。

（2）使用 bridge-domain bd-id 命令，创建 BD，并进入 BD 视图。默认情况下，未创建 BD。

（3）（可选）使用 description description 命令，配置 BD 的描述信息。默认情况下，未配置 BD 的描述信息。

（4）使用 quit 命令，退出 BD 视图，返回系统视图。

（5）配置业务接入点接入业务。

基于 VLAN 方式的命令如下。

① 使用 vlan vlan-id 命令，创建 VLAN 并进入 VLAN 视图。

② 使用 quit 命令，退出 VLAN 视图，返回系统视图。

③ 使用 bridge-domain bd-id 命令，进入已经创建的 BD 视图。

④ 使用 l2 binding vlan vlan-id 命令，将指定 VLAN 与 BD 相关联，实现数据报文在 BD 内的转发。默认情况下，VLAN 与 BD 无关联。

基于报文流封装类型的命令如下。

① 使用 interface interface-type interface-number.subnum mode l2 命令，创建二层子端口，并进入二层子端口视图。默认情况下，未创建二层子端口。

② 使用 encapsulation { dot1q { vid pe-vid } | default | untag | qinq { vid vlanvidce-vidce-vid } } 命令，配置二层子端口允许通过的流封装类型，使不同的端口接入不同的数据报文。默认情况下，二层子端口未配置允许通过的流封装类型。

③ 使用 bridge-domain bd-id 命令，将指定二层子端口与 BD 相关联，实现数据报文在 BD 内的转发。默认情况下，二层子端口与 BD 无关联。

2. 配置 VXLAN 隧道

在设备上部署 VXLAN 时，需要对上行 VXLAN 隧道建立部署。

VXLAN 隧道是通过隧道两端的 VTEP 的 IP 地址建立的，通过 BGP EVPN 方式动态建立 VXLAN 隧道，就是在两端 VTEP 之间建立 BGP EVPN 对等体，对等体之间利用 BGP EVPN 路由来互相传递 VNI 和 VTEP IP 地址信息，从而动态建立 VXLAN 隧道。

（1）配置 BGP EVPN 对等体关系。

① 使用 system-view 命令，进入系统视图。

② 使用 bgp as-number 命令，进入 BGP 视图。

③ 使用 peer ipv4-address as-number { as-number-plain | as-number-dot } 命令，将对端 PE 配置为对等体。

④ 使用 peer ipv4-address connect-interface loopback interface-number 命令，指定 BGP 建立 TCP 连接的端口。

⑤ 使用 l2vpn-family evpn 命令，使能并进入 BGP-EVPN 地址族视图。

⑥ 使用 peer { ipv4-address | group-name } enable 命令，使能对等体或组间交换 EVPN 路由信息的能力，以互相传递 VNI 和 VTEP IP 地址信息，从而动态建立 VXLAN 隧道。

⑦ 使用 quit 命令，退出 BGP-EVPN 地址族视图，返回 BGP 视图。

⑧ 使用 quit 命令，退出 BGP 视图，返回系统视图。

（2）配置源端 VTEP 的 IP 地址。

① 使用 interface nve nve-number 命令，创建 NVE 端口，并进入 NVE 端口视图。默认情况下，未创建 NVE 端口。

② 使用 source ip-address 命令，配置 VXLAN 隧道源端 VTEP 的 IP 地址。默认情况下，源端 VTEP 未配置 IP 地址。

3. 配置 VXLAN 三层网关

VBDIF 端口在 VXLAN 三层网关上配置，用于跨网段报文的转发，因此同网段通信时不需要配置此端口。

VXLAN 为站点内的租户提供三层业务时，需要在网络中部署 VXLAN 三层网关，以便站点内的租户通过 VXLAN 三层网关与外界网络或其他 VXLAN 网络内的租户进行三层通信。

基于 BD 可创建三层逻辑端口——VBDIF 端口，每个 BD 对应一个 VBDIF 端口，在为 VBDIF 端口配置 IP 地址后，该端口即可作为本 BD 内租户的网关，对需要进行通信的报文进行基于 IP 地址的三层转发。为了实现不同网段的终端用户互通，终端用户的默认网关地址必须对应 VXLAN 三层网关上的 VBDIF 端口的 IP 地址。

（1）配置 VPN 实例。

① 使用 system-view 命令，进入系统视图。

② 使用 ip vpn-instance vpn-instance-name 命令，创建 VPN 实例，并进入 VPN 实例视图。

③ 使用 ipv4-family 命令，使能 VPN 实例 IPv4 地址族，并进入 VPN 实例 IPv4 地址族视图。默认情况下，未使能 VPN 实例的 IPv4 地址族。

④ 使用 route-distinguisher route-distinguisher 命令，配置 VPN 实例的路由区分符（Route Distinguisher，RD）。默认情况下，VPN 实例地址族未配置 RD。

⑤ 使用 vpn-target vpn-target &<1-8> [both | export-extcommunity | importextcommunity] evpn 命令，为 VPN 实例配置 EVPN-VPN-Target 扩展团体属性。

EVPN-VPN-Target 是 BGP 的扩展团体属性，用来控制 EVPN 路由信息的接收和发布。一条 vpn-target evpn 命令最多可以配置 8 个 EVPN-VPN-Target。如果希望在 EVPN 实例地址族下配置更多的 EVPN-VPN-Target，则可以多次使用 vpn-target evpn 命令进行配置。

⑥ 使用 quit 命令，退出 VPN 实例 IPv4 地址族视图，返回 VPN 视图。

⑦ 使用 vxlan vni vni-id 命令，将 VNI 与 VPN 实例绑定。默认情况下，VNI 没有与 VPN 实例绑定。

⑧ 使用 quit 命令，退出 VPN 实例视图，返回系统视图。

（2）配置三层网关并绑定 VPN 实例。

① 使用 interface vbdif bd-id 命令，创建 VBDIF 端口，并进入 VBDIF 端口视图。VBDIF 端口的编号必须对应一个已经存在的 ID 的 BD。

② 使用 ip binding vpn-instance vpn-instance-name 命令，将 VBDIF 端口与 VPN 实例绑定。使用 ip binding vpn-instance 命令删除端口上已经配置的 IP 地址、路由协议等三层特性（包括 IPv4 和 IPv6），如果需要可重新配置，端口不能与未使能任何地址族的 VPN 实例绑定。

③ 使用 ip address ip-address { mask | mask-length } [sub] 命令，配置 VBDIF 端口的 IP 地址，实现三层互通。默认情况下，在 VBDIF 端口未配置 IP 地址。

④（可选）使用 mac-address mac-address 命令，配置 VBDIF 端口的 MAC 地址。默认情况下，VBDIF 端口的 MAC 地址是系统的 MAC 地址。

⑤ 使用 quit 命令，退出 VBDIF 端口视图，返回系统视图。

（3）配置 VXLAN 网关之间发布 IP 前缀类型的路由。

① 使用 bgp as-number 命令，进入 BGP 视图。

② 使用 ipv4-family vpn-instance vpn-instance-name 命令，进入 BGP-VPN 实例 IPv4 地址族视图。

③ 使用 import-route protocol [process-id] [med med | route-policy route-policyname] * 命令，将其他协议路由引入当前 BGP-VPN 实例 IPv4 地址族视图中。

如果要发布主机 IP 路由，则只需配置引入直连路由即可。如果要发布主机所在的网段路由，则需要先利用动态路由协议（如 OSPF 协议等）发布该网段路由，再配置引入相应动态协议的路由。

④ 使用 advertise l2vpn evpn 命令，使能 VPN 实例向 BGP-EVPN 地址族发布 IP 路由功能。默认情况下，未使能 VPN 实例向 BGP-EVPN 地址族发布 IP 路由功能。

（4）查看配置结果。

完成 VXLAN 业务接入点和 VXLAN 隧道配置后，可以使用以下命令查看配置信息。

① 使用 display bridge-domain [bd-id [brief | verbose]] 命令，可以查看 BD 的配置信息。

② 使用 display vxlan tunnel [tunnel-id] [verbose] 命令，可以查看 VXLAN 隧道的信息。

③ 使用 display vxlan vni [vni-id [verbose]]命令，可以查看 VNI 的 VXLAN 的配置信息。
④ 使用 display bgp evpn routing-table 命令，可以查看 EVPN 路由信息。

13.3 项目实施

13.3.1 VXLAN 隧道实现同网段用户互通配置实例

如图 13.7 所示，某公司在不同地域部署了自己的业务部门，不同的业务部门的数据中心都拥有自己的 VM，服务器 Server1 上的 VM1 属于 VLAN 10，服务器 Server2 上的 VM1 属于 VLAN 20，全网运行 OSPF 协议。由于这两个部门间有相同的业务诉求，因此可以将其规划到相同网段中，现在需要通过 VXLAN 隧道实现各业务部门的终端用户的相互通信，配置相关端口与 IP 地址，进行网络拓扑连接。

图 13.7 VXLAN 隧道实现同网段用户互通配置实例

V13-1 VXLAN 隧道实现同网段用户互通配置实例——基础配置

1. 配置思路

采用如下思路使同网段用户通过 VXLAN 隧道互通。

（1）分别在 CPE1、CE2、CE3 上配置路由协议，保证网络三层互通。

（2）分别在 CE2、CE3 上配置 VXLAN 接入业务选择部署方式。

（3）分别在 CE2、CE3 上配置 VXLAN 隧道。

V13-2 VXLAN 隧道实现同网段用户互通配置实例——VXLAN 配置与结果测试

2. 配置步骤

（1）配置交换机各端口的 IP 地址。

配置交换机 CPE1 各端口的 IP 地址。

```
<HUAWEI>system-view
[~HUAWEI]sysname CPE1
[~HUAWEI]commit                    //提交配置
[~CPE1]interface GE 1/0/1
[~CPE1-GE1/0/1]undo shutdown
[~CPE1-GE1/0/1]undo portswitch
[~CPE1-GE1/0/1]ip address 10.1.1.2 24
[~CPE1-GE1/0/1]quit
[~CPE1]interface GE 1/0/2
[~CPE1-GE1/0/2]undo shutdown       //默认端口是关闭的，这里需开启端口
[~CPE1-GE1/0/2]undo portswitch     //关闭二层端口，开启三层端口
[~CPE1-GE1/0/2]ip address 20.1.1.2 24
[~CPE1-GE1/0/2]quit
[~CPE1]interface LoopBack 1
[~CPE1-LoopBack1]ip address 1.1.1.1 32
[~CPE1-LoopBack1]quit
[~CPE1]commit
```

配置交换机 CE2 各端口的 IP 地址。

```
<HUAWEI>system-view
[~HUAWEI]sysname CE2
[~HUAWEI]commit
[~CE2]interface GE 1/0/1
[~CE2-GE1/0/1]undo shutdown
[~CE2-GE1/0/1]undo portswitch
[~CE2-GE1/0/1]ip address 10.1.1.1 24
[~CE2-GE1/0/1]quit
[~CE2]interface LoopBack 1
[~CE2-LoopBack1]ip address 2.2.2.2 32
[~CE2-LoopBack1]quit
[~CE2]commit
```

配置交换机 CE3 各端口的 IP 地址。

```
<HUAWEI>system-view
[~HUAWEI]sysname CE3
[~HUAWEI]commit
[~CE3]interface GE 1/0/2
[~CE3-GE1/0/2]undo shutdown
[~CE3-GE1/0/2]undo portswitch
[~CE3-GE1/0/2]ip address 20.1.1.1 24
```

[~CE3-GE1/0/2]quit
[~CE3]interface LoopBack 1
[~CE3-LoopBack1]ip address 3.3.3.3 32
[~CE3-LoopBack1]quit
[~CE3]commit

配置交换机 LSW1 的各端口。

<Huawei>system-view
[Huawei]sysname LSW1
[LSW1]vlan 10
[LSW1-vlan10]quit
[LSW1]interface GigabitEthernet 0/0/1
[LSW1-GigabitEthernet0/0/1]port link-type access
[LSW1-GigabitEthernet0/0/1]port default vlan 10
[LSW1-GigabitEthernet0/0/1]quit
[LSW1]interface GigabitEthernet 0/0/24
[LSW1-GigabitEthernet0/0/24]port link-type trunk
[LSW1-GigabitEthernet0/0/24]port trunk allow-pass vlan 10
[LSW1-GigabitEthernet0/0/24]quit

配置交换机 LSW2 的各端口。

<Huawei>system-view
[Huawei]sysname LSW2
[LSW2]vlan 20
[LSW2-vlan10]quit
[LSW2]interface GigabitEthernet 0/0/1
[LSW2-GigabitEthernet0/0/1]port link-type access
[LSW2-GigabitEthernet0/0/1]port default vlan 20
[LSW2-GigabitEthernet0/0/1]quit
[LSW2]interface GigabitEthernet 0/0/24
[LSW2-GigabitEthernet0/0/24]port link-type trunk
[LSW2-GigabitEthernet0/0/24]port trunk allow-pass vlan 20
[LSW2-GigabitEthernet0/0/24]quit

（2）配置 OSPF 协议，注意，需要发布设备上的 32 位 Loopback 端口地址。

配置交换机 CPE1 的 OSPF 协议。

[~CPE1]ospf 1
[~CPE1-ospf-1]area 0
[~CPE1-ospf-1-area-0.0.0.0]network 1.1.1.1 0.0.0.0
[~CPE1-ospf-1-area-0.0.0.0]network 10.1.1.0 0.0.0.255
[~CPE1-ospf-1-area-0.0.0.0]network 20.1.1.0 0.0.0.255
[~CPE1-ospf-1-area-0.0.0.0]quit
[~CPE1-ospf-1]quit
[~CPE1]commit

配置交换机 CE2 的 OSPF 协议。

[~CE2]ospf 1
[~CE2-ospf-1]area 0
[~CE2-ospf-1-area-0.0.0.0]network 2.2.2.2 0.0.0.0
[~CE2-ospf-1-area-0.0.0.0]network 10.1.1.0 0.0.0.255
[~CE2-ospf-1-area-0.0.0.0]quit

```
[~CE2-ospf-1]quit
[~CE2]commit
```
配置交换机 CE3 的 OSPF 协议。
```
[~CE3]ospf 1
[~CE3-ospf-1]area 0
[~CE3-ospf-1-area-0.0.0.0]network 3.3.3.3 0.0.0.0
[~CE3-ospf-1-area-0.0.0.0]network 20.1.1.0 0.0.0.255
[~CE3-ospf-1-area-0.0.0.0]quit
[~CE3-ospf-1]quit
[~CE3]commit
```
（3）OSPF 协议配置成功后，交换机之间可通过 OSPF 协议发现对方的 Loopback 端口的 IP 地址，并能互相连通，这里以 CE2 连通 CE3 为例进行介绍。
```
[~CE2]ping –a 2.2.2.2 3.3.3.3
  PING 3.3.3.3: 56    data bytes, press CTRL_C to break
    Reply from 3.3.3.3: bytes=56 Sequence=1 ttl=254 time=8 ms
    Reply from 3.3.3.3: bytes=56 Sequence=2 ttl=254 time=8 ms
    Reply from 3.3.3.3: bytes=56 Sequence=3 ttl=254 time=8 ms
    Reply from 3.3.3.3: bytes=56 Sequence=4 ttl=254 time=8 ms
    Reply from 3.3.3.3: bytes=56 Sequence=5 ttl=254 time=4 ms
  --- 3.3.3.3 ping statistics ---
    5 packet(s) transmitted
    5 packet(s) received
    0.00% packet loss
    round-trip min/avg/max = 4/7/8 ms
[~CE2]
```
（4）分别在交换机 CE2、CE3 上配置业务接入点。
配置交换机 CE2 的 GE 1/0/0 端口。
```
[~CE2]bridge-domain ?
    INTEGER<1-16777215>    Bridge-domain ID      //取值为 1~16777215
[~CE2]bridge-domain 10                           //创建 BD，并进入 BD 视图
[*CE2-bd10]quit
[*CE2]interface GE 1/0/0.?                       //取值为 1~99999999
    <1-99999999>
[*CE2]interface GE 1/0/0.1 mode l2               //创建二层子端口，并进入二层子端口视图
[*CE2-GE1/0/0.1]encapsulation dot1q vid 10
                        //配置二层子端口允许通过的流封装类型，允许 VLAN 10 的流量通过
[*CE2-GE1/0/0.1]bridge-domain 10
                        //将指定二层子端口与 BD 相关联，实现数据报文在 BD 内的转发
[*CE2-GE1/0/0.1] undo shutdown                   //开启端口
[*CE2-GE1/0/0.1]quit
[*CE2]interface GE 1/0/0
[~CE2-GE1/0/0]undo shutdown                      //开启端口
[~CE2-GE1/0/0]port link-type trunk
[~CE2-GE1/0/0]quit
[*CE2]
```
配置交换机 CE3 的 GE 1/0/0 端口。
```
[~CE3]bridge-domain 10                           //创建 BD，并进入 BD 视图
```

```
[*CE3-bd10]quit
[*CE3]interface GE 1/0/0.1 mode l2              //创建二层子端口，并进入二层子端口视图
[*CE3-GE1/0/0.1]encapsulation dot1q vid 20
                                                //配置二层子端口允许通过的流封装类型，允许 VLAN 20 的流量通过
[*CE3-GE1/0/0.1]bridge-domain 10
                                                //将指定二层子端口与 BD 相关联，实现数据报文在 BD 内的转发
[*CE3-GE1/0/0.1]undo shutdown                   //开启端口
[*CE3-GE1/0/0.1]quit
[*CE3]interface GE 1/0/0
[~CE3-GE1/0/0]undo shutdown                     //开启端口
[~CE3-GE1/0/0]port link-type trunk
[~CE3-GE1/0/0]quit
[*CE3]
```

（5）分别在交换机 CE2、CE3 上配置 VXLAN 隧道。

在交换机 CE2 上配置 VXLAN 隧道。

```
[~CE2]bridge-domain 10
[~CE2-bd10]vxlan vni ?                          //取值为 1～16777215
    INTEGER<1-16777215>   Value of VXLAN network identifier
[~CE2-bd10]vxlan vni 2000
[*CE2-bd10]quit
[*CE2]interface nve ?
    <1-2>   Nve interface number                //取值为 1 和 2
[*CE2]interface nve 1
[*CE2-Nve1]source 2.2.2.2                       //配置 VXLAN 隧道源端 VTEP 的 IP 地址
[*CE2-Nve1]vni 2000 head-end peer-list 3.3.3.3  //配置头端复制列表
[*CE2-Nve1]quit
[*CE2]commit
[~CE2]
```

在交换机 CE3 上配置 VXLAN 隧道。

```
[~CE3]bridge-domain 10
[~CE3-bd10]vxlan vni 2000
[*CE3-bd10]quit
[*CE3]interface nve 1
[*CE3-Nve1]source 3.3.3.3                       //配置 VXLAN 隧道源端 VTEP 的 IP 地址
[*CE3-Nve1]vni 2000 head-end peer-list 2.2.2.2  //配置头端复制列表
[*CE3-Nve1]quit
[*CE3]commit
[~CE3]
```

（6）验证配置结果，上述内容配置成功后，在交换机 CE2、CE3 上使用 display vxlan vni 命令，可以看到 VNI 的状态为 Up；使用 display vxlan tunnel 命令，可以看到 VXLAN 隧道的信息。

在交换机 CE2 上使用 display vxlan vni 命令，可以看到 VNI 的状态为 Up。

```
[~CE2]display vxlan vni
Number of vxlan vni : 1
VNI             BD-ID           State
---------------------------------------
2000            10              up
[~CE2]
```

在交换机 CE2 上使用 display vxlan tunnel 命令，可以看到 VXLAN 隧道的信息。

```
[~CE2]display vxlan tunnel
Number of vxlan tunnel : 1
Tunnel ID      Source          Destination        State    Type     Uptime
--------------------------------------------------------------------------------
4026531841    2.2.2.2         3.3.3.3            up       static   00:05:34
[~CE2]
```

在交换机 CE3 上使用 display vxlan vni 命令，可以看到 VNI 的状态为 Up。

```
[~CE3]display vxlan vni
Number of vxlan vni : 1
VNI            BD-ID           State
----------------------------------------
2000           10              up
[~CE3]
```

在交换机 CE3 上使用 display vxlan tunnel 命令，可以看到 VXLAN 隧道的信息。

```
[~CE3]display vxlan tunnel
Number of vxlan tunnel : 1
Tunnel ID      Source          Destination        State    Type     Uptime
--------------------------------------------------------------------------------
4026531841    3.3.3.3         2.2.2.2            up       static   00:06:40
[~CE3]
```

（7）查看交换机 CPE1、CE2、CE3 的配置信息，这里以交换机 CE2 为例进行介绍，其主要配置实例代码如下。

```
<CE2>display current-configuration
#
sysname CE2
#
router id 2.2.2.2
#
bridge-domain 10
 vxlan vni 2000
#
interface GE1/0/0.1 mode l2
 encapsulation dot1q vid 10
 bridge-domain 10
#
interface GE1/0/1
 undo portswitch
 undo shutdown
 ip address 10.1.1.1 255.255.255.0
#
interface LoopBack1
 ip address 2.2.2.2 255.255.255.255
#
interface Nve1
 source 2.2.2.2
 vni 2000 head-end peer-list 3.3.3.3
```

```
#
ospf 1
 area 0.0.0.0
  network 2.2.2.2 0.0.0.0
  network 10.1.1.0 0.0.0.255
#
return
<CE2>
```

（8）查看交换机 LSW1、LSW2 的配置信息，这里以交换机 LSW1 为例进行介绍，其主要配置实例代码如下。

```
<LSW1>display current-configuration
#
sysname LSW1
#
vlan batch 10
#
interface GigabitEthernet0/0/1
 port link-type access
 port default vlan 10
#
interface GigabitEthernet0/0/24
 port link-type trunk
 port trunk allow-pass vlan 10
#
return
<LSW1>
```

（9）配置完成后，验证配置结果，使用 ping 命令，服务器 Server1 访问服务器 Server2 的 IP 地址，其测试结果如图 13.8 所示。

图 13.8　服务器 Server1 访问服务器 Server2 的 IP 地址的测试结果

13.3.2　VXLAN 三层网关实现不同网段用户互通配置实例

如图 13.9 所示，有某一已经建成且比较成熟的网络，但是其没有专用的数据中心网络，所有的服务器都分布在不同的部门，并且不具备集中放置的条件，现在企业希望在已有的网络中构建一个虚拟网络，需要将散落在不同部门的服务器构建成一个虚拟网络，实现资源整合和业务灵活部署。各服务器上部署着大量的 VM，不同业务的服务器之间需要实现三层互通，企业在不同的位置拥有自己的 VM，服务器 Server1 属于 VLAN 10，服务器 Server2 属于 VLAN 20，且属于不同的网段，现需要通过 VXLAN 隧道实现不同业务的服务器之间的三层互通，配置相关端口与 IP 地址，进行网络拓扑连接。

图 13.9 VXLAN 三层网关实现不同网段用户互通配置实例

V13-3 VXLAN 三层网关实现不同网段用户互通配置实例——基础配置

1. 配置思路
采用如下思路使不同网段用户之间通过 VXLAN 三层网关实现互通。
（1）分别在 CPE1、CE2、CE3 上配置路由协议，保证网络三层互通。
（2）分别在 CE2、CE3 上配置 VXLAN 接入业务部署方式。
（3）分别在 CPE1、CE2 和 CE3 上配置 VXLAN 隧道。
（4）在 CPE1 上配置三层网关。

V13-4 VXLAN 三层网关实现不同网段用户互通配置实例——VXLAN 配置与结果测试

2. 配置步骤与方法
（1）配置交换机各端口的 IP 地址。
配置交换机 CPE1 各端口的 IP 地址。

```
<HUAWEI>system-view
```

```
[~HUAWEI]sysname CPE1
[~HUAWEI]commit                              //提交配置
[~CPE1]interface GE 1/0/1
[~CPE1-GE1/0/1]undo shutdown
[~CPE1-GE1/0/1]undo portswitch
[~CPE1-GE1/0/1]ip address 10.1.1.2 24
[~CPE1-GE1/0/1]quit
[~CPE1]interface GE 1/0/2
[~CPE1-GE1/0/2]undo shutdown          //默认端口是关闭的，这里需开启端口
[~CPE1-GE1/0/2]undo portswitch        //关闭二层端口，开启三层端口
[~CPE1-GE1/0/2]ip address 20.1.1.2 24
[~CPE1-GE1/0/2]quit
[~CPE1]interface LoopBack 1
[~CPE1-LoopBack1]ip address 1.1.1.1 32
[~CPE1-LoopBack1]quit
[~CPE1]commit
```

配置交换机 CE2 各端口的 IP 地址。

```
<HUAWEI>system-view
[~HUAWEI]sysname CE2
[~HUAWEI]commit
[~CE2]interface GE 1/0/1
[~CE2-GE1/0/1]undo shutdown
[~CE2-GE1/0/1]undo portswitch
[~CE2-GE1/0/1]ip address 10.1.1.1 24
[~CE2-GE1/0/1]quit
[~CE2]interface LoopBack 1
[~CE2-LoopBack1]ip address 2.2.2.2 32
[~CE2-LoopBack1]quit
[~CE2]commit
```

配置交换机 CE3 各端口的 IP 地址。

```
<HUAWEI>system-view
[~HUAWEI]sysname CE3
[~HUAWEI]commit
[~CE3]interface GE 1/0/2
[~CE3-GE1/0/2]undo shutdown
[~CE3-GE1/0/2]undo portswitch
[~CE3-GE1/0/2]ip address 20.1.1.1 24
[~CE3-GE1/0/2]quit
[~CE3]interface LoopBack 1
[~CE3-LoopBack1]ip address 3.3.3.3 32
[~CE3-LoopBack1]quit
[~CE3]commit
```

配置交换机 LSW1 的端口。

```
<Huawei>system-view
[Huawei]sysname LSW1
[LSW1]vlan 10
[LSW1-vlan10]quit
```

```
[LSW1]interface GigabitEthernet 0/0/1
[LSW1-GigabitEthernet0/0/1]port link-type access
[LSW1-GigabitEthernet0/0/1]port default vlan 10
[LSW1-GigabitEthernet0/0/1]quit
[LSW1]interface GigabitEthernet 0/0/24
[LSW1-GigabitEthernet0/0/24]port link-type trunk
[LSW1-GigabitEthernet0/0/24]port trunk allow-pass vlan 10
[LSW1-GigabitEthernet0/0/24]quit
```

配置交换机 LSW2 的端口。

```
<Huawei>system-view
[Huawei]sysname LSW2
[LSW2]vlan 20
[LSW2-vlan10]quit
[LSW2]interface GigabitEthernet 0/0/1
[LSW2-GigabitEthernet0/0/1]port link-type access
[LSW2-GigabitEthernet0/0/1]port default vlan 20
[LSW2-GigabitEthernet0/0/1]quit
[LSW2]interface GigabitEthernet 0/0/24
[LSW2-GigabitEthernet0/0/24]port link-type trunk
[LSW2-GigabitEthernet0/0/24]port trunk allow-pass vlan 20
[LSW2-GigabitEthernet0/0/24]quit
```

（2）配置 OSPF 协议，注意，需要发布设备上的 32 位 Loopback 端口 IP 地址。

配置交换机 CPE1 的 OSPF 协议。

```
[~CPE1]ospf 1
[~CPE1-ospf-1]area 0
[~CPE1-ospf-1-area-0.0.0.0]network 1.1.1.1 0.0.0.0
[~CPE1-ospf-1-area-0.0.0.0]network 10.1.1.0 0.0.0.255
[~CPE1-ospf-1-area-0.0.0.0]network 20.1.1.0 0.0.0.255
[~CPE1-ospf-1-area-0.0.0.0]quit
[~CPE1-ospf-1]quit
[~CPE1]commit
```

配置交换机 CE2 的 OSPF 协议。

```
[~CE2]ospf 1
[~CE2-ospf-1]area 0
[~CE2-ospf-1-area-0.0.0.0]network 2.2.2.2 0.0.0.0
[~CE2-ospf-1-area-0.0.0.0]network 10.1.1.0 0.0.0.255
[~CE2-ospf-1-area-0.0.0.0]quit
[~CE2-ospf-1]quit
[~CE2]commit
```

配置交换机 CE3 的 OSPF 协议。

```
[~CE3]ospf 1
[~CE3-ospf-1]area 0
[~CE3-ospf-1-area-0.0.0.0]network 3.3.3.3 0.0.0.0
[~CE3-ospf-1-area-0.0.0.0]network 20.1.1.0 0.0.0.255
[~CE3-ospf-1-area-0.0.0.0]quit
[~CE3-ospf-1]quit
[~CE3]commit
```

（3）OSPF 协议配置成功后，交换机之间可通过 OSPF 协议发现对方的 Loopback 端口的 IP 地址，并能互相连通，这里以 CE2 连通 CE3 为例进行介绍。

```
[~CE2]ping -a 2.2.2.2 3.3.3.3
  PING 3.3.3.3: 56  data bytes, press CTRL_C to break
    Reply from 3.3.3.3: bytes=56 Sequence=1 ttl=254 time=8 ms
    Reply from 3.3.3.3: bytes=56 Sequence=2 ttl=254 time=8 ms
    Reply from 3.3.3.3: bytes=56 Sequence=3 ttl=254 time=8 ms
    Reply from 3.3.3.3: bytes=56 Sequence=4 ttl=254 time=8 ms
    Reply from 3.3.3.3: bytes=56 Sequence=5 ttl=254 time=4 ms
  --- 3.3.3.3 ping statistics ---
    5 packet(s) transmitted
    5 packet(s) received
    0.00% packet loss
    round-trip min/avg/max = 4/7/8 ms
[~CE2]
```

（4）分别在交换机 CE2、CE3 上配置业务接入点。

配置交换机 CE2 的 GE 1/0/0 端口。

```
[~CE2]bridge-domain 10                    //创建 BD，并进入 BD 视图
[*CE2-bd10]quit
[*CE2]interface GE 1/0/0.1 mode l2        //创建二层子端口，并进入二层子端口视图
[*CE2-GE1/0/0.1]encapsulation dot1q vid 10
                                          //配置二层子端口允许通过的流封装类型，允许 VLAN 10 的流量通过
[*CE2-GE1/0/0.1]bridge-domain 10
                                          //将指定二层子端口与 BD 相关联，实现数据报文在 BD 内的转发
[*CE2-GE1/0/0.1] undo shutdown            //开启端口
[*CE2-GE1/0/0.1]quit
[*CE2]interface GE 1/0/0
[~CE2-GE1/0/0]undo shutdown               //开启端口
[~CE2-GE1/0/0]port link-type trunk
[~CE2-GE1/0/0]quit
[*CE2]
```

配置交换机 CE3 的 GE 1/0/0 端口。

```
[~CE3]bridge-domain 20                    //创建 BD，并进入 BD 视图
[*CE3-bd10]quit
[*CE3]interface GE 1/0/0.1 mode l2        //创建二层子端口，并进入二层子端口视图
[*CE3-GE1/0/0.1]encapsulation dot1q vid 20
                                          //配置二层子端口允许通过的流封装类型，允许 VLAN 20 的流量通过
[*CE3-GE1/0/0.1]bridge-domain 20
                                          //将指定二层子端口与 BD 相关联，实现数据报文在 BD 内的转发
[*CE3-GE1/0/0.1]undo shutdown             //开启端口
[*CE3-GE1/0/0.1]quit
[*CE3]interface GE 1/0/0
[~CE3-GE1/0/0]undo shutdown               //开启端口
[~CE3-GE1/0/0]port link-type trunk
[~CE3-GE1/0/0]quit
[*CE3]
```

（5）分别在交换机 CE2、CE3 上配置 VXLAN 隧道。

在交换机 CE1 上配置 VXLAN 隧道。

```
[*CPE1]bridge-domain 10
[*CPE1-bd10]vxlan vni 5
[*CPE1-bd10]quit
[*CPE1]interface nve 1
[*CPE1-Nve1]source 1.1.1.1
[*CPE1-Nve1]vni 5 head-end peer-list 2.2.2.2
[*CPE1-Nve1]quit
[*CPE1]bridge-domain 20
[*CPE1-bd20]vxlan vni 8
[*CPE1-bd20]quit
[*CPE1]interface nve 1
[*CPE1-Nve1]source 1.1.1.1
[*CPE1-Nve1]vni 8 head-end peer-list 3.3.3.3
[*CPE1-Nve1]quit
[*CPE1]
```

在交换机 CE2 上配置 VXLAN 隧道。

```
[~CE2]bridge-domain 10
[~CE2-bd10]vxlan vni 5
[*CE2-bd10]quit
[*CE2]interface nve 1
[*CE2-Nve1]source 2.2.2.2                    //配置 VXLAN 隧道源端 VTEP 的 IP 地址
[*CE2-Nve1]vni 5 head-end peer-list 1.1.1.1  //配置头端复制列表
[*CE2-Nve1]quit
[*CE2]commit
[~CE2]
```

在交换机 CE3 上配置 VXLAN 隧道。

```
[~CE3]bridge-domain 20
[~CE3-bd10]vxlan vni 8
[*CE3-bd10]quit
[*CE3]interface nve 1
[*CE3-Nve1]source 3.3.3.3                    //配置 VXLAN 隧道源端 VTEP 的 IP 地址
[*CE3-Nve1]vni 8 head-end peer-list 1.1.1.1  //配置头端复制列表
[*CE3-Nve1]quit
[*CE3]commit
[~CE3]
```

(6)在 CPE1 上配置 VXLAN 三层网关。

```
[*CPE1]interface vbdif 10
[*CPE1-Vbdif10]ip add 192.168.10.254 24
[*CPE1-Vbdif10]quit
[*CPE1]interface vbdif 20
[*CPE1-Vbdif20]ip address 192.168.20.254 24
[*CPE1-Vbdif20]quit
[*CPE1]
```

(7)验证配置结果,上述内容配置成功后,在交换机 CE2、CE3 上使用 display vxlan vni 命令,可以看到 VNI 的状态为 Up;使用 display vxlan tunnel 命令,可以看到 VXLAN 隧道的信息。

在交换机 CE1 上使用 display vxlan vni 命令,可以看到 VNI 的状态为 Up。

```
[~CPE1]display vxlan vni
Number of vxlan vni : 2
VNI            BD-ID           State
--------------------------------------------
5              10              up
8              20              up
[~CPE1]
```

在交换机 CE1 上使用 display vxlan tunnel 命令，可以看到 VXLAN 隧道的信息。

```
[~CPE1]display vxlan tunnel
Number of vxlan tunnel : 2
Tunnel ID      Source          Destination       State    Type      Uptime
--------------------------------------------------------------------------------
4026531841     1.1.1.1         2.2.2.2           up       static    01:07:59
4026531843     1.1.1.1         3.3.3.3           up       static    00:42:31
[~CPE1]
```

在交换机 CE2 上使用 display vxlan vni 命令，可以看到 VNI 的状态为 Up。

```
[~CE2]display vxlan vni
Number of vxlan vni : 1
VNI            BD-ID           State
--------------------------------------------
5              10              up
[~CE2]
```

在交换机 CE2 上使用 display vxlan tunnel 命令，可以看到 VXLAN 隧道的信息。

```
[~CE2]display vxlan tunnel
Number of vxlan tunnel : 1
Tunnel ID      Source          Destination       State    Type      Uptime
--------------------------------------------------------------------------------
4026531842     2.2.2.2         1.1.1.1           up       static    01:18:37
[~CE2]
```

在交换机 CE3 上使用 display vxlan vni 命令，可以看到 VNI 的状态为 Up。

```
[~CE3]display vxlan vni
Number of vxlan vni : 1
VNI            BD-ID           State
--------------------------------------------
8              20              up
[~CE3]
```

在交换机 CE3 上使用 display vxlan tunnel 命令，可以看到 VXLAN 隧道的信息。

```
[~CE3]display vxlan tunnel
Number of vxlan tunnel : 1
Tunnel ID      Source          Destination       State    Type      Uptime
--------------------------------------------------------------------------------
4026531842     3.3.3.3         1.1.1.1           up       static    01:19:00
[~CE3]
```

（8）显示交换机 CPE1 的配置信息，主要配置实例代码如下。

```
<CPE1>display current-configuration
#
sysname CPE1
```

```
#
router id 1.1.1.1
#
bridge-domain 10
 vxlan vni 5
#
bridge-domain 20
 vxlan vni 8
#
interface Vbdif10
 ip address 192.168.10.254 255.255.255.0
#
interface Vbdif20
 ip address 192.168.20.254 255.255.255.0
#
interface GE1/0/1
 undo portswitch
 undo shutdown
 ip address 10.1.1.2 255.255.255.0
#
interface GE1/0/2
 undo portswitch
 undo shutdown
 ip address 20.1.1.2 255.255.255.0
#
interface Nve1
 source 1.1.1.1
 vni 5 head-end peer-list 2.2.2.2
 vni 8 head-end peer-list 3.3.3.3
#
ospf 1
 area 0.0.0.0
  network 1.1.1.1 0.0.0.0
  network 10.1.1.0 0.0.0.255
  network 20.1.1.0 0.0.0.255
#
return
<CPE1>
```

（9）显示交换机 CE2 的配置信息，主要配置实例代码如下。

```
<CE2>display current-configuration
#
sysname CE2
#
router id 2.2.2.2
#
bridge-domain 10
 vxlan vni 5
#
```

```
interface GE1/0/0
 undo shutdown
 port link-type trunk
#
interface GE1/0/0.1 mode l2
 encapsulation dot1q vid 10
 bridge-domain 10
#
interface GE1/0/1
 undo portswitch
 undo shutdown
 ip address 10.1.1.1 255.255.255.0
#
interface LoopBack1
 ip address 2.2.2.2 255.255.255.255
#
interface Nve1
 source 2.2.2.2
 vni 5 head-end peer-list 1.1.1.1
#
ospf 1
 area 0.0.0.0
  network 2.2.2.2 0.0.0.0
  network 10.1.1.0 0.0.0.255
#
return
<CE2>
```

（10）显示交换机 CE3 的配置信息，主要配置实例代码如下。

```
<CE3>display current-configuration
#
sysname CE3
#
router id 3.3.3.3
#
bridge-domain 20
 vxlan vni 8
#
interface GE1/0/0
 undo shutdown
 port link-type trunk
#
interface GE1/0/0.1 mode l2
 encapsulation dot1q vid 20
 bridge-domain 20
#
interface GE1/0/2
 undo portswitch
 undo shutdown
```

```
  ip address 20.1.1.1 255.255.255.0
 #
 interface LoopBack1
  ip address 3.3.3.3 255.255.255.255
 #
 interface Nve1
  source 3.3.3.3
  vni 8 head-end peer-list 1.1.1.1
 #
 ospf 1
  area 0.0.0.0
   network 3.3.3.3 0.0.0.0
   network 20.1.1.0 0.0.0.255
 #
 return
 <CE3>
```

课后习题

1. 选择题

（1）下列关于 VXLAN 技术的阐述中，正确的是（　　）。

    A. VXLAN 用于替代传统的 VLAN

    B. VXLAN 是一种叠加在三层网络之上的二层网络，用于解决 VM 在 VLAN 中只能在二层网络下迁移的局限性

    C. 经过 VXLAN 封装后的数据包无法跨二层网络传递

    D. 经过 VXLAN 封装后会导致可用带宽不足

（2）【多选】VXLAN 的主要作用包括（　　）。

    A. 突破二层网络边界的限制

    B. 解决 VLAN 数量不足的限制

    C. 适用于多租户场景

    D. 解决了虚拟化带来的挑战，简化了对网络的要求

2. 简答题

（1）简述 VXLAN 的优点。

（2）简述 VXLAN 的运行机制。

（3）简述 VXLAN 的集中式网关和分布式网关的区别。

（4）简述 VXLAN 的配置方法。